Intelligent Systems Reference Library

Volume 208

The aim of this series is to publish a Reference Library, including novel advances and developments in all aspects of Intelligent Systems in an easily accessible and well structured form. The series includes reference works, handbooks, compendia, textbooks, well-structured monographs, dictionaries, and encyclopedias. It contains well integrated knowledge and current information in the field of Intelligent Systems. The series covers the theory, applications, and design methods of Intelligent Systems. Virtually all disciplines such as engineering, computer science, avionics, business, e-commerce, environment, healthcare, physics and life science are included. The list of topics spans all the areas of modern intelligent systems such as: Ambient intelligence, Computational intelligence, Social intelligence, Computational neuroscience, Artificial life, Virtual society, Cognitive systems, DNA and immunity-based systems, e-Learning and teaching, Human-centred computing and Machine ethics, Intelligent control, Intelligent data analysis, Knowledge-based paradigms, Knowledge management, Intelligent agents, Intelligent decision making, Intelligent network security, Interactive entertainment, Learning paradigms, Recommender systems, Robotics and Mechatronics including human-machine teaming, Self-organizing and adaptive systems, Soft computing including Neural systems, Fuzzy systems, Evolutionary computing and the Fusion of these paradigms, Perception and Vision, Web intelligence and Multimedia.

Indexed by SCOPUS, DBLP, zbMATH, SCImago.

All books published in the series are submitted for consideration in Web of Science.

More information about this series at http://www.springer.com/series/8578

Javidan Kazemi Kordestani ·
Mehdi Razapoor Mirsaleh ·
Alireza Rezvanian · Mohammad Reza Meybodi
Editors

Advances in Learning Automata and Intelligent Optimization

Springer

Editors
Javidan Kazemi Kordestani
Department of Computer Engineering
Science and Research Branch
Islamic Azad University
Tehran, Iran

Mehdi Razapoor Mirsaleh
Department of Computer Engineering
and Information Technology
Payame Noor University (PNU)
Tehran, Iran

Alireza Rezvanian
Department of Computer Engineering
University of Science and Culture
Tehran, Iran

Mohammad Reza Meybodi
Department of Computer Engineering
Amirkabir University of Technology
(Tehran Polytechnic)
Tehran, Iran

ISSN 1868-4394 ISSN 1868-4408 (electronic)
Intelligent Systems Reference Library
ISBN 978-3-030-76290-2 ISBN 978-3-030-76291-9 (eBook)
https://doi.org/10.1007/978-3-030-76291-9

This Springer imprint is published by the registered company Springer Nature Switzerland AG
The registered company address is: Gewerbestrasse 11, 6330 Cham, Switzerland

Dedication

To my beloved wife, for her love, her invaluable support, her patience, and understanding during the time, it has taken me to complete this book

Javidan

To my family

Mehdi

To my family for their lovely support

Alireza

To my family

Mohammad Reza

Preface

This book is written for computer scientists, graduate students, and researchers studying artificial intelligence, machine learning, reinforcement learning, learning automata techniques, and engineers working on real-world problem-solving in engineering domains. In particular, the reader is assumed already familiar with basic mathematics, statistics, probability, and algorithm. Prior exposure to mathematics, stochastic process, and learning automata is helpful but not necessary. The book in detail describes verities of learning automaton models and their recent developments of applications in solving real-world problems and optimization with detailed mathematical and theoretical perspectives.

This book consists of nine chapters devoted to the theory of learning automata and cellular learning automata models for optimization. Chapter 1 gives a preliminary introduction and an overview of various learning automata models and static and dynamic optimization concepts. Chapter 2 provides a bibliometric analysis of the research studies on learning automata and optimization as a systematic review. Chapter 3 is dedicated to describing the recent hybrid algorithms with the aid of cellular learning automata. Chapter 4 is devoted to learning automata for behavior control in evolutionary computation in local and global optimization. In Chapter 5, applications of a memetic model of learning automata for solving NP-hard problems are discussed. Chapter 6 provides object migration automata for solving graph and network problems. Chapter 7 gives an overview of multi-population methods for dynamic environments. Chapter 8 describes learning automata for online function evaluation management in evolutionary multi-population methods for dynamic optimization problems. Finally, Chapter 9 provides a detailed discussion on function management in multi-population methods with a variable number of populations using a learning automaton approach.

The authors would like to thank Dr. Thomas Ditzinger, Springer, Editorial Director & Interdisciplinary Applied Sciences, Holger Schaepe, Senior Editorial Assistant, Springer-Verlag Heidelberg in Engineering Editorial, Silvia Schneider, and Ms. Varsha Prabakaran, Project Coordinator & Books Production administrator of Springer Nature, for the editorial assistance, cooperative collaboration, excellent support, and Saranya Kalidoss for providing continuous assistance and advice whenever needed to produce this important scientific work. We hope that readers will share our pleasure to present this book on the theory of learning automata and optimization will find it useful in their research.

Acknowledgment

We are grateful to many people who have contributed to the work presented here and offered critical reviews of prior publication. We thank Springer for its assistance in publishing the book. We are also grateful to our academic supervisor, family, parents, and friends for their love and support.

March 2021

Javidan Kazemi Kordestani
Mehdi Razapoor Mirsaleh
Alireza Rezvanian
Mohammad Reza Meybodi

Contents

1 An Introduction to Learning Automata and Optimization 1
Javidan Kazemi Kordestani, Mehdi Razapoor Mirsaleh,
Alireza Rezvanian, and Mohammad Reza Meybodi
1.1 Introduction 1
1.2 Learning Automata 3
1.2.1 Learning Automata Variants 5
1.2.2 Recent Applications of Learning Automata 23
1.3 Optimization 23
1.3.1 Evolutionary Algorithms and Swarm Intelligence 27
1.4 Reinforcement Learning and Optimization Methods 29
1.4.1 Static Optimization 31
1.4.2 Dynamic Optimization 34
1.5 LA and Optimization Timeline 35
1.6 Chapter Map 37
1.7 Conclusion 37
References 37

2 Learning Automaton and Its Variants for Optimization: A Bibliometric Analysis 51
Javidan Kazemi Kordestani, Mehdi Razapoor Mirsaleh,
Alireza Rezvanian, and Mohammad Reza Meybodi
2.1 Introduction 51
2.2 Learning Automata Models and Optimization 52
2.3 Material and Method 54
2.3.1 Data Collection and Initial Results 54
2.3.2 Refining the Initial Results 55
2.4 Analyzing the Results 55
2.4.1 Initial Result Statistics 56
2.4.2 Top Journals 56
2.4.3 Top Researchers 57

2.4.4 Top Papers 60
2.4.5 Top Affiliations 62
2.4.6 Top Keywords 67
2.5 Conclusion 70
References 71

3 Cellular Automata, Learning Automata, and Cellular Learning Automata for Optimization 75
Javidan Kazemi Kordestani, Mehdi Razapoor Mirsaleh, Alireza Rezvanian, and Mohammad Reza Meybodi
3.1 Introduction 75
3.2 Preliminaries 77
3.2.1 Cellular Automata 77
3.2.2 Learning Automata 78
3.2.3 Cellular Learning Automata 78
3.3 CA, CLA, and LA Models for Optimization 82
3.3.1 Cellular Learning Automata-Based Evolutionary Computing (CLA-EC) 82
3.3.2 Cooperative Cellular Learning Automata-Based Evolutionary Computing (CLA-EC) 84
3.3.3 Recombinative Cellular Learning Automata-Based Evolutionary Computing (RCLA-EC) 84
3.3.4 CLA-EC with Extremal Optimization (CLA-EC-EO) 88
3.3.5 Cellular Learning Automata-Based Differential Evolution (CLA-DE) 89
3.3.6 Cellular Particle Swarm Optimization (Cellular PSO) 90
3.3.7 Firefly Algorithm Based on Cellular Learning Automata (CLA-FA) 93
3.3.8 Harmony Search Algorithm Based on Learning Automata (LAHS) 94
3.3.9 Learning Automata Based Butterfly Optimization Algorithm (LABOA) 95
3.3.10 Grey Wolf Optimizer Based on Learning Automata (GWO-LA) 97
3.3.11 Learning Automata Models with Multiple Reinforcements (MLA) 99
3.3.12 Cellular Learning Automata Models with Multiple Reinforcements (MCLA) 106
3.3.13 Multi-reinforcement CLA with the Maximum Expected Rewards (MCLA) 107
3.3.14 Gravitational Search Algorithm Based on Learning Automata (GSA-LA) 114
3.4 Conclusion 115
References 116

4 Learning Automata for Behavior Control in Evolutionary Computation 127
Javidan Kazemi Kordestani, Mehdi Razapoor Mirsaleh, Alireza Rezvanian, and Mohammad Reza Meybodi
4.1 Introduction 128
4.2 Types of Parameter Adjustment in EC Community 128
4.2.1 EC with Constant Parameters 128
4.2.2 EC with Time-Varying Parameters 129
4.3 Differential Evolution 135
4.3.1 Initialization 136
4.3.2 Difference-Vector Based Mutation 136
4.3.3 Repair Operator 137
4.3.4 Crossover 137
4.3.5 Selection 137
4.4 Learning Automata for Adaptive Control of Behavior in Differential Evolution 138
4.4.1 Behavior Control in DE with Variable-Structure Learning Automaton 138
4.4.2 Behavior Control in DE with Fixed-Structure Learning Automaton 142
4.5 Experimental Setup 143
4.5.1 Benchmark Functions 143
4.5.2 Algorithm's Configuration 145
4.5.3 Simulation Settings and Results 146
4.5.4 Experimental Results 147
4.6 Conclusion 154
References 155

5 A Memetic Model Based on Fixed Structure Learning Automata for Solving NP-Hard Problems 159
Javidan Kazemi Kordestani, Mehdi Razapoor Mirsaleh, Alireza Rezvanian, and Mohammad Reza Meybodi
5.1 Introduction 160
5.2 Fixed Structure Learning Automata and Object Migrating Automata 161
5.2.1 Fixed Structure Learning Automata 161
5.2.2 Object Migration Automata 164
5.3 GALA 165
5.3.1 Global Search in GALA 165
5.3.2 Crossover Operator 167
5.3.3 Mutation Operator 169
5.3.4 Local Learning in GALA 170
5.3.5 Applications of GALA 173

5.4 The New Memetic Model Based on Fixed Structure Learning Automata . . . 175
5.4.1 Hybrid Fitness Function . . . 175
5.4.2 Mutation Operators . . . 176
5.4.3 Crossover Operators . . . 179
5.5 The OneMax Problem . . . 183
5.5.1 Local Search for OneMax . . . 185
5.5.2 Experimental Results . . . 185
5.6 Conclusion . . . 190
References . . . 190

6 The Applications of Object Migration Automaton (OMA)-Memetic Algorithm for Solving NP-Hard Problems . . . 195
Javidan Kazemi Kordestani, Mehdi Razapoor Mirsaleh, Alireza Rezvanian, and Mohammad Reza Meybodi
6.1 Introduction . . . 195
6.2 The Equipartitioning Problem . . . 196
6.2.1 Local Search for EPP . . . 197
6.2.2 Experimental Results . . . 197
6.3 The Graph Isomorphism Problem . . . 212
6.3.1 The Local Search in the Graph Isomorphism Problem . . . 212
6.3.2 Experimental Results . . . 217
6.4 Assignment of Cells to Switches Problem (ACTSP) in Cellular Mobile Network . . . 227
6.4.1 Background and Related Work . . . 231
6.4.2 The OMA-MA for Assignment of Cells to Switches Problem . . . 233
6.4.3 The Framework of the OMA-MA Algorithm . . . 236
6.4.4 Experimental Result . . . 239
6.5 Conclusion . . . 251
References . . . 251

7 An Overview of Multi-population Methods for Dynamic Environments . . . 253
Javidan Kazemi Kordestani, Mehdi Razapoor Mirsaleh, Alireza Rezvanian, and Mohammad Reza Meybodi
7.1 Introduction . . . 254
7.2 Moving Peaks Benchmark . . . 256
7.2.1 Extended Versions of MPB . . . 257
7.3 Performance Measurement . . . 259
7.4 Types of Multi-population Methods . . . 260
7.4.1 Methods with a Fixed Number of Populations . . . 260
7.4.2 Methods with a Variable Number of Populations . . . 271

7.4.3 Methods Based on Population Clustering 276
7.4.4 Self-adapting the Number of Populations 279
7.5 Numerical Results 280
7.6 Conclusions 283
References 283

8 Learning Automata for Online Function Evaluation Management in Evolutionary Multi-population Methods for Dynamic Optimization Problems 287
Javidan Kazemi Kordestani, Mehdi Razapoor Mirsaleh, Alireza Rezvanian, and Mohammad Reza Meybodi
8.1 Introduction 288
8.2 Preliminaries 289
8.2.1 Waste of FEs Due to Change Detection 289
8.2.2 Waste of FEs Due to the Excessive Number of Sub-populations 289
8.2.3 Waste of FEs Due to Overcrowding of Subpopulations in the Same Area of the Search Space 290
8.2.4 Waste of FEs Due to Exclusion Operator 290
8.2.5 Allocation of FEs to Unproductive Populations 290
8.2.6 Unsuitable Parameter Configuration of the EC Methods . . . 291
8.2.7 Equal Distribution of FEs Among Sub-populations 291
8.3 Theory of Learning Automata 292
8.3.1 Fixed Structure Learning Automata 293
8.3.2 Variable Structure Learning Automata 294
8.4 EC Techniques under Study 294
8.4.1 Particle Swarm Optimization 294
8.4.2 Firefly Algorithm 295
8.4.3 Jaya 296
8.5 LA-Based FE Management Model for MP Evolutionary Dynamic Optimization 297
8.5.1 Initialization of Sub-populations 299
8.5.2 Detection and Response to Environmental Changes 300
8.5.3 Choose a Sub-population for Execution 300
8.5.4 Evaluate the Search Progress of Populations and Generate the Reinforcement Signal 300
8.5.5 Exclusion 301
8.6 FE-Management in MP Method with a Fixed Number of Populations 302
8.6.1 VSLA-Based FE Management Strategy 302
8.6.2 FSLA-Based FE Management Strategies 303
8.7 Experimental Study 307
8.7.1 Experimental Setup 307
8.7.2 Experimental Results and Discussion 310

8.8 Conclusion 318
References 319

9 Function Management in Multi-population Methods with a Variable Number of Populations: A Variable Action Learning Automaton Approach 323
Javidan Kazemi Kordestani, Mehdi Razapoor Mirsaleh, Alireza Rezvanian, and Mohammad Reza Meybodi
9.1 Introduction 324
9.2 Main Framework of Clustering Particle Swarm Optimization 325
9.2.1 Creating Multiple Sub-swarms from the Cradle Swarm . . . 325
9.2.2 Local Search by PSO 327
9.2.3 Status of Sub-swarms 328
9.2.4 Detection and Response to Environmental Changes 329
9.3 Variable Action-Set Learning Automata 330
9.4 FEM in MP Methods with a Variable Number of Populations 330
9.5 Experimental Study 332
9.5.1 Dynamic Test Function 332
9.5.2 Performance Measure 335
9.5.3 Experimental Settings 335
9.5.4 Experimental Results 336
9.6 Conclusions 339
References 339

About the Authors

Javidan Kazemi Kordestani received the B.Sc. in computer engineering (software engineering) from the Islamic Azad University of Karaj, Iran, in 2008, and his M.Sc. in computer engineering (artificial intelligence) from Islamic Azad University of Qazvin, Iran, in 2012. He also received the Ph.D. degree in computer engineering (artificial intelligence) at the Department of Computer Engineering, Science and Research Branch, Islamic Azad University, Tehran, Iran. He has authored or co-authored numerous research publications in reputable peer-reviewed journals of Elsevier, Springer, Taylor & Francis, and Wiley. He has also acted as a reviewer for several prestigious international journals. His current research interests include evolutionary computation, dynamic optimization problems, learning systems, and real-world applications.

Mehdi Rezapoor Mirsaleh received the B.Sc. in computer engineering from Kharazmi University, Tehran, Iran, in 2000. He also received the M.Sc. and Ph.D. degrees from Amirkabir University of Technology (Tehran Polytechnic), Tehran, Iran, in 2003 and 2016, respectively, in computer engineering. Currently, he is Assistant Professor at the Department of Computer Engineering and Information Technology, Payame Noor University (PNU), Tehran, Iran. His research interests include learning systems, machine learning, social networks, and soft computing.

Alireza Rezvanian received the B.Sc. degree from Bu-Ali Sina University of Hamedan, Iran, in 2007, the M.Sc. degree in computer engineering with honors from Islamic Azad University of Qazvin, Iran, in 2010, and the Ph.D. degree in computer engineering at the Computer Engineering Department from Amirkabir University of Technology (Tehran Polytechnic), Tehran, Iran, in 2016. Currently, he is Assistant Professor at the Department of Computer Engineering, University of Science and Culture, Tehran, Iran. He worked from 2016 to 2020 as a researcher at the School of Computer Science from the Institute for Research in Fundamental Sciences (IPM), Tehran, Iran. He has authored or co-authored more than 70 research publications in reputable peer-reviewed journals and conferences, including IEEE, Elsevier, Springer, Wiley, and Taylor & Francis. He has been a guest editor of the special issue on new applications of learning automata-based techniques in real-world environments for the journal of computational science (Elsevier). He is an editorial board member and one of the associate editors of human-centric computing and information sciences (Springer), CAAI Transactions on Intelligence Technology (IET), The Journal of Engineering (IET), and Data in Brief (Elsevier). His research activities include soft computing, learning automata, complex networks, social network analysis, data mining, data science, machine learning, and evolutionary algorithms.

Mohammad Reza Meybodi received the B.S. and M.S. degrees in economics from the Shahid Beheshti University in Iran in 1973 and 1977, respectively. He also received the M.S. and Ph.D. degrees from Oklahoma University, USA, in 1980 and 1983, respectively, in computer science. Currently, he is Full Professor in the Computer Engineering Department, Amirkabir University of Technology, Tehran, Iran. Prior to the current position, he worked from 1983 to 1985 as Assistant Professor at the Western Michigan University and from 1985 to 1991 as Associate Professor at Ohio University, USA. His current research interests include learning systems, cloud computing, soft computing, and social networks.

Abbreviations

ACO	Ant colony optimization
ACPSO	Adaptive cooperative particle swarm optimizer
AFSA-CLA	Artificial fish swarm algorithm based on cellular learning automata
AIS	Artificial immune system
ACLA	Asynchronous cellular learning automata
ADCLA	Asynchronous dynamic cellular learning automata
ACS	Adaptive cuckoo search
BBExpPSO	Exploiting barebones particle swarm optimization
BBPSO	Barebones particle swarm optimization
BCPSO	Basic clustering particle swarm optimizer
BLA	Bayesian learning automata
BOMA-MA	Baldwinian object migration automation-based memetic algorithm
CA	Cellular automata
CADCLA	Closed asynchronous dynamic cellular learning automata
CALA	Continuous action-set learning automata
CARLA	Continuous action-set reinforcement learning automata
CCGA+LA	Learning automata-based co-evolutionary genetic algorithm
CCLA	Cooperative cellular learning automata
CCLA-EC	Cooperative cellular learning automata-based evolutionary computing
CCPSO	Competitive clustering particle swarm optimizer
CCS	Converged chromosomes set
CDE	Crowding-based differential evolution
CI	Computational intelligence
CLA	Cellular learning automata
CLA-AIS	Cellular learning automata-based artificial immune system
CLA-BBPSO	Cellular learning automata-based barebones particle swarm optimization

CLA-BBPSO-R	Cellular learning automata-based barebones PSO with rotated mutations
CLA-DE	Cellular learning automata-based differential evolution
CLA-EC	Cellular learning automata-based evolutionary computing
CLA-EC-EO	Cellular learning automata-based evolutionary computing with extremal optimization
CLA-FA	Cellular learning automata-based firefly algorithm
CLA-PSO	Cellular learning automata-based particle swarm optimization
CLAMA	Cellular learning automata-based memetic algorithm
CLAMS	Cellular learning automata-based multi-swarm
CLA-MPD	Cellular learning automata-based multi-population
CMA	Canonical memetic algorithm
CPSO	Clustering particle swarm optimization
CPSOLA	Cooperative particle swarm optimization based on learning automata
CS	Cuckoo search
DCLA-PSO	Discrete cellular learning automata-based particle swarm optimization
DE	Differential evolution
DGPA	Discrete generalist pursuit algorithm
DICLA	Dynamic irregular cellular learning automata
DLA	Distributed learning automata
DOP	Dynamic optimization problem
DRLA	Dimensionality ranking in learning automata
EA	Evolutionary algorithm
EC	Evolutionary computation
EDA	Estimation of distribution algorithm
EDLA	Extended distributed learning automata
EO	Extremal optimization
EPP	Equipartitioning problem
ES	Evolutionary strategies
FA	Firefly algorithm
FE	Fitness evaluation
FEM	Fitness evaluation management
FALA	Finite action-set learning automata
FLA	Finite learning automata
FSLA	Fixed structure learning automata
GA	Genetic algorithm
GALA	Genetic algorithm based on learning automata
GIP	Graph isomorphism problem
GLA	Game of learning automata
GSA	Gravitational search algorithm
GSA-LA	Gravitational search algorithm based on learning automata
GSO	Group search optimizer
GWO-LA	Gray wolf optimizer based on learning automata

HS	Harmony search
HOMA-MA	Hybrid (Baldwinian–Lamarckian) object migration automaton-based memetic algorithm
HSLA	Hierarchical structure of learning automata
IADE	Individual-based adaptive differential evolution
IAPSO	Independent adaptive particle swarm optimization
ICA-LA	Imperialist competitive-based learning automata
ICLA	Irregular cellular learning automata
ICLA-EC	Irregular cellular learning automata-based evolutionary computing
ISADE	Independent strategy adaptive differential evolution
LA	Learning automaton
LA-AIN	Learning automata-based artificial immune network
LABSO	Brain storm optimization based on learning automata
LACAIS	Learning automata-based cooperative artificial immune system
LADE	Learning automata-based differential evolution
LAHS	Harmony search algorithm based on learning automata
LA-MA	Learning automata-based memetic algorithm
LBA	Learning bee colony
LOMA–MA	Lamarckian object migration automaton-based memetic algorithm
LS	Local search
MP	Multi-population
MPB	Moving peaks benchmark
MCLA	Multi-reinforcement cellular learning automata
M-CLA-PSO	Memetic cellular learning automata-based particle swarm optimization
MLATI	Multi-reinforcement learning automaton type I
MLATII	Multi-reinforcement learning automaton type II
MLATIII	Multi-reinforcement learning automaton type III
MLAMA	Michigan memetic learning automata
MMLA	Multi-reinforcement learning automata
MNLA	Multi-reinforcement N-tuple learning automata
MOLA	Multi-objective learning automata
NCC	Non-converged chromosome
NLA	Network of learning automata
NSGA-II	Non-dominated sorting genetic algorithm II
OE	Offline error
OMA	Object migration automata
OMA-MA	Object migration automaton-based memetic algorithm
OCLA	Open cellular learning automata
PADE	Population-based adaptive differential evolution
PGAM	Parallel genetic algorithms with migration
PGAM-SA	Parallel genetic algorithms with migrations and simulated annealing

PGAM-TS	Parallel genetic algorithms with migrations and Tabu search
PLA	Pursuit learning automata
PSADE	Population strategy adaptive differential evolution
PSO	Particle swarm optimization
PSO-LA	Particle swarm optimization based on learning automata
QPSO	Quantum-behaved particle swarm optimization
RCLA-EC	Recombinative cellular learning automata-based evolutionary computing
RL	Reinforcement learning
RLMPSO	Reinforcement learning-based memetic particle swarm optimizer
SI	Swarm intelligence
SLA	Stochastic learning automata
TS	Tabu search
TDL	Temporal-difference learning
UAPSO	Unified adaptive particle swarm optimization
VALA	Variable-action-set learning automaton
VSLA	Variable structure learning automata
WCLA	Wavefront cellular learning automata
WoS	Web of Science

Chapter 1
An Introduction to Learning Automata and Optimization

Javidan Kazemi Kordestani, Mehdi Razapoor Mirsaleh, Alireza Rezvanian, and Mohammad Reza Meybodi

Abstract Learning automaton (LA) is one of the reinforcement learning techniques in artificial intelligence. Learning automata's learning ability in unknown environments is a useful technique for modeling, controlling, and solving many real problems in the distributed and decentralized environments. In this chapter, first, we provide an overview of LA concepts and recent variants of LA models. Then, we present a brief description of the recent reinforcement learning mechanisms for solving optimization problems. Finally, the evolution of the recent LA models for optimization is presented.

1.1 Introduction

One of the goals of artificial intelligence is to make a machine thinks/acts like human behavior or thinks/acts rationally (Stuart and Peter 2002). This synthetic machine needs to learn to solve problems and adapt to environments. From a psychology perspective, learning is considered any systematic change in the system's performance with a specific goal (Fu 1970). Learning is defined as any relatively permanent change in behavior resulting from experience. The learning system is characterized

J. Kazemi Kordestani
Department of Computer Engineering, Science and Research Branch, Islamic Azad University, Tehran, Iran

M. Razapoor Mirsaleh
Department of Computer Engineering and Information Technology, Payame Noor University (PNU), P.O. BOX 19395-3697, Tehran, Iran
e-mail: mrezapoorm@aut.ac.ir

A. Rezvanian (✉)
Department of Computer Engineering, University of Science and Culture, Tehran, Iran
e-mail: rezvanian@usc.ac.ir

M. R. Meybodi
Computer Engineering Department, Amirkabir University of Technology (Tehran Polytechnic), Tehran, Iran
e-mail: mmeybodi@aut.ac.ir

J. Kazemi Kordestani et al. (eds.), *Advances in Learning Automata and Intelligent Optimization*, Intelligent Systems Reference Library 208, https://doi.org/10.1007/978-3-030-76291-9_1

by its ability to improve its behavior with time, in some sense tending towards an ultimate goal (Narendra and Thathachar 1974). The concept of learning makes it possible to design systems that can gradually improve their performance during actual operation through learning from past experiences. Every learning task consists of two parts: a learning system and environment. The learning system must learn to act in the environment.

Among the machine learning methods, reinforcement learning as a leading method for unknown environments is learning from positive and negative rewards (Montague 1999). In standard reinforcement learning, the system is connected to its environment via perception and action. The learning system receives an input at each instant, which indicates the current state of the environment. Then the learning system chooses an action to generate as output. This action changes the state of the environment, and the new state of the environment is communicated to the learning system with a scalar reinforcement signal. The reinforcement signal changes the learning system's behavior, and the learning system can learn to choose the best action over time through systematic trial and error. Reinforcement learning tasks can be divided naturally into two types: sequential and non-sequential tasks. In non-sequential tasks, the objective is to learn a mapping from situations to actions that maximize the expected immediate payoff. Non-sequential reinforcement learning has been studied extensively in learning automata (Kumpati and Narendra 1989). In sequential tasks, the objective is to maximize the expected long-term payoffs. Sequential tasks are difficult because the action selected may influence future situations and thus future payoffs. The sequential tasks are studied in reinforcement learning techniques based on dynamic programming, which approximates dynamic programming (Kaelbling et al. 1996).

The theory of learning automaton (LA) (Rezvanian et al. 2018a, c) as a promising field of artificial intelligence is a powerful source of computational technique that could be exploited for solving many real-world problems. LA is a self-adaptive decision-making unit that interacts with an unknown stochastic environment and can progressively make optimal decisions, even if provided with probabilistic wrong hints. Learning automata is especially suitable for modeling, learning, controlling, and solving complex real-world problems where the available information is incomplete; the environment is either noisy or highly uncertain. Thus, LA has made a significant impact in many areas of computer science and engineering problems. In the last decade, a wide range of learning automata theories, models, paradigms, simulations, and applications have been published by researchers.

This chapter's primary focus is to overview the varieties of learning automata (LA) models. This chapter is organized into the following subsections. We give a brief introduction to learning automata. Then we give a brief description of optimization problems. Next, we provide a comparative description regarding reinforcement learning and optimization algorithms. Finally, we will give a summarization of the recent successful applications of learning automata.

1.2 Learning Automata

Learning automaton (LA) as one of the computational intelligence techniques has been found a very useful tool to solve many complex and real-world problems where a large amount of uncertainty or lacking the information about the environment exists (Thathachar and Sastry 2002; Rezvanian et al. 2018a, 2019a). A learning automaton is a stochastic model operating in a reinforcement learning framework (Narendra and Thathachar 1974; Thathachar and Sastry 2004; Rezvanian et al. 2019b). This model can be classified under the reinforcement learning schemes in the temporal-difference (TD) learning methods category. *TD* learning is a combination of Monte Carlo ideas and dynamic programming ideas. Like Monte Carlo methods, *TD* methods can learn directly from raw experience without a model of the environment's dynamics. Like the dynamic programming, *TD* methods update estimates based on the other learned estimates without waiting for an outcome (Sutton and Barto 1998a). Sarsa (Rummery and Niranjan 1994), Q-learning (Watkins 1989), Actor-Critic methods (Barto et al. 1983), and R-learning (Schwartz 1993) are other samples of *TD* methods. Learning automata differs from other TD methods in representing the internal states and updating the internal states' updating method.

The automata approach to learning can be considered to determine an optimal action from a set of actions. A learning automaton can be regarded as an abstract object that has a finite number of actions. It selects an action from its finite set of actions and applies it to an environment. The environment evaluates the applied action and sends a reinforcement signal to the learning automaton, as shown in Fig. 1.1. The reinforcement signal provided by the environment is used to update the internal state of the learning automaton. By continuing this process, the learning automaton gradually learns to select the optimal action, which leads to favorable responses from the environment.

Formally, a learning automaton is a quintuple $\langle \alpha, \Phi, \beta, F, G\rangle$, where $\alpha = \{\alpha_1, \alpha_2, \ldots, \alpha_r\}$ is the set of actions that it must choose from, $\Phi = (\Phi_1, \Phi_2, \ldots, \Phi_s)$ is the set of states, $\beta = \{\beta_1, \beta_2, \ldots, \beta_q\}$ is the set of inputs, G: $\Phi \to \alpha$ is the output map and determines the action taken by the learning automaton if it is in the state Φ_j, and F: $\Phi \times \beta \to \Phi$ is the transition map that defines the transition of the state of the learning automaton upon receiving an input from the environment.

The selected action at the time instant k, denoted by $\alpha(k)$, serves as the input to the environment, which in turn emits a stochastic response, $\beta(k)$, at the time instant k, which is considered as the response of the environment to the learning automaton. Based upon the nature of β, environments could be classified into three classes: P-, Q-, and S-models. The output of a P-model environment has two elements of success or failure. Usually, in P-model environments, a failure (or unfavorable response) is denoted by one, while a success (or a favorable response) is denoted by 0. In Q-model environments, β can take a finite number of values in the interval [0, 1] while in S-model environments β lies in the interval [0, 1]. Based on the response $\beta(k)$, the state of the learning automaton $\Phi(k)$ is updated, and a new action is chosen at the time instant $(k + 1)$. Learning automata can be classified into two main classes:

fixed structure learning automata and variable structure learning automata (VSLA) (Kumpati and Narendra 1989). A simple pseudo-code for the behavior of an r-action learning automaton within a stationary environment with $\beta \in \{0, 1\}$ is presented in Fig. 1.2.

Learning automata have several excellent features, which make them suitable for use in many applications. The main features of learning automata are stated below.

1. Learning automata can be used without any prior information about the underlying application.
2. Learning automata are very useful for applications with a large amount of uncertainty.

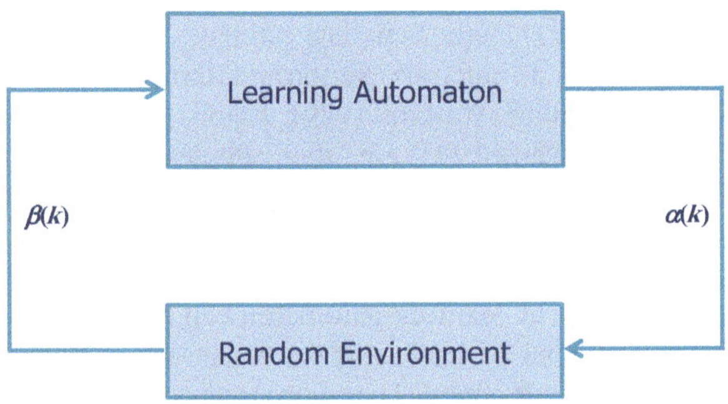

Fig. 1.1 The interaction of a learning automaton and its environment

```
Algorithm 1-1. Learning automaton
Input: Action-set α
Output: Action probability vector p
Assumptions
    Initialize r-dimensional action-set: α ={α1, α2, ..., αr} with r actions
    Initialize r-dimensional action probability vector p(k) = {p1, p2, ..., pr} = {1/r, 1/r, ..., 1/r} at instant k
    Let i denotes the selected action by the automaton
    Let j denotes the current checking action
Begin
  While (automaton does not converge to any action)
      The learning automaton selects an action based on the action probability vector p.
      The environment evaluates the selected action and gives the reinforcement signal β∈{0, 1} to the learning
         automaton.
      //The learning automaton updates its probability vector p(k) using the provided reinforcement signal β
      For each action j∈{1, ..., r} Do
         If (β=0) // favorable response
             The selected action by learning automaton is rewarded according to equation (1-1)
         Else If (β=1) // unfavorable response
             The selected action by learning automaton is penalized according to equation (1-2)
         End If
      End For
   End While
End Algorithm
```

Fig. 1.2 Pseudo-code of a learning automaton (LA)

3. Unlike traditional hill-climbing algorithms, hill-climbing in learning automata is done in the usual sense in a probability space.
4. Learning automata require very little and straightforward feedback from their environment.
5. Learning automata are very useful in multi-agent and distributed systems with limited intercommunication and incomplete information.
6. Learning automata are elementary in structure and can be implemented easily in software or hardware.
7. The action set of learning automata can be a set of symbolic or numeric values.
8. Optimization algorithms based on learning automata do not need the objective function to be an analytical function of adjustable parameters.
9. Learning automata can be analyzed by powerful mathematical methodologies. It has been shown that learning automata are optimal in single, hierarchical, and distributed structures.
10. Learning automata requires a few mathematical operations at each iteration to be used in real-time applications.
11. Learning automata have the flexibility and analytical tractability needed for most applications.

1.2.1 Learning Automata Variants

Stochastic learning automata can be categorized into two main classes, namely, finite action-set learning automata (FALA) and continuous action-set learning automata (CALA) (Thathachar and Sastry 2004). The learning automaton is called FALA if it has a finite set of actions, and it is called CALA otherwise. For a FALA with r actions, the action probability distribution is an r-dimensional probability vector. FALA can be categorized into two main families: variable structure learning automata (VSLA), in which the transition and output functions vary in time, and otherwise, fixed structure learning automata (FSLA) (Kumpati and Narendra 1989). Also, learning automata algorithms can be divided into two groups: non-estimator and estimator learning algorithms.

In many applications, there is a need to learn a real-valued parameter. In this situation, the actions of the automaton can be possible values of that parameter. To use a FALA for such an optimization problem, we have to discretize the parameter's value space to obtain a finite number of actions. However, a fine discretization increases the number of actions, which decreases the automaton's convergence speed. A natural solution to this problem would be to employ an automaton with a continuous space of actions. Such a model of LA is called continuous action-set learning automata (CALA).

In the finite action-set learning automaton (FALA), learning algorithms can update action probability vectors in discrete or continuous steps. The former is called the discretized learning algorithm, while the latter is called a continuous learning algorithm. The learning algorithms can be divided into two groups: non-estimator and estimator learning algorithms, briefly described in the following subsections.

1.2.1.1 Fixed-Structure Learning Automata (FSLA)

The learning automaton is called fixed-structure; either the probability of the transition from one state to another state or the action probability of any state action is fixed. An FSLA is a quintuple $\langle \alpha, \beta, \Phi, \mathcal{F}, \mathcal{G} \rangle$, where $\alpha = \{\alpha_1, \alpha_2, \ldots \alpha_r\}$ is the set of the actions that automaton chooses from, $\beta = \{0, 1\}$ is automaton's set of inputs where it receives a penalty if $\beta = 1$ and receives a reward otherwise, $\Phi = \{\phi_1, \phi_2, \ldots \phi_{rN}\}$ indicates its set of states, where N is called the depth of memory of the automaton, $\mathcal{F} : \Phi \times \beta \rightarrow \Phi$ illustrates the transition of the state of the automaton on receiving an input from the environment. $\mathcal{F}$ can be stochastic, $\mathcal{G} : \Phi \rightarrow \alpha$ is the output function of the automaton. The action taken by the automaton is determined according to its current state. This means that the automaton selects action α_i if it is in any of the states $\left\{\phi_{(i-1)N+1}, \ \phi_{(i-1)N+2}, \ldots, \phi_{iN}\right\}$. The state $\phi_{(i-1)N+1}$ is considered to be the most internal state, and ϕ_{iN} is considered to be the boundary state of action α_i, indicating that the automaton has the most and the least certainty in performing the action α_i, respectively.

The action chosen by the automaton is applied to the environment, which in turn emits a reinforcement signal β. Based on the received signal β, the automaton's state is updated, and the new action is chosen according to the functions $\mathcal{F}$ and $\mathcal{G}$ respectively. There exist different types of FSLA based on the state transition function $\mathcal{F}$ and the output function $\mathcal{G}$. $L_{2N,2}$, $G_{2N,2}$, and Krinsky (Tsetlin 1962) are essential FSLA types.

1.2.1.1.1 The Tsetlin Automaton ($L_{2N,2}$)

The Tsetlin automaton (Tsetlin 1962) is an extension of the more simple $L_{2,2}$ automaton. This automaton has $2N$ states, that are denoted by $\phi_1, \phi_2, \ldots \phi_{2N}$, and two actions α_1 and α_2. These states keep track of the automaton's prior behavior and its received feedback. The automaton chooses an action based on the current state it resides in it. That is, action α_1 is chosen by the automaton if it is in the states $\phi_1, \phi_2, \ldots \phi_N$, while if the current state is $\phi_{N+1}, \phi_{N+2}, \ldots \phi_{2N}$ action α_2 is chosen. The automaton moves towards its most internal state whenever it receives a reward, and conversely, on receiving a penalty, it moves towards its boundary state or to the boundary state of the other action. For instance, consider that the automaton is in state ϕ_N. The automaton moves to state ϕ_{N-1} if it has been rewarded and it moves to state ϕ_{2N} in the case of punishment. However, if the current state is either ϕ_1 or ϕ_{N+1} the automaton remains in the mentioned states until it keeps on being rewarded. The state transition graph of $L_{2N,2}$ automaton is depicted in Fig. 1.3.

As mentioned before, the state transition function $\mathcal{F}$ can be considered a stochastic function (Tsetlin 1962). In such a case, on receiving a signal from the environment, the automaton's transition among its states is not deterministic. For instance, when an

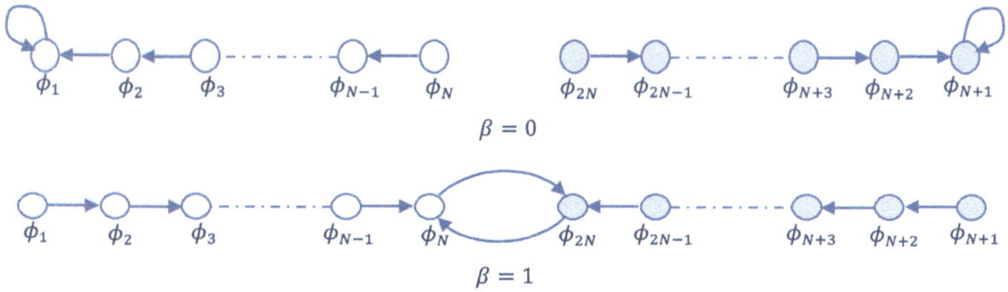

Fig. 1.3 The state transition graph for $L_{2N,2}$ automaton

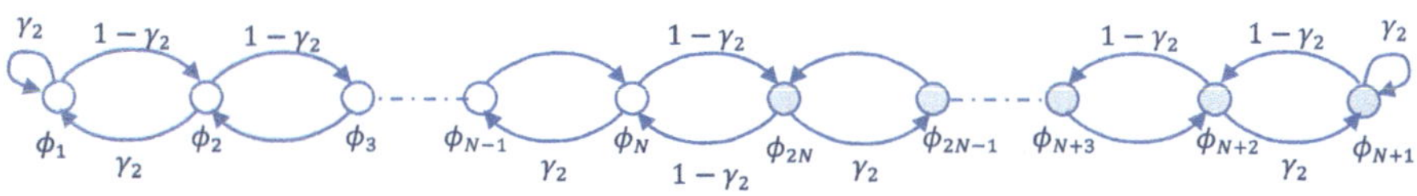

Fig. 1.4 The state transition graph for $L_{2N,2}$ in case of punishment

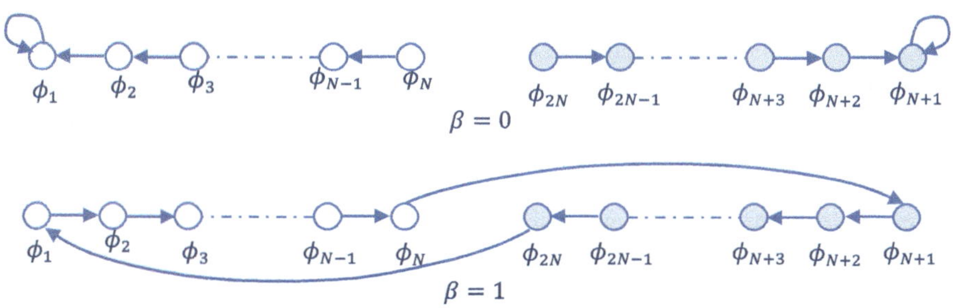

Fig. 1.5 The state transition graph for $G_{2N,2}$ automaton

action results in a reward, the automaton may move one state towards the boundary state with probability $\gamma_1 \in [0, 1)$, and move one state towards its most internal state with probability $1 - \gamma_1$, and reverse this procedure using probabilities γ_2 and $1 - \gamma_2$ when the response of the environment is a penalty. In some situations where rewarding a favorable action may be preferred more than penalizing an unfavorable action, one can set $\gamma_1 = 0$ and $\gamma_2 \in [0, 1)$. These settings result in state transitions of $L_{2N,2}$ automaton become deterministic when the automaton receives a reward ($\beta = 0$), as shown in Fig. 1.3. But, in the case of punishment, the $L_{2N,2}$ automaton will transit among its states stochastically, as shown in Fig. 1.4, by considering $\gamma_1 = 0$, the automaton will be expedient for all values of γ_2 in interval [0, 1) (Thathachar and Sastry 2002).

1.2.1.1.2 The TsetlinG Automaton ($G_{2N,2}$)

The TsetlinG automaton (Tsetlin 1962) is another type of the FSLA. This automaton behaves the same as the $L_{2N,2}$ automaton, except for the times when it is being

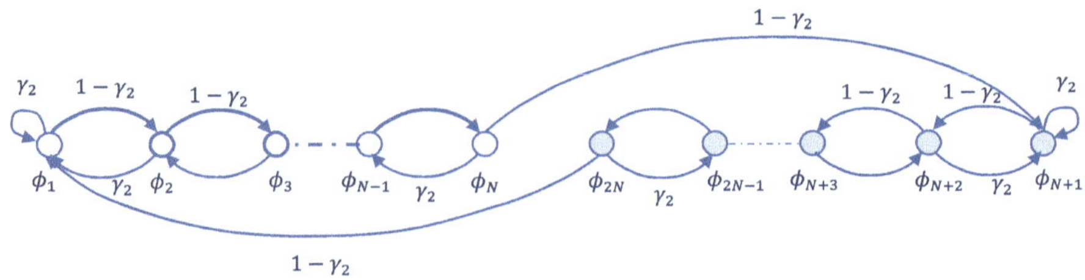

Fig. 1.6 The state transition graph for $G_{2N,2}$ in case of punishment

punished where the automaton moves from state ϕ_N to ϕ_{N+1} and from state ϕ_{2N} to state ϕ_1. The state transition graph of $G_{2N,2}$ automaton is illustrated in Fig. 1.5. When the state transition function of $G_{2N,2}$ automaton is stochastic and $\gamma_1 = 0$ and $\gamma_2 \in [0, 1)$, the automaton transits among its states on receiving a reward as shown in Fig. 1.5 and the case of punishment as shown in Fig. 1.6.

1.2.1.1.3 The Krinsky Automaton

This subsection briefly describes another type of fixed structure learning automata: the Krinsky automaton (Tsetlin 1962). When this automaton's chosen action results in a penalty, it acts exactly like the $L_{2N,2}$ automaton. However, in situations where the automaton is rewarded, any of the states $\phi_1, \phi_2, \ldots \phi_N$ pass to the state ϕ_1 and any of the states $\phi_{N+1}, \phi_{N+2}, \ldots \phi_{2N}$ transit to the state ϕ_{N+1}. Therefore, N successive penalties are needed for the automaton to switch from its current action to the other. One can note the state transition graph of the introduced automaton in Fig. 1.7. In the case of a stochastic state transition function and for settings $\gamma_1 = 0$ and $\gamma_2 \in [0, 1)$, the automaton behaves the same as before deterministically when it is rewarded (see Fig. 1.7), and on receiving a penalty, the behavior of the Krinsky automaton is identical to the $L_{2N,2}$ automaton as shown in Fig. 1.7.

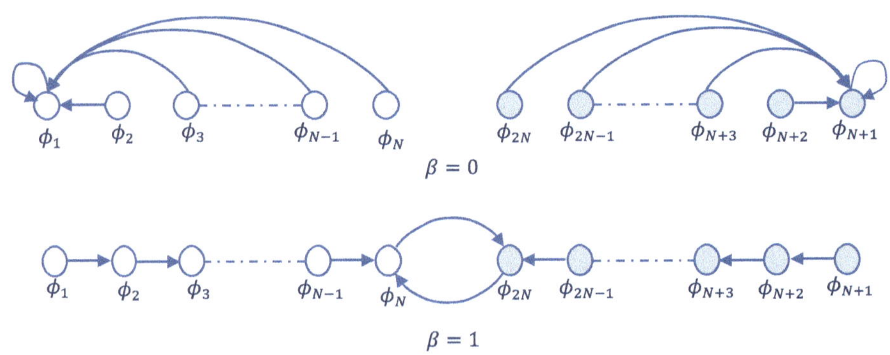

Fig. 1.7 The state transition graph for Krinsky automaton

1.2.1.1.4 Object Migration Automata (OMA)

Object migration automata were first introduced by Oommen and Ma (Oommen and Ma 1988). OMAs are a type of fixed structure learning automata and are defined by a quintuple <$\underline{\alpha}$, $\underline{\Phi}$, $\underline{\beta}$, F, G>.$\underline{\alpha} = \{\alpha_1, \ldots \alpha_r\}$. is the set of allowed actions for the automaton. For each action α_k, there is a set of states $\{\varphi_{(k-1)N+1}, \ldots, \varphi_{kN}\}$, where N is the depth of memory. The states $\varphi_{(k-1)N+1}$ and φ_{kN} are the most internal state and the boundary state of action α_k, respectively. The set of all states is represented by $\underline{\Phi} = \{\varphi_1, \ldots, \varphi_s\}$, where $s = N * r$. $\underline{\beta} = \{0, 1\}$ is the set of inputs, where 1 represents an unfavorable response, and 0 represents a favorable response. $F : \underline{\Phi} * \underline{\beta} \to \underline{\Phi}$ is a function that maps the current state and current input into the next state, and $G : \underline{\Phi} \to \underline{\alpha}$ is a function that maps the current state into the current output. In other words, G determines the action taken by the automaton. W objects are assigned to actions in an OMA and moved around the automaton states instead of general learning automata, in which the automaton can move from one action to another by the environmental response. The state of objects is changed based on the feedback response from the environment. If the object w_i is assigned to action α_k (i.e., w_i is in state ξ_i, where $\xi_i \in \{\varphi_{(k-1)N+1}, \ldots, \varphi_{KN}\}$), and the feedback response from the environment is 0, α_k is rewarded, and w_i is moved toward the most internal state ($\varphi_{(k-1)N+1}$) of that action. If the feedback from the environment is 1, then α_k is penalized, and w_i is moved toward the boundary state (φ_{kN}) of action α_k. The variable γ_k denotes the reverse of the state number of the object assigned to action α_k (i.e., degree of association between action α_k and its assigned object). By rewarding an action, the degree of association between that action and its assigned object will be increased. Conversely, penalizing an action causes the degree of association between that action and its assigned object to be decreased. An object associated with state $\varphi_{(k-1)N+1}$ has the most degree of association with action α_k, and an object associated with state φ_{kN} has the least degree of association with action α_k.

1.2.1.2 Variable-Structure Learning Automata (VSLA)

VSLA can be represented by a quadruple $\langle \alpha, \beta, p, T \rangle$, where $\alpha = \{\alpha_1, \alpha_2, \ldots, \alpha_r\}$ indicates the action set from which the automaton chooses, $\beta = \{\beta_1, \beta_2, \ldots, \beta_k\}$ indicates the set of inputs to the automaton, $p = \{p_1, p_2, \ldots, p_r\}$ indicates the action probability vector, such that p_i is the probability of selecting the action α_i, T indicates the learning algorithm that is used to update the action probability vector of the automaton in terms of the environment's response, i.e., $p(t+1) = T[\alpha(t), \beta(t), p(t)]$, where the inputs are the chosen action $\alpha(t)$, the response of the environment $\beta(t)$ and the action probability vector $p(t)$ at time.

Let $\alpha_i(t)$ be the action selected by the automaton at time t. The action probability vector $p(t)$ is updated as given in Eq. (1.1), if the environment's response is the reward, and $p(t)$ is updated according to Eq. (1.2), if the response of the environment is the penalty.

$$p_j(t+1) = \begin{cases} p_j(t) + a\left[1 - p_j(t)\right] & j = i \\ (1-a)p_j(t) & \forall j \neq i \end{cases} \tag{1.1}$$

$$p_j(t+1) = \begin{cases} (1-b)p_j(t) & j = i \\ \left(\frac{b}{r-1}\right) + (1-b)p_j(t) & \forall j \neq i \end{cases} \tag{1.2}$$

where r denotes the number of actions taken by the automaton, and a and b are the reward and penalty parameters that determine the amount of increases and decreases of the action probabilities, respectively. If $a = b$, the learning algorithm is a linear reward–penalty (L_{R-P}) algorithm, if $a \gg b$, it is a linear reward–$\in$ penalty ($L_{R-\in P}$) algorithm, and finally if $b = 0$, it is a linear reward–Inaction (L_{R-I}) algorithm in which the action probability vector remains unchanged when the environment penalizes the taken action. The reward and penalty parameters a and b influence convergence speed and how closely the automaton approaches optimal behavior (the convergence accuracy) (Thathachar and Sastry 2002). If a is too small, the learning process is too slow. In contrary, if a is too large, the increments in the action probabilities are too high and the automaton's accuracy in perceiving the optimal behavior becomes low. By choosing the parameter a to be sufficiently small, the probability of convergence to the optimal behavior may be made as close to 1 as desired (Thathachar and Sastry 2002). In the $L_{R-\in P}$ learning algorithm, the penalty parameter b is considered to be small in comparison with the reward parameter a ($b = \in a$, where $0 < \in \ll 1$). In this algorithm, the action probability distribution $p(t)$ of the automaton converges in distribution to a random variable p^* which can be made as close to the optimal vector as desired by choosing $\in$ sufficiently small (Thathachar and Ramachandran 1984).

To investigate a learning automaton's learning ability, a pure-chance automaton that always selects its available actions with equal probabilities is used as the standard for comparison (Thathachar and Sastry 2002). Any automaton that is said to learn must perform at least better than such a pure-chance automaton. To make this comparison, one measure can be the average penalty for a given action probability vector. For a stationary random environment with the penalty probability vector $c = \{c_1, c_2, \ldots, c_r\}$, the average penalty probability $M(t)$ received by an automaton is equal to

$$M(t) = E[\beta(t)|p(t)] = \sum\nolimits_{\alpha_j \in \alpha} c_i p_i(t) \tag{1.3}$$

For a pure-chance automaton, $M(t)$ is a constant and is defined as

$$M(0) = \frac{1}{r} \sum\nolimits_{\alpha_i \in \alpha} c_i \tag{1.4}$$

An automaton that does better than pure chance must have the average penalty $M(t)$ less than $M(0)$ at least asymptotically as $t \to \infty$. Since $p(t)$ and consequently $M(t)$ are random variables in general, the expected value $E[M(t)]$ is compared with $M(0)$.

Definition 1.1. A learning automaton interacting with a P-, Q-, or S-model environment is said to be *expedient* if

$$\lim_{t \to \infty} E[M(t)] < M(0) \tag{1.5}$$

Expediency means that the average penalty probability decreases when the automaton updates its action probability function. It would be more interesting in determining an updating procedure that would result in $E[M(t)]$ attaining its minimum value. In such a case, the automaton is called *optimal.*

Definition 1.2. A learning automaton interacting with a P-, Q-, or S-model environment is said to be *optimal*

$$\lim_{t \to \infty} E[M(t)] = c_{\ell} \tag{1.6}$$

where $c_{\ell} = \min_{i} c_i$. Optimality implies that asymptotically the action α_{ℓ} with the minimum penalty probability c_{ℓ} is chosen with probability one. While optimality is a very desirable property in stationary environments, it may not be possible to achieve it in a given situation. In such a case, one might aim at a suboptimal performance, which is represented by *ε-optimality.*

Definition 1.3. A learning automaton is interacting with a P-, Q-, or S-model environment is said to be *ε-optimal* if the following equation can be obtained for any $\varepsilon > 0$ by a proper choice of the parameters of the learning automaton.

$$\lim_{t \to \infty} E[M(t)] < c_{\ell} + \varepsilon \tag{1.7}$$

ε-optimality implies that the performance of the automaton can be made as close to the optimal as desired.

1.2.1.3 Variable Action Set Learning Automata (VASLA)

A variable action set learning automaton (also known as a learning automaton with a variable number of actions) is defined as an automaton in which the number of available actions at each instant varies over time (Thathachar and Harita 1987). Such a learning automaton has a finite set of r actions, $\alpha = \{\alpha_1, \alpha_2, \ldots, \alpha_r\}$. At each instant t, the action subset $\hat{\alpha}(t) \subseteq \alpha$ is available for the learning automaton to choose from. Selecting the elements of $\hat{\alpha}(t)$ is made randomly by an external agency. The procedure of choosing an action and updating the action probability vector in this automaton is done as follows.

Let $K(t) = \sum_{\alpha_i \in \hat{\alpha}(t)} p_i(t)$ presents the sum of the probabilities of the available actions in subset $\hat{\alpha}(t)$. Before choosing an action, the available actions probability vector is scaled, as given below.

$$\hat{p}_i(t) = \frac{p_i(t)}{k(t)} \quad \forall \alpha_i \in \hat{\alpha}(t) \tag{1.8}$$

Then, the automaton randomly chooses one of its available actions according to the scaled action probability vector $\hat{p}(t)$. Depending on the reinforcement signal received from the environment, the automaton updates the vector $\hat{p}(t)$. Finally, the available actions probability vector $\hat{p}(t)$ is rescaled according to Eq. (1.9). ε-optimality of this type of LA have been proved in Thathachar and Harita (1987).

$$p_i(t+1) = \hat{p}_i(t+1).K(t) \quad \forall_{\alpha_i} \in \hat{\alpha}_i(t) \tag{1.9}$$

The pseudo-code of the behavior of a variable action set learning automaton is shown in Fig. 1.8.

Algorithm 1-2. Variable action-set learning automata
Input: Action-set α
Output: Action probability vector p
Assumptions
  Initialize r-dimensional action-set: $\alpha=\{\alpha_1, \alpha_2, \ldots, \alpha_r\}$ with r actions
  Initialize r-dimensional action probability vector $p(k) = \{p_1, p_2, \ldots, p_r\} = \left\{\frac{1}{r}, \frac{1}{r}, \ldots, \frac{1}{r}\right\}$ at instant k
  Let i denotes the selected action by the automaton
  Let j denotes the current checking action
Begin
  While (automaton does not converge to any action)
    Calculate available actions of a learning automaton
    Calculate the sum of the probability of available actions
    For each action $j\in\{1, \ldots, r\}$ **Do**
      If (α_j is available action)
        Scale action probability vector $\hat{p}_i(k)$ according to equation (1-8)
      End if
    End for
    Generate map function between available actions and all actions of a learning automaton
    The learning automaton selects an action according to the probability vector of available actions $\hat{p}(k)$
    The environment evaluates the selected action and gives the reinforcement signal $\beta\in\{0, 1\}$ to the learning automaton.
    For each available action $j\in\{1, \ldots, m\}$ **Do**
      If (β=0) // favorable response
        The selected action by learning automaton is rewarded according to equation (1-1)
      Else If (β=1) // unfavorable response
        The selected action by learning automaton is rewarded according to equation (1-2)
      End if
    End for
    For each action $j\in[1, \ldots, r]$ do
      If (α_j is available action)
        Rescale the probability vector of selected available action by equation (1-9)
      End
    End for
  End While
End Algorithm

Fig. 1.8 Pseudo-code of the behavior of a variable action-set learning automaton

1.2.1.4 Continuous Action-Set Learning Automata (CALA)

In a continuous action-set learning automaton (CALA), the action-set is defined as a continuous interval over the real numbers. This means that each automaton chooses its actions from the real line. In such a learning automaton, the action probability of the possible actions is defined as a probability distribution function. All actions are initially selected with the same probability. That is, the probability distribution function under which the actions are initially selected has a uniform distribution. The probability distribution function is updated depending upon the responses received from the environment.

A continuous action-set learning automaton (CALA) (Thathachar and Sastry 2004) is an automaton whose action-set is the real line, and its action probability distribution is considered to be a normal distribution with mean $\mu(t)$ and standard deviation $\sigma(t)$. At each time instant t, the CALA chooses a real number α at random based on the current action probability distribution $N(\mu(t), \sigma(t))$. The two actions $\alpha(t)$ and $\mu(t)$ are served as the inputs to the random environment. The CALA receives the reinforcement signals $\beta_{\alpha(t)}$ and $\beta_{\mu(t)}$ from the environment for both actions. At last, $\mu(t)$ and $\sigma(t)$ are updated as

$$\mu(t+1) = \mu(t) + \lambda \frac{\left(\beta_{\alpha(t)} - \beta_{\mu(t)}\right)}{\phi(\sigma(t))} \frac{(\alpha(t) - \mu(t))}{\phi(\sigma(t))} \tag{1.10}$$

$$\sigma(t+1) = \sigma(t) + \lambda \frac{\left(\beta_{\alpha(t)} - \beta_{\mu(t)}\right)}{\phi(\sigma(t))} \left[\left(\frac{(\alpha(t) - \mu(t))}{\phi(\sigma(t))} \right)^2 - 1 \right] - \lambda K (\sigma(t) - \sigma_L) \tag{1.11}$$

where,

$$\phi(\sigma(t)) = \begin{cases} \sigma_L & for\ \sigma(t) \le \sigma_L \\ \sigma(t) & for\ \sigma(t) > \sigma_L \end{cases} \tag{1.12}$$

and $0 < \lambda < 1$ denotes the learning parameter, $K > 0$ is a large positive constant controlling the shrinking of $\sigma(t)$ and σ_L is a sufficiently small lower bound on $\sigma(t)$. Since the updating given for $\sigma(t)$ does not automatically ensure that $\sigma(t) \ge \sigma_L$, the function ϕ provides a projected version of $\sigma(t)$, denoted by $\phi(\sigma(t))$. The interaction with the environment continue until $\mu(t)$ does not change noticeably and $\sigma(t)$ converges close to σ_L.

The objective of CALA is to learn the value of α for which $E\left[\beta_{\alpha(t)}\right]$ attains a minimum. That is, the objective is to make $N(\mu(t), \sigma(t))$ converge to $N(\alpha^*, 0)$, where α^* is a minimum of $E\left[\beta_{\alpha(t)}\right]$. However, we cannot let $\sigma(t)$ converge to zero since we want the asymptotic behavior of the algorithm to be analytically tractable. Hence, the lower bound $\sigma_L > 0$ is used, and the objective of learning is kept as $\sigma(t)$ converging to σ_L and $\mu(t)$ converging to α^*. By choosing σ_L and λ sufficiently small and K sufficiently large, $\mu(t)$ of the CALA algorithm will be close to a minimum of $E\left[\beta_{\alpha(t)}\right]$ with probability close to unity after a long enough time (Thathachar and Sastry 2004).

1.2.1.5 Non-estimator Learning Algorithms

In non-estimator learning algorithms, the reinforcement signal's current value is the only parameter used to update the action probability vector. Let us assume a finite action-set learning automaton (FALA) r actions operating in a stationary P-model environment. α_i (for $i = 1, 2, \ldots, r$) denotes the action taken by this automaton at instant n (i.e., $\alpha(n) = \alpha_i$), and $\underline{\beta} = \{0, 1\}$ is the response of the environment to this action. The following is a general learning algorithm for updating the action probability vector presented in Aso and Kimura (1979).

$$p_j n + 1 = \begin{cases} p_j(n) - g_{ij}\left[\underline{p}(n)\right] \text{ if } \beta(n) = 0 \\ p_j(n) - h_{ij}\left[\underline{p}(n)\right] \text{ if } \beta(n) = 1 \end{cases} \tag{1.13}$$

for all $j \neq i$. For preserving probability measure, we must have $\sum_{j=1}^{r} p_j(n) = 1$, so we obtain

$$p_i(n+1) = \begin{cases} p_i(n) - g_{ii}\left[\underline{p}_n\right] \; if \; \beta(n) = 0 \\ p_i(n) - h_{ii}\left[\underline{p}_n\right] \; if \; \beta(n) = 1 \end{cases} \tag{1.14}$$

where $g_{ii}\left(\underline{p}\right) = -\sum_{\substack{j=1 \\ j\neq i}}^{r} g_{ij}\left[\underline{p}(n)\right]$ and $h_{ii}\left(\underline{p}\right) = -\sum_{\substack{j=1 \\ j\neq i}}^{r} h_{ij}\left[\underline{p}(n)\right]$. Functions g_{ij} and h_{ii} (for $i, j = 1, 2, \cdots, r$) are called as reward and penalty functions, respectively.

Non-estimator learning algorithms are divided as linear (Linear Reward-Penalty ($L_{R\text{-}P}$), Linear Reward-Penalty ($L_{R\text{-}P}$), Linear Reward Inaction ($L_{R\text{-}I}$), Linear Inaction-Penalty ($L_{I\text{-}P}$) Linear Reward-Epsilon Penalty ($L_{R\text{-}\varepsilon P}$), Linear Reward-Reward ($L_{R\text{-}R}$)), nonlinear (Meybodi and Lakshmivarahan 1982) and hybrid learning algorithms (Friedman and Shenker 1996).

1.2.1.6 Estimator Learning Algorithms

One of the main difficulties in using LA in many practical applications is the slow rate of convergence. Although several attempts are conducted to increase the rate of convergence, these are not enough. An improvement in the convergence rate could be to determine the environment's characteristics as the learning proceeds. This additional information is used when the action probabilities are updated. These algorithms are called *estimator learning algorithms* (Thathachar and Sastry 1985a).

Non-estimator learning algorithms update their action probability vectors based on the current response from the environment. In contrast, estimator learning algorithms maintain a running estimate of reward strengths for each action. The action probability vector will then be updated based on the environment's current response and the running estimate of reward strengths.

The state of an automaton operating under a P-model environment which is equipped with an estimator learning algorithm is defined as $\langle p(n), \hat{d}(n)\rangle$ at instance n, where $p(n)$ denotes the action probability vector, $\hat{d}(n) = \left[\hat{d}_1(n), \hat{d}_2(n), \ldots, \hat{d}_r(n)\right]^T$ represents the set of reward estimations and $\hat{d}_i(n)$ denotes the estimation of the reward probability d_j at instant n. The estimation of reward probability d_j is defined as the ratio of the number of times that action α_i is rewarded to the number of times α_i is selected (Thathachar and Sastry 1985a).

An estimator learning algorithm operates as follows. The learning automaton initially chooses an action (e.g., α_i). The selected action is applied to the environment. The random environment generates response $\beta(n)$ after it evaluates the chosen action. Learning automaton updates the action probability vector by using $\hat{d}(n)$ and $\beta(n)$. $\hat{d}(n)$ is finally updated based on the current response by the following rules.

$$R_i(n+1) = R_i(n) + [1 - \beta(n)]$$

$$R_j(n+1) = R_j(n) \qquad j \neq i \tag{1.15}$$

$$Z_i(n+1) = Z_i(n) + 1$$

$$Z_j(n+1) = Z_j(n) \qquad j \neq i \tag{1.16}$$

$$\hat{d}(n) = \frac{R_k(n)}{Z_k(n)} \qquad k = 1, 2, \ldots, r, \tag{1.17}$$

where $R_j(n)$ denotes the number of times action α_i is rewarded and $Z_i(n)$ denotes the number of times action α_i is selected. Several well-known estimator learning algorithms are presented in the literature, including TS stochastic estimator (TSE), Discretized TS estimator algorithm (DTSE) (Lanctot and Oommen 1992), relative strength learning automata (Simha and Kurose 1989), Stochastic estimator learning automata (SELA) (Papadimitriou et al. 1991), S-model ergodic discretized estimator learning algorithm (SEDEL) (Vasilakos and Paximadis 1994), and absorbing stochastic estimator learning algorithm (ASELA) (Papadimitriou et al. 2002).

1.2.1.7 Pursuit Algorithms

From the name of this class of finite action-set learning automata, it can be concluded that in these algorithms, the action probability vector chases after the action, which is most recently estimated as the optimal action. A pursuit learning algorithm is a simplified version of the estimator algorithms inheriting their main characteristics. In pursuit learning algorithms, the choice probability of the action with the maximum rewarding estimation is increased. By this updating method, the learning algorithm always pursues the optimal action. In the following subsections, several well-known pursuit learning algorithms are suggested by researchers, including Pursuit Reward-Penalty ($P_{R\text{-}P}$) (Thathachar and Sastry 1986) and Pursuit Reward-Inaction ($P_{R\text{-}I}$) (John Oommen and Agache 2001).

1.2.1.8 Generalized Pursuit Algorithm

The main disadvantage of the above-mentioned pursuit algorithms is that the choice probability of the action with the highest rewarding estimation has to be increased at each iteration of the algorithm. This results in the movement of the action probability vector toward the action with the maximum rewarding estimation. This means that the learning automaton may converge to a wrong (non-optimal) action when the action with the highest rewarding estimation is not the action with the minimum penalty probability. To avoid such a wrong convergence problem, the generalized pursuit algorithm (GP) was introduced in Agache and Oommen (2002). In this algorithm, a set of possible actions with higher estimations than the currently selected action can be pursued at each instant. In Agache and Oommen (2002), it has been shown that this algorithm is ε-optimal in all stationary environments. Let $K(n)$ denotes the number of actions that have higher estimations than the action selected at instant n. Equation (1.18) shows the updating rule of the generalized pursuit algorithm.

$$
\begin{array}{ll}
p_j(n+1) = p_j(n)(1-a) + \frac{a}{K(n)} & \forall j(j \neq m) \text{ such that } \hat{d}_j > \hat{d}_i \\
p_j(n+1) = p_j(n)(1-a) & \forall j(j \neq m) \text{ such that } \hat{d}_j \leq \hat{d}_i \\
p_m(n+1) = 1 - \sum_{j \neq m} p_j(n+1) &
\end{array}
\tag{1.18}
$$

where α_m denotes the action with the highest rewarding estimation.

A discretized generalized pursuit algorithm (DGP) (Agache and Oommen 2002) is a particular case of the generalized pursuit algorithm in which the action probability vector is updated in discrete steps. This algorithm is called pseudo-discretized since the step size of the algorithm varies in different steps. In this algorithm, the choice probability of all the actions with higher rewarding estimations (than the selected action) increases with amount $\frac{\Delta}{K(n)}$, and that of the other actions decreases with amount $\frac{\Delta}{r-K(n)}$. The ε-*optimality* of the discretized generalized pursuit algorithm in all stationary environments has been shown in Agache and Oommen (2002) DGP updates the action probability vector by the following updating rule.

$$
\begin{aligned}
&p_j(n+1) = \min\left\{p_j(n) + \tfrac{\Delta}{K(n)}, 1\right\} \quad \forall j (j \neq m) \text{ such that } \hat{d}_j > \hat{d}_i \\
&p_j(n+1) = \max\left\{p_j(n) - \tfrac{\Delta}{r-K(n)}, 0\right\} \forall j (j \neq m) \text{ such that } \hat{d}_j \leq \hat{d}_i \\
&p_m(n+1) = 1 - \textstyle\sum_{j \neq m} p_j(n+1)
\end{aligned}
\tag{1.19}
$$

The pursuit algorithms are ranked in decreasing order of performance as *DGP*, DP_{R-I}, *GP*, DP_{R-P}, P_{R-I}, and P_{R-P}.

1.2.1.9 Interconnected Learning Automata

It seems that the full potential of learning automaton is realized when multiple automata interact with each other. It is shown that a set of interconnected learning automata can describe the behavior of an ant colony capable of finding the shortest path from their nest to food sources and back (Verbeeck et al. 2002). In this section, we study the interconnected learning automata. The interconnected learning automata techniques based on activation type of learning automata for taking action can be classified into three classes: synchronous, sequential, and asynchronous.

1.2.1.10 Hierarchical Structure Learning Automata

When the number of actions for a learning automaton becomes large (more than ten actions), the action probability vector's time to converges also increases. Under such circumstances, a hierarchical structure of learning automata (HSLA) can be used. A hierarchical system of automata is a tree structure with a depth of M where each node corresponds to an automaton, and the arcs emanating from that node corresponds to that automaton's actions. In HSLA, an automaton with r actions is in the first level (the root of the tree), and the kth level has r^{k-1} automata, each with r actions. The root node corresponds to an automaton, referred to as the first-level or top-level automaton. The selection of each action of this automaton leads to activate an automaton at the second level. In this way, the structure can be extended to an arbitrary number of levels.

1.2.1.11 Network of Learning Automata (NLA)

A network of learning automata (Willianms 1988) is a collection of LAs connected as a hierarchical feed-forward layered structure. In this structure, the outgoing link of the LAs in the other layers is the input of the LAs of the succeeding layers. In this model, the LAs (and consequently the layers) are classified into three separate groups. The first group includes the LAs located at the first level of the network, called the input LAs (input layer). The second group comprises the LAs located at the last level of the network, called the output LAs (output layer). The third group includes the LAs located between the first and the last layers, called the hidden LAs (hidden layer). In a network of LAs, the input LAs receive the context vectors as external inputs from the environment, and the output LAs apply the output of the network to the environment. The difference between the feed-forward neural networks and NLA is that units of neural networks are deterministic.

In contrast, NLA units are stochastic, and the learning algorithms used in the two networks are different. Since units are stochastic, the output of a particular unit i is drawn from a distribution depending on its input weight vector and the units' output in the other layers. This model operates as follows. The context vector is applied to the input LAs. Each input LA selects one of its possible actions based on its action probability vector and the input signals it receives from the environment. The chosen action activates the LAs of the next level, which are connected to this LA. Each activated LA selects one of its actions, as stated before. The actions selected by the output LAs are applied to the random environment. The environment evaluates the output action in comparison with the desired output and generates the reinforcement signal. All LAs then use this reinforcement signal for updating their states.

1.2.1.12 Distributed Learning Automata (DLA)

The hierarchical structure learning automata has a tree structure, in which there exists a unique path between the root of the tree and each of its leaves. However, in some applications, such as routing in computer networks, there may be multiple paths between the source and destination nodes. This system is a generalization of HSLA, referred to as distributed learning automata (DLA). A Distributed learning automata (DLA) (Beigy and Meybodi 2006a) is a network of interconnected learning automata that collectively cooperate to solve a particular problem. The number of actions for a particular LA in DLA is equal to the number of LA's that are connected to this LA. The selection of an action by a LA in DLA activates another LA, which corresponds to this action. Formally, a DLA can be defined by a quadruple $\langle A, E, T, A_0\rangle$, where $A = \{A_1, A_2, \ldots, A_n\}$ is the set of learning automata, $E \subset A \times A$ is the set of edges where edge (v_i, v_j) corresponds to action α_j^i of automaton A_i, T is the set of learning algorithms with which learning automata update their action probability vectors, and A_I is the root automaton of DLA at which activation of DLA starts. The DLA was used for solving the several stochastic graph problems (Akbari Torkestani and Meybodi 2010, 2012; Mollakhalili Meybodi and Meybodi 2014a, b; Mostafaei 2015; Rezvanian and Meybodi 2015a, b), cognitive radio networks (Moradabadi and Meybodi 2017a; Fahimi and Ghasemi 2017) and social network analytics problems (Khomami et al. 2016b; Ghavipour and Meybodi 2018b).

1.2.1.13 Extended Distributed Learning Automata (*e*DLA)

An extended distributed learning automata (*e*DLA) (Mollakhalili Meybodi and Meybodi 2014b) is a new extension of DLA supervised by a set of rules governing the operation of the LAs. *Mollakhalili-Meybodi* et al. presented a framework based on *e*DLA for solving stochastic graph optimization problems such as stochastic shortest path problems and stochastic minimum spanning tree problems. The eDLA was also applied for solving several social network analytics problems (Ghamgosar et al. 2017).

1.2.1.14 Cellular Learning Automata (CLA)

Cellular learning automaton (CLA) (Beigy and Meybodi 2004; Rezvanian et al. 2018f), as depicted in Fig. 1.9, is a combination of cellular automata (Wolfram 1986) and learning automata (Kumpati and Narendra 1989; Rezvanian et al. 2018a). In this model, each cell of a cellular automaton (CA) contains one or more learning automata (LA). The LA or LAs residing in a particular cell define the state of the cell. Like CA, a CLA rule controls the behavior of each cell. At each time step, the local rule defines the reinforcement signal for a particular LA based on its selected action and the actions chosen by its neighboring LAs.

Consequently, the neighborhood of each LA is considered to be its local environment. A formal definition of CLA is provided by Beigy and Meybodi (Beigy and Meybodi 2004). The authors also investigated the asymptotic behavior of the model and provided some theoretical analysis on its convergence.

A CLA structure can be represented by a graph, where each vertex denotes a CLA cell. Each edge of this graph defines a neighborhood relation between its two incident nodes. We represent the action set of the ith LA by $A_i = \{a_{i1}, \ldots, a_{im_i}\}$, where m_i is the number of actions of the LA. The probability vector of the ith LA, which determines its actions' selection probability, is denoted by p_i. This probability vector defines the LA's internal state, and the probability vectors of all learning automata in the CLA define the configuration of the CLA at each step k. The configuration of a CLA at step k is denoted by $P(k) = (p_1(\text{k}), p_2(\text{k}), \ldots, p_n(k))^{\text{T}}$, where n is the number of cells. A set of local rules governs the operation of a CLA. At each step k, the local rule of the ith cell determines the reinforcement signal to the learning automaton in the cell as follows:

$$\begin{aligned} &F_i : \varphi_i \rightarrow \underline{\beta} \\ &with \\ &\varphi_i = \prod_{j=0}^{n_i} A_{N_i(j)} \end{aligned} \tag{1.20}$$

where N_i is the neighborhood function of the ith cell, and $N_i(j)$ returns the index of the jth neighboring cell to cell i with $N_i(0)$ defined as $N_i(0) = i$. n_i denotes the number of cells in the neighborhood of the ith cell. Finally, $\underline{\beta}$ is the set of values that the reinforcement signal can take. Each LA's expected reward at each step can be obtained based on its associated local rule and the CLA configuration. The expected reward of action like $a_r \in A_i$ in configuration P is denoted by $d_{ir}(P)$. A configuration P is called compatible if the following condition holds for any configurations Q and

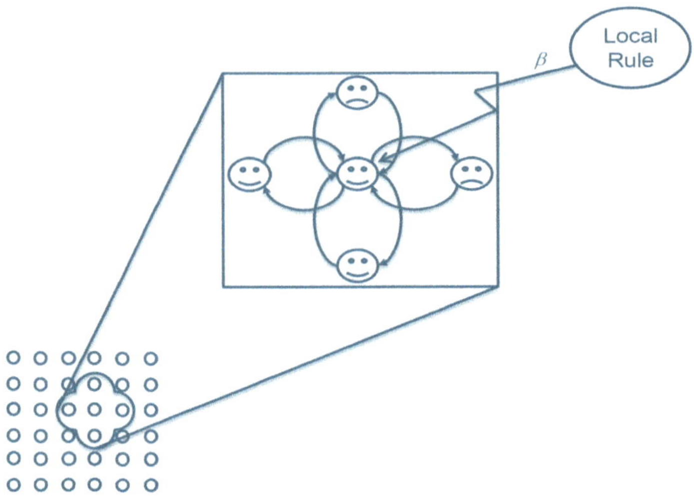

Fig. 1.9 Cellular learning automaton (Beigy and Meybodi 2004)

any cell i in the CLA.

$$\sum_r d_{ir}(P) \times p_{ir} \geq \sum_r d_{ir}(Q) \times q_{ir} \tag{1.21}$$

A CLA starts from some initial state (it is the internal state of every cell); at each stage, each automaton selects an action according to its probability vector and performs it in its local environment. Next, the CLA rule determines the reinforcement signals (the responses) to the LAs residing in its cells, and based on the received signals, each LA updates its internal probability vector. This procedure continues until a termination condition is satisfied.

1.2.1.14.1 Open Cellular Learning Automata

The basic CLA can be considered closed since it does not consider the CLA interaction and the external environment. In closed models, the behavior of each cell is affected only by its neighboring cells. However, CLA can easily model some applications if an external environment is also taken into account. This external environment also influences the behavior of the cells. Based on this idea, Beigy and Meybodi introduced open CLA (Beigy and Meybodi 2007), in which, in addition to the neighboring environment of a cell, two new environments are also considered: global environment and group environment. The global environment affects all of the LAs of the CLA, and the group environment affects a group of LAs in the CLA.

1.2.1.14.2 Asynchronous Cellular Learning Automata

Most CLA models are synchronous. In synchronous models, all cells receive their reinforcement signals simultaneously and, based on the received signals, update their internal action probabilities, also, at the same time. The synchronous models can be extended into asynchronous ones. In asynchronous models, the LAs can be updated at different times. In these models, cells can be updated in a time-driven or step-driven manner. In step driven manner, cells are updated in some fixed or random orders (Beigy and Meybodi 2008), while in a time-driven manner, each cell possesses an internal clock that determines its activation times. Asynchronous CLA has been employed in applications such as data mining and peer-to-peer networks.

1.2.1.14.3 Irregular Cellular Learning Automata

In basic cellular learning automaton (CLA), cells are arranged in regular forms like girds or rings. However, there exist some applications which require irregular arrangements of the cells. Esnaashari and Meybodi have presented an irregular CLA (Esnaashari and Meybodi 2018) for clustering the nodes in a wireless sensor network. Irregular CLA is modeled as an undirected graph in which each vertex represents a cell of the CLA, and its adjacent vertices determine its neighborhood. Up to now, various applications of Irregular CLA are reported in the literature. The application areas of this model include, for instance, (but not limited to) wireless sensor networks (Rezvanian et al. 2018e), graph problems (Vahidipour et al. 2017b), cloud computing (Morshedlou and Meybodi 2017), complex networks (Khomami et al. 2018), and social network analysis (Ghavipour and Meybodi 2017).

1.2.1.14.4 Dynamic Cellular Learning Automata

In a dynamic CLA, one of its aspects, such as structure, local rule, attributes, or neighborhood radius, may change over time. It should be noted that CLA can have various categories. For instance, it can be jointly categorized as irregular, asynchronous, and open. In this regard, dynamic CLA can be either closed or open. Even it can be synchronous or asynchronous. Several models of dynamic CLA have been investigated in the literature. Esnaashari and Meybodi introduced an interest-based dynamic irregular CLA in which two cells are considered to be neighbors if they share similar interests. In this regard, a set of interests is defined for dynamic irregular CLA. Each cell represents its tendency for each interest using a vector called a tendency vector. Two cells are considered to be neighbors if their tendency vectors are close enough. In addition to reinforcement signals, their presented dynamic irregular CLA uses a second kind of signal called restructuring signal. This latter signal is used to update the cells' tendency vectors, changing the neighborhood structures. Later, Saghiri and Meybodi have introduced other dynamic CLA models, such as asynchronous dynamic CLA and asynchronous dynamic CLA with varying numbers of LAs in each cell. These later CLA models have proven to be quite successful in landmark clustering in peer-to-peer networks and adaptive super-peer selection.

1.2.1.14.5 Wavefront Cellular Learning Automata

Wavefront CLA (WCLA) (Moradabadi and Meybodi 2018b; Rezvanian et al. 2019c) is an asynchronous CLA with a diffusion property. The activation sequence of cells in WCLA is controlled through waves. Each LA in WCLA receiving a wave is activated and selects an action according to its probability vector. If the LA's newly chosen action is different from its previous action, the LA propagates the wave to its neighbors. A wave propagates through cells in WCLA for some time until one of the following conditions holds: all LAs that have newly received the wave choose similar actions as their previous actions, or the energy of the wave drops to zero. The energy of each wave determines the wave's ability to propagate itself. This energy decreases as it moves through the network until it reaches zero. WCLA has been utilized for solving online social network problems such as prediction methods and sampling.

1.2.1.14.6 Associative Cellular Learning Automata

In associative CLA (Ahangaran et al. 2017), each cell receives two inputs: an input vector from the global environment of the CLA and a reinforcement signal for its performed actions. During each iteration, each cell receives an input from the environment. Then, it selects an action and applies it to the environment. Based on the performed actions, the local rule of the CLA determines the reinforcement signal to each LA. Associative CLA has been applied to applications such as clustering and image segmentation. For instance, clustering can be performed in the following manner. During each iteration, sample data are chosen by the environment and are given to each LA. According to the received input, each LA selects an action using its decision function. The defined action selection works based on the cell's current state and the input vector's distance. An LA is rewarded if its chosen action is smaller than those selected in its neighborhood. The proposed LA learning algorithm is defined so that an LA updates its state upon receiving the reward and becomes nearer to the input data while the receiving penalty does not affect the cell's state. Accordingly, the smallest action is chosen by the nearest cell to the input data. So, it is expected that, after some iterations, the state-vectors of cells lie on the centers of the clusters. Followings have briefly reviewed the classifications of CLA:

- **Static CLA vs. Dynamic CLA**: in static CLAs, the cellular structure of the CLA remains fixed during the evolution of the CLA, while in dynamic CLAs, one of its aspects such as structure, local rule, or neighborhood radius may vary with time (Esnaashari and Meybodi 2011, 2013)
- **Open CLA vs. Close CLA**: in close CLAs, the action of each LA depends on the neighboring cells, whereas in open CLAs, the action of each LA depends on the neighboring cells, a global environment that influences all cells, and an exclusive environment for each particular cell (Beigy and Meybodi 2007; Saghiri and Meybodi 2017a).
- **Asynchronous CLA vs. Synchronous CLA**: In synchronous CLA, all cells use their local rules simultaneously. This model assumes that there is an external clock that triggers synchronous events for the cells. In asynchronous CLA, only some cells are activated at a given time, and the state of the rest of the cells remains unchanged (Beigy and Meybodi 2007). The LAs may be activated in either time-driven where each cell is assumed to have an internal clock that wakes up the LA associated with that cell, or in step-driven where a cell is selected in a fixed or random sequence.

- **Regular CLA vs. Irregular CLA**: in regular CLAs, the structure of CLA is represented as a lattice of d-tuples of integer numbers, while in Irregular CLA (ICLA), the structure regularity assumption is replaced with an undirected graph (Ghavipour and Meybodi 2017; Vahidipour et al. 2017a; Esnaashari and Meybodi 2018).
- **CLAs with one LAs in each cell vs. CLAs with multiple LAs in each cell**: in conventional CLA with one LA in each cell, each cell is equipped with one LA, while in CLAs with multiple LAs in each cell, each cell is equipped with multiple LAs (Beigy and Meybodi 2010).
- **CLAs with a fixed number of LAs in each cell vs. CLAs with varying number of LAs in each cell**: in conventional CALAs, the number of LAs in each cell remains fixed during the evolution of CLA, while in CLAs with varying number of LAs in each cell, the number of LAs of each cell changes over time (Saghiri and Meybodi 2017a).
- **CLAs with fixed structure LAs vs. CLAs with variable structure LAs**: since LAs can be classified into two leading families; fixed and variable structure (Kumpati and Narendra 1989; Thathachar and Sastry 2004). In CLAs with fixed structure LAs, constituting LAs are of fixed structure type, whereas in CLAs with variable structure LAs, LAs are of variable structure type.

Up to now, various CLA models (Rezvanian et al. 2018f) such as open CLA (Beigy and Meybodi 2007), asynchronous CLA (Beigy and Meybodi 2008), irregular CLA (Esnaashari and Meybodi 2008), associative CLA (Ahangaran et al. 2017), dynamic irregular CLA (Esnaashari and Meybodi 2018), asynchronous dynamic CLA (Saghiri and Meybodi 2018b), and Wavefront CLA (Rezvanian et al. 2019b) are developed and successfully applied on different application domains.

1.2.2 Recent Applications of Learning Automata

In recent years, learning automata as one of the powerful computational intelligence techniques have been found a very useful technique to solve many real, complex, and dynamic environments where a large amount of uncertainty or lacking information about the environment exists (Rezvanian et al. 2018a; 2018d). Table 1.1 summarized some recent applications of learning automata.

1.3 Optimization

Various scientific and engineering problems can be modeled as optimization tasks. Roughly speaking, optimization problems can be categorized as static problems and dynamic problems. The problem attributes are considered to be constant in a static optimization problem. Accordingly, the optima of the problem do not change during

Table 1.1 Summary of recent applications of learning automata models

Applications	Learning automata model
5G networks	CLA (Qureshi et al. 2019)
Big Data	LA (Irandoost et al. 2019b)
Bioinformatics	CLA (Vafaee Sharbaf et al. 2016)
Biomedical	ACO-CLA (Boveiri et al. 2020)
Business Process Management System	CLA (Saraeian et al. 2019)
Cellular networks	CLA (Beigy and Meybodi 2010), LA (Rezapoor Mirsaleh and Meybodi 2018b)
Channel Assignment	CLA (Vafashoar and Meybodi 2019a)
Cloud computing	DLA (Hasanzadeh and Meybodi 2014), LA (Jobava et al. 2018), LA (Rahmanian et al. 2018), ICLA (Morshedlou and Meybodi 2018), LA (Morshedlou and Meybodi 2014), FALA (Velusamy and Lent 2018), LA (Qavami et al. 2017), LA (Rasouli et al. 2020), CLA-EC (Jalali Moghaddam et al. 2020), ICLA (Morshedlou and Meybodi 2017), CLA (Kheradmand and Meybodi 2014), CLA-EC (Jalali Moghaddam et al. 2020)
Community detection	ICLA (Zhao et al. 2015; Khomami et al. 2018), ACLA (Motiee and Meybodi 2009), CLA (Daliri Khomami et al. 2020a, b; Khomami et al. 2020)
Cognitive radio network	BLA (Mahmoudi et al. 2020)
Cyber-Physical Systems	LA (Ren et al. 2018)
Dynamic optimization	LA (Kazemi Kordestani et al. 2020)
Data mining	FALA (Hasanzadeh-Mofrad and Rezvanian 2018), ACLA (Ahangaran et al. 2017), CLA (Sohrabi and Roshani 2017), LA (Savargiv et al. 2020), DLA (Goodwin and Yazidi 2020), LA (Hasanzadeh-Mofrad and Rezvanian 2018), CLA-EC (Rastegar et al. 2005)
Graph problems	CLA (Vahidipour et al. 2017b), LA (Rezapoor Mirsaleh and Meybodi 2018c), LA (Mousavian et al. 2013), ICLA (Mousavian et al. 2014), DLA (Soleimani-Pouri et al. 2012), LA (Khomami et al. 2016b), FALA (Vahidipour et al. 2019), DLA (Vahidipour et al. 2019), ICAL (Vahidipour et al. 2019), (Daliri Khomami et al. 2017), GAPN-LA (Vahidipour et al. 2019), DLA (Rezvanian and Meybodi 2015a), DLA (Rezvanian and Meybodi 2015b), WCLA (Moradabadi and Meybodi 2018b)
Graph sampling	DLA (Rezvanian et al. 2014), DLA (Rezvanian and Meybodi 2017a), EDLA (Rezvanian and Meybodi 2017b), EDLA (Rezvanian and Meybodi 2017a), ICLA (Ghavipour and Meybodi 2017), VSLA (Rezvanian and Meybodi 2017a), FSLA (Ghavipour and Meybodi 2018c), FALA (Khadangi et al. 2016)
Hardware design	CLA (Hariri et al. 2005), CLA (Zamani et al. 2003)
Image processing	CLA (Hasanzadeh Mofrad et al. 2015), LA (Damerchilu et al. 2016), LA (Kumar et al. 2015b), CLA (Adinehvand et al. 2017)
Influence maximization	NLA (Daliri Khomami et al. 2018), DGCPA (Ge et al. 2017), DL_{ri} (Huang et al. 2018), CLA (Aldrees and Ykhlef 2014), LA (Khomami et al. 2021)
Intrusion detection	ICLA (FathiNavid and Aghababa 2012), CLA (Aghababa et al. 2012)
Internet of Things (IoT)	LA (Di et al. 2018), LA (Sikeridis et al. 2018), LA (Saleem et al. 2020), LA (Deng et al. 2020)
Link prediction	FALA (Moradabadi and Meybodi 2018c), FALA (Moradabadi and Meybodi 2017b), CALA (Moradabadi and Meybodi 2018a), DLA (Moradabadi and Meybodi 2017a), CALA (Moradabadi and Meybodi 2016), WCLA (Rezvanian et al. 2019h), ICLA-EC (Manshad et al. 2021)
Machine vision	LA (Betka et al. 2020)
Marketing	OCLA (Aldrees and Ykhlef 2014)
MapReduce	GLA (Irandoost et al. 2019a)
Medical imaging	CLA (Hadavi et al. 2014)
Multi-agent systems	CLA (Khani et al. 2017), CLA (Saraeian et al. 2019), CLA (Xue et al. 2019)

(continued)

Table 1.1 (continued)

Applications	Learning automata model
Network security	LA (Krishna et al. 2014), FALA (Di et al. 2019), LA (Hasan Farsi; Reza Nasiripour; Sajjad Mohammadzadeh 2018), CALA (Kahani and Fallah 2018), FALA (Su et al. 2018), $DP_{R\text{-}P}$ (Seyyedi and Minaei-Bidgoli 2018)
Operations research	LA (Nesi et al. 2020)
Opportunistic networks	CLA (Zhang et al. 2016)
Optimization	FALA (Rezvanian and Meybodi 2010b), FALA (Mahdaviani et al. 2015b), VSLA (Kordestani et al. 2018), CLA (Vafashoar and Meybodi 2016), LA (Rezapoor Mirsaleh and Meybodi 2015), LA (Li et al. 2018), LA (Rezapoor Mirsaleh and Meybodi 2018a), LA (Zarei and Meybodi 2020), PLA (Yazidi et al. 2020), LA (Alirezanejad et al. 2020), CLA-DE (Vafashoar et al. 2012), CLA (Vafashoar and Meybodi 2016), CLA (Vafashoar and Meybodi 2018), CLA (Mozafari et al. 2015), CLA (Vafashoar and Meybodi 2020)
Peer-to-peer networks	LA (Saghiri and Meybodi 2016, 2007a), CLA (Saghiri and Meybodi 2018a), (Amirazodi et al. 2018), ICLA (Rezvanian et al. 2018d), LA (Safara et al. 2020)
Recommender systems	FALA (Krishna et al. 2013), CALA (Ghavipour and Meybodi 2016), CLA (Toozandehjani et al. 2014)
Robotics	LA (Ghosh et al. 2019), CLA (Santoso et al. 2016)
Scheduling	CLA (Abdolzadeh and Rashidi 2010), ACO-CLA (Boveiri et al. 2020)
Social network analysis	LA (Amiri et al. 2013), DLA (Khomami et al. 2016b), CALA (Moradabadi and Meybodi 2016), FSLA (Ghavipour and Meybodi 2018c), ICLA (Ghavipour and Meybodi 2017), ICLA (Khomami et al. 2018), ICLA (Zhao et al. 2015), CLA (Aldrees and Ykhlef 2014), FSLA (Roohollahi et al. 2020), LA (Rezvanian et al. 2019f), ICLA (Rezvanian et al. 2019d), LA (Rezvanian et al. 2019c), NLA (Rezvanian and Meybodi 2016a)
Software-defined networks	ICLA (Thakur and Khatua 2019)
Stochastic social networks	DLA (Rezvanian and Meybodi 2016b), LA (Moradabadi and Meybodi 2018c), DLA (Vahidipour et al. 2017b), ICLA (Vahidipour et al. 2019), EDLA (Rezvanian and Meybodi 2017a)
Transportation	DCLA (Ruan et al. 2019), CLA (Chen et al. 2018)
Trust management	DLA (Ghavipour and Meybodi 2018b), DLA (Ghavipour and Meybodi 2018c), FALA (Lingam et al. 2018), CLA (Bushehrian and Nejad 2017), LA (Rezvanian et al. 2019e)
Vehicular environments	LA (Misra et al. 2014; Kumar et al. 2015a), LA (Toffolo et al. 2018)
Wireless mesh networks	LA (Parvanak et al. 2018; Beheshtifard and Meybodi 2018)
Wireless sensor networks	LA (Han and Li 2019), ICLA (Rezvanian et al. 2018e), FALA (Javadi et al. 2018), DLA (Mostafaei 2018), ICLA (Mostafaei and Obaidat 2018b), DLA (Mostafaei and Obaidat 2018a), GLA (Rahmani et al. 2018), CLA (Rezvanian et al. 2018e)

the optimization procedure. However, many optimization problems often contain several uncertain and dynamic factors. Typical examples include traffic-aware routing in wireless networks and air traffic scheduling. Such problems are commonly referred to as dynamic optimization problems. Due to the time-varying characteristics of a dynamic optimization problem, its fitness landscape changes with time. Accordingly, the locations of its optima may change with time as well.

Among different optimization approaches, nature-inspired methods such as evolutionary algorithms (Yu and Gen 2010; Eiben and Smith 2015) have attracted considerable interest in recent years. A nature-inspired algorithm is an iterative process. During each iteration, the algorithm aims at generating new and better candidate solutions to a given problem from the set of current candidate solutions. To generate new candidate solutions, nature-inspired optimization algorithms incorporate schemes or operators such as mutation, crossover, and selection. Evolutionary algorithms (EAs) as optimization algorithms are based on the survival of the fittest principle. Imperialist Competitive Algorithm (ICA) (Yas et al. 2014) consists of random points called countries where their fitness measure by their power. Countries divide the problem space into two types of empires, including colonies and imperialists. The powerful countries become Imperialist and start to take control of weak ones (colonies), Harmony Search (HS) (Estahbanati 2014) inspired by extemporaneously music composition. In the HS algorithm, each decision variable (musician) generates (plays) a value (note) for finding a global optimum (best harmony).

Particle Swarm Optimization (PSO) (Nabizadeh et al. 2012; Hasanzadeh et al. 2013; Rezaee Jordehi and Jasni 2013; Kordestani et al. 2014a) is a heuristic-based iterative technique that uses a population of particles. Each particle of PSO represents a feasible solution to problem space. PSO keeps track of each individual's best values and the entire population to optimize the problem. Cuckoo Search (CS) Algorithm (Kavousi-Fard and Kavousi-Fard 2013) inspired cuckoo's unusual breeding behavior. The cuckoos (population) are laying their eggs (new solutions) in other species nests. Firefly Algorithm (FA) (Kamarian et al. 2014) inspired by the flashing behavior of fireflies. All the fireflies (population) attract other fireflies through the associated light intensity.

Artificial Immune System (AIS) (Rezvanian and Meybodi 2010d) consists of several optimization techniques inspired by human immune mechanisms. The immune cells (population) try to protect the body by immune mechanisms (operators). Genetic Algorithm (GA) (Manurung et al. 2012) is a search meta-heuristic that emulates the process of natural selection. Moreover, Non-dominated Sorting Genetic Algorithm II (NSGA II) (Javadi et al. 2014) is a multiobjective GA covering the Pareto-optimal front. Finally, Ant Colony Optimization (ACO) (Shyu et al. 2003; Soleimani-pouri et al. 2014) was inspired by ants' colonial behavior. ACO is a probabilistic approach for finding the shortest paths through a graph. The Group Search Optimizer (GSO) (He et al. 2009; Hasanzadeh et al. 2016) is a population-based optimization heuristic that adopts the Producer-Scrounger (PS) model (Barnard and Sibly 1981). The PS model is a group living methodology with two strategies: (1) producing, e.g., searching for food, and (2) joining (scrounging), e.g., joining resources uncovered by others. Besides producer and scrounger members, the GSO algorithm population also contains ranger members who perform random walks to avoid getting trapped in pseudo-minima. Moreover, GSO inspires from the concept of animal foraging behavior to search the most promising areas of problem space.

Much effort has been put into the development of versatile nature-inspired optimization methods. However, it is well accepted that a general-purpose universal optimization algorithm is impossible: no strategy or scheme can outperform the others on all possible optimization problems (Wu et al. 2019). Additionally, a single evolutionary scheme or operator may always follow similar trajectories. Therefore, it would be better to use diverse evolutionary operators to increase the chance of finding optima. Moreover, the parameter and strategy configuration of an optimization algorithm can significantly affect its performance. On a specific problem, proper configurations can result in high-quality solutions and high convergence speeds.

Considering the discussed issues, the utilization of multiple strategies or techniques and a suitable adaptation mechanism can significantly enhance a nature-inspired optimization method. Some recently developed nature-inspired optimization methods incorporate various strategies in some of their algorithmic elements, such as mutation, crossover, or neighborhood structure. Such collections are referred to as ensemble components in some literature, such as Wu et al. (2019). In what follows, we will show how learning automata (LA) and reinforcement learning (RL) can be utilized for the adaptation in nature-inspired optimization algorithms.

1.3.1 Evolutionary Algorithms and Swarm Intelligence

Biological and natural principles have long been a source of inspiration for human inventions. Natural evolution can be viewed as an optimization process of spices based on learning how to adapt to the environment. These Evolutionary algorithms generally have three main characteristics (Yu and Gen 2010):

- Population-based: EAs maintain a group of individuals, termed as a population. Each individual is a candidate solution to the given optimization problem.
- Fitness-oriented: Each individual has a gen representation. Also, EAs have mechanisms to determine the quality of a candidate solution. The quality of an individual is called its fitness value. During their optimization procedure, EAs prefer to preserve high-quality individuals.
- Variation-driven: To search the solution space, EAs incorporates various operations that mimic the genetic gene changes.

Inspired by the natural evolution process, several EAs like a genetic algorithm (GA), evolutionary strategies (ES), differential evolution (DE) are introduced in the literature (Yu and Gen 2010; Eiben and Smith 2015). There are other similar phenomena in nature that can be viewed as an optimization and learning process. A swarm of not smart insects sometimes demonstrates intelligent behavior. This swarm-level intelligent behavior has inspired the design of algorithms such as particle swarm optimization (PSO). In what follows, we briefly review two well-known nature-inspired optimization algorithms including, DE and PSO.

1.3.1.1 Differential Evolution

DE evolves a population of *NP* candidate solutions, called individuals, through a process consisting of mutation, crossover, and selection. Each individual $X_i(G) = \{x_i^1(G),\ldots, x_i^D(G)\}\ \forall i = 1,\ldots,NP$ at generation G represents a D dimensional real vector within the problem search space. The search space is defined using its upper and lower bounds: $X_{min} = \{x_{min1}, x_{min2}, \ldots, x_{minD}\}$ and $X_{max} = \{x_{max1}, x_{max2}, \ldots, x_{maxD}\}$. The DE population is initially generated using uniform distribution within these bounds:

$$x_{i,0}^{j} = x_{min}^{j\,\mathrm{rand}_{max}^{j_{min}^{j\ and}}}, \tag{1.22}$$

At each generation G, one mutant vector $V_{i,G}$ is generated for each individual using a mutation strategy. Several mutation strategies have been developed in the DE literature; three of the most common ones are listed below:

$$\begin{aligned} &DE/rand/1: \\ &V_{i,G} = X_{r1,G} + F.\left(X_{r2,G} - X_{r3,G}\right), \end{aligned} \tag{1.23}$$

$$\begin{aligned} &DE/best/1: \\ &V_{i,G} = X_{best,G} + F.\left(X_{r1,G} - X_{r2,G}\right), \end{aligned} \tag{1.24}$$

$$\begin{aligned} &DE/rand - to - best/1: \\ &V_{i,G} = X_{i,G} + F.\left(X_{best,G} - X_{i,G}\right) + F.\left(X_{r1,G} - X_{r2,G}\right), \end{aligned} \tag{1.25}$$

where r_1, r_2, and r_3 are three mutually different random integer numbers uniformly taken from the interval [1, *NP*]. F is a constant control parameter that lies in the range [0,2] and scales the differential vectors. $X_{best,G}$ represents the fittest individual of the Gth generation.

After the generation of mutant vectors, a crossover operator is applied to each mutant vector $V_{i,G}$ in order to generate a trail vector $U_i(G) = \{u_i^{\,1}(G),\ldots, u_i^{\,D}(G)\}$ $\forall i = 1,\ldots, NP$. DE algorithms use two types of crossover operators: exponential and binomial. The binomial operator is the most preferred crossover operator in the DE literature and is defined as follows:

$$U_{i,G}^{j} = \begin{cases} V_{i,G}^{j} & \text{if rand(0, 1)} < CR \ \text{ or } j = irand \\ X_{i,G}^{j} & \text{otherwise} \end{cases} \quad j = 1, \ldots, D, \tag{1.26}$$

where *CR* is a constant parameter controlling the portion of the offspring chosen from the mutant vector. *irand* is a random integer generated uniformly from the interval [1, *N*] for each trial vector to ensure that at least one of its components is chosen from the mutant vector. rand(0,1) is a uniform random real number generator from the interval [0,1].

Each generated trial vector is evaluated using a fitness function; then, either the parent vector (also called target vector) or its corresponding trial vector is selected for the next generation by using a greedy selection scheme:

$$X_{i,G+1} = \begin{cases} U_{i,G} \text{ if } f\left(U_{i,G}\right) < f\left(X_{i,G}\right) \\ X_{i,G} \text{ otherwise} \end{cases}. \tag{1.27}$$

1.3.1.2 Particle Swarm Optimization

PSO executes its search in a specified space through the accumulation of velocity and position information. In the beginning, each particle is initiated randomly within a D dimensional search space. It keeps the information about its best-visited position known as *pbest*, swarm best position known as *gbest*, and its current flying velocity. Canonical PSO maintains a swarm of particles; each particle X_i in the swarm maintains three D-dimensional vectors: $x_i = [x_{i1}, x_{i2}, \ldots, x_{iD}]$ representing its current position, $p_i = [p_{i1}, p_{i2}, \ldots, p_{iD}]$ representing the best location previously experienced by the particle, and $v_i = [v_{i1}, v_{i2}, \ldots, v_{iD}]$ representing its flying velocity. The whole swarm keeps the information about its best-experienced position in a D dimensional vector like $p_g = [p_{g1}, p_{g2}, \ldots, p_{gD}]$. During the search, at step k, the velocity of the particle X_i and its current position is updated according to the following equations:

$$\begin{aligned} v_{id}(k+1) &\leftarrow \omega v_{id}(k) + c_1 r_1 (p_{id}) - x_{id}(k) + c_2 r_2 \big(p_{gd} - x_{id}(k)\big) \\ x_{id}(k+1) &\leftarrow x_{id}(k) + v_{id}(k+1), \end{aligned} \tag{1.28}$$

where $v_{id}(k)$ is the dth dimension of the velocity vector of the particle in step k; $x_{id}(k)$ and $p_{id}(k)$ are respectively the dth dimension of its position and historical best position vectors; $p_{gd}(k)$ represents the dth dimension of the historical best position of the whole swarm in step k; ω is the inertia weight, which was introduced to bring a balance between the exploration and exploitation characteristics (Shi and Eberhart 1998); c_1 and c_2 are acceleration constants and represent cognitive and social learning weights; and, finally, r_1 and r_2 are two random numbers from the uniform distribution $u(0, 1)$. After acquiring the particles' new positions, each particle's historical best position is updated, affecting the swarm's historical global best position.

1.4 Reinforcement Learning and Optimization Methods

Up to now, various hybrid models based on reinforcement learning schemes and nature-inspired methods have been introduced for solving a variety of problems (Table 1.2). A great deal of these methods uses a reinforcement learning scheme as a control unit to control an optimization method comprising components. This section briefly reviews some of these approaches. Although in most of the works reviewed in this section, reinforcement learning is utilized to improve a nature-inspired method's performance, some works are based on a different idea. Iima and Kuroe introduced some swarm reinforcement learning algorithms (Iima and Kuroe 2008). In these algorithms, several Q-learning agents learn concurrently based on two learning strategies: independent learning and cooperative learning. In the independent learning phase, each agent learns according to the standard Q-learning algorithm. In the cooperative phase, agents exchange their obtained knowledge according to swarm updating rules of PSO.

Table 1.2 Some applications of reinforcement learning in population-based optimization methods

Brief description	Application	Heuristic	Reinforcement learning mechanism	Ref.
Parameter/strategy adaptation	Numerical optimization	PSO	LA	Hashemi and Meybodi (2011), Hasanzadeh et al. (2012,2014); Zhang et al. (2018)
Parameter/strategy adaptation	Numerical optimization, economic dispatch problem	Quantum-behaved PSO	Q-learning	Sheng and Xu (2015)
Parameter/strategy adaptation	Numerical optimization	Memetic PSO	Q-learning	Samma et al. (2016)
Parameter/strategy adaptation	Numerical optimization	DE	LA	Sengupta et al. (2012; Kordestani et al. (2014a), Mahdaviani et al. (2015b)
Parameter/strategy adaptation	Numerical optimization	GA	LA	Ali and Brohi (2013)
Parameter/strategy adaptation	Numerical optimization	Butterfly algorithm	LA	Arora and Anand (2018)
Parameter/strategy adaptation	Numerical optimization	Harmony search	LA	Enayatifar et al. (2013a)
Evaluation budget control	Numerical optimization in noisy environments	PSO	LA	Zhang et al. (2014), Zhang et al. (2015)
Optimization entity/local search	Graph isomorphism problem, OneMax problem, object partitioning	Memetic GA	LA, object migration automaton	Rezapoor Mirsaleh and Meybodi (2015, 2018c)
Optimization entity/genotype	Numerical optimization	GA	LA	Howell et al. (2002)
Optimization entity/quantization orthogonal crossover	Numerical multi-objective optimization	Orthogonal EA	LA	Dai et al. (2016)
Optimization entity/genotype	Traveling salesman problem	GA	Q-learning	Alipour et al. (2018)

(continued)

Table 1.2 (continued)

Brief description	Application	Heuristic	Reinforcement learning mechanism	Ref.
Upper-level heuristic/strategy adaptation	Numerical multi-objective optimization, vehicle crashworthiness problem	Hyper-heuristic	LA	Li et al. (2018)
Optimization entity/genotype	Numerical optimization	DE, GA	CLA	Rastegar and Meybodi (2004, Vafashoar et al. (2012, Mozafari et al. (2015)
Scheduler	Dynamic numerical optimization	Multi-population optimizer	LA	Kordestani et al. (2019)
Parameter/strategy adaptation	Dynamic numerical optimization	Cuckoo search	LA	(Kordestani et al. (2018)
Parameter/strategy adaptation	Dynamic numerical optimization	Artificial immune system	LA	Rezvanian and Meybodi 2010a, b, d),
Parameter/strategy adaptation	Dynamic numerical optimization	Firefly algorithm	LA	Abshouri et al. (2011)
Parameter/strategy adaptation	Dynamic numerical optimization	PSO	LA	Geshlag and Sheykhzadeh (2012)
Parameter/strategy adaptation	Dynamic software project scheduling	Multi-objective memetic algorithm	Q-learning	Shen et al. (2018)
Parameter/strategy adaptation	Dynamic multi-objective optimization, controlling traffic lights	PSO	Q-learning	El Hatri and Boumhidi (2016)

1.4.1 Static Optimization

One of the earliest uses of a reinforcement learning mechanism in adaptive parameter control of a nature-inspired method is reported by Hashemi and Meybodi (Hashemi and Meybodi 2011). Unified adaptive PSO (UAPSO) and independent adaptive PSO (IAPSO) algorithms use LA-based mechanisms for parameter adaption during the search procedure. During each iteration of UAPSO, a set of learning automata determine the parameter values of the algorithm. In contrast, IAPSO associates a set of learning automata with each particle, and each set determines the parameter values for its associated particle. After the evolution of particles, the learning automata are updated to choose better parameter values in future generations (Hashemi and Meybodi 2011). Another reinforcement learning parameter adaptation mechanism is presented by Sheng and Xu, which adaptively adjusts quantum-behaved PSO parameter values using the Q-learning approach (Sheng and Xu 2015). The presented algorithm is employed for solving the economic dispatch problem and demonstrated some beneficial characteristics.

The reinforcement learning-based memetic particle swarm optimizer (RLMPSO) model uses five well-designed operations consisting of exploration, convergence, high-jump, low-jump, and fine-tuning update each particle (Samma et al. 2016). To control the application sequence of these five operations, RLMPSO utilizes a Q-learning mechanism. Each state of the Q-learner entity is associated with one specific operation. After executing a specific operation on a particle, the Q-learner decides the next operation, which best suits the currently applied operation. The LAPSO algorithm presented by Zhang et al. uses two updating strategies for particles (Zhang et al. 2018). During each generation, an equipped LA chooses the updating mechanism of all particles. Then, using the chosen mechanism, the particles get updated. Next, the probability vector of the LA is updated based on the number of successful position updates.

Similarly, cooperative particle swarm optimization based on learning automata (CPSOLA) hybridizes cooperative and canonical PSO methods into one algorithm to inherit both methods' beneficial characteristics. It uses a learning automaton to switch between two methods based on the algorithm's performance (Hasanzadeh et al. 2014). The authors also presented another adaptive cooperative particle swarm optimizer in Hasanzadeh et al. (2013). In this algorithm, a set of learning automata is associated with the dimensions of the problem. These learning automata aim to find the correlated variables of the search space. By considering the existent dependencies among variables, the proposed PSO can solve complex optimization problems efficiently.

They are historically learning automata are decision-making units designed for unknown, random, and noisy environments. Consequently, they can be well suited for optimization in such environments. One common approach for dealing with noise corrupted fitness values is based on re-evaluation. After re-evaluating a particular candidate solution several times, its fitness can be defined as the mean value of the obtained noise corrupted fitness values. PSOLA and LAPSO, introduced by Zhang et al., utilize LA to allocate re-evaluation budgets to promising particles in an intelligent manner (Zhang et al. 2014, 2015).

Learning automata is also utilized successfully for adaptively controlling the parameter settings and the mutation strategies of differential evolution algorithm (Kordestani et al. 2014a; Mahdaviani et al. 2015). In the approach presented in Sengupta et al. (2012), a population of *NP* candidate solutions is equipped with a set of *NP* learning automata. During each generation, the individuals are ranked according to their fitness values. Then, the ith learning automaton chooses the mutation parameter value of the ith ranked individual. Next, the individual is updated according to the chosen settings and is evaluated by the fitness function. Based on this evaluation, the probability vector of the corresponding learning automaton is updated.

Reinforcement learning has also integrated, in a similar manner as the ones described before, into approaches such as genetic algorithm, harmony search, butterfly algorithm, and firefly algorithm. Ali and Brohi have used fast learning automata to adaptively control the application of three mutation operators, including Cauchy, Gaussian, and Levy, in their presented genetic algorithm (Ali and Brohi 2013). Similarly, Arora and Anand have used an LA agent to control butterflies' behavior in the butterfly optimization algorithm. In their approach, a learning automaton is associated with the whole population, which decides whether a butterfly should perform a random local search or move towards the best butterfly (Arora and Anand 2018). The authors have also investigated the applicability of their introduced model on some classical engineering design problems. Another approach for controlling parameter values through a reinforcement learning mechanism is presented in Enayatifar et al. (2013). This method learns a set of proper parameter values for a variant of the harmony search algorithm.

Rezapoor Mirsaleh and Meybodi (2015) introduced a Baldwinian memetic algorithm called MGALA by combining genetic algorithms with learning automata. In this method, LA provides the local search function of the memetic algorithm. Each chromosome of MGALA is an object migration automaton whose states keep information about the local search process's history. The critical aspect of MGALA is the computation of the fitness function, which is based on the history of the local search and the chromosome's objective value. The same authors suggested an optimization method composed of two parts: a genetic part and a memetic part. They demonstrated how to use learning automata to create a balance between exploration performed by evolution and exploitation performed by local search (Rezapoor Mirsaleh and Meybodi 2018c).

Howell et al. used probability strings to represent individuals (Howell et al. 2002). The value of the ith component of a probability string defines the probability of the allele value in a corresponding bit string being one at that position. These probability strings can be regarded as genotypes. At each generation, a population of phonotypes is generated by sampling the probability strings. The fitness function then evaluates these generated bit strings, and the probability vectors are updated accordingly.

Additionally, the probability strings can be affected by crossover and mutation operators with the hope of producing better genotypes. An interesting orthogonal evolutionary algorithm based on learning automata is presented by Dai et al. for multi-objective optimization (Dai et al. 2016). In this algorithm, learning automata provide mechanisms for mutation and grouping decision variables for quantization orthogonal crossover.

Q-learning has been an efficient approach for solving the traveling salesman problem (Alipour et al. 2018). In the approach presented in Alipour et al. (2018), each city is modeled as a Q-learner state. There are several traveling salesmen considered traversing the cities to obtain the best solution. During each iteration, each salesman starts from a random city. Then, it selects an action from the action set of the corresponding state. It follows the action and moves to the next city. This process is repeated until the tour is complete. The best tour fond by all salesmen is used to generate reinforcement signals for the Q-learning system. According, the salesmen can choose better tours as the algorithm proceeds.

The reinforcement learning paradigm can well suit the hyper-heuristic framework of optimization. Several works reported employing learning automata to control the application of a set of low-level heuristics. An excellent example of this hyper-heuristics category has been presented by Li et al. for multi-objective optimization (Li et al. 2018). The method is applied for solving vehicle crashworthiness problems; however, it can be used as a general framework for hyper-heuristic design. In this method, the reinforcement learning scheme sits at the core of the meta-heuristic selection process. For each meta-heuristic, it learns the most appropriate meta-heuristic to be applied next. By doing this, a transition matrix is learned by the reinforcement learning mechanism, which governs the application sequence of low-level meta-heuristics during the searching process.

1.4.1.1 Static Optimization Based on CLA

Several works have been reported using CLA to solve optimization problems (Rastegar and Meybodi 2004; Vafashoar et al. 2012; Mozafari et al. 2015). In some of these works, in a similar way (Howell et al. 2002), CLA models the solution space. Usually, a set of learning automata is considered in each cell of the CLA, and each combination of their actions corresponds to a candidate solution to the given problem. Each cell's learning automata give their combination of chosen actions (a sampled candidate solution) to the environment. The environment decides the favorability of the received candidate solution or its comprising components and generates reinforcement signals accordingly. The LAs use these reinforcement feedbacks to update their internal probability distributions and generate better candidate solutions in future generations (Rastegar and Meybodi 2004; Vafashoar et al. 2012; Mozafari et al. 2015).

1.4.2 Dynamic Optimization

A general framework for scheduling subpopulations in multi-population dynamic optimization methods is presented in Kordestani et al. (2019). In this approach, a learning automaton schedules the evolutionary process of subpopulations. The general idea of the presented approach is to allocate more function evaluations to the best-performing subpopulations. In another dynamic optimization approach, LA is used to balance the exploration and exploitation characteristics of cuckoos in the cuckoo search algorithm (Kordestani et al. 2018). The presented method uses two types of Levy distributions with two different β values: one for long jumps and the other for short ones. During each iteration, the search strategy distribution for each egg is adaptively determined through a learning automaton. In a similar approach,

Rezvanian and Meybodi presented to control the mutation probability adaptively in artificial immune systems by learning automata (Rezvanian and Meybodi 2010a, b). Learning automata have been employed for parameter adjustment in the firefly algorithm (Abshouri et al. 2011). In the dynamic optimization approach presented in Abshouri et al. (2011), each firefly is equipped with three learning automata, and each learning automaton adjusts a specific parameter for its associated firefly. Another approach using learning automata to provide adaptive strategy selection for PSO is presented in Geshlag and Sheykhzadeh (2012) for dynamic optimization. In this approach, a particle can be update based on two rules. Each particle is associated with a particular LA, and this LA controls which strategy to be utilized by the particle.

Software project scheduling, which allocates employees to tasks in a software project, can be viewed as a dynamic multi-objective optimization problem (Shen et al. 2018). A dynamic multi-objective memetic algorithm based on Q-learning is presented by Shen et al. for solving this scheduling/rescheduling problem (Shen et al. 2018). In this algorithm, Q-learner learns the most appropriate global and local search methods for different software project environment states. A motivating multi-objective optimization approach is presented in El Hatri and Boumhidi (2016), which can be utilized in static and dynamic environments. The algorithm is tested as a system for controlling traffic lights. It uses Q-learning to adaptively control the parameter settings of each PSO particle based on its state. A particle's state is defined as its distance from its previous best position and the global best position. Additionally, Q-learning is modified to meet the requirements of multi-objective optimization problems.

1.5 LA and Optimization Timeline

We depict the attempts of researchers on the hybrid learning automata models for optimization in which LA commonly is applied for adaptively tuning of parameters to improving the performance of each algorithm including: CLA-EC (Rastegar and Meybodi 2004), CALA (Beigy and Meybodi 20006), PSO-LA (Sheybani and Meybodi 2017a), CLA-PSO (Sheybani and Meybodi 2007b), DCLA-PSO (Jafarpour et al. 2007), RCLA-EC (Jafarpour and Meybodi 2007), CCGA+LA (Abtahi et al. 2008), CCLA-EC (Zanganeh et al. 2010), CLA-AIS (Javadzadeh et al. 2010), AFSA-CLA (Yazdani et al. 2010), LA-AIN (Rezvanian and Meybodi 2010a), Lopt-aiNet: (Rezvanian and Meybodi 2010b), LACAIS (Rezvanian and Meybodi 2010c), M-CLA-PSO (Akhtari and Meybodi 2011), LBA (Aghazadeh and Meybodi 2011), ICA-LA Improving Imperialist Competitive (Khezri and Meybodi 2011), CLA-DE (Vafashoar et al. 2012), CLA-FA (Hassanzadeh and Meybodi 2012), LA-MA (Rezapoor Mirsaleh and Meybodi 2013), ACPSO (Hasanzadeh et al. 2013), LAHS (Enayatifar et al. 2013), LADE: differential evolution (Mahdaviani et al. 2015c), EDLA (Mollakhalili Meybodi and Meybodi 2014a), CCLA (Mozafari et al. 2015), CLA-BBPSO (Vafashoar and Meybodi 2016), CLAMA (Rezapoor Mirsaleh and Meybodi 2016), ACS (Abedi Firouzjaee et al. 2017), CLAMS (Vafashoar and Meybodi 2018),

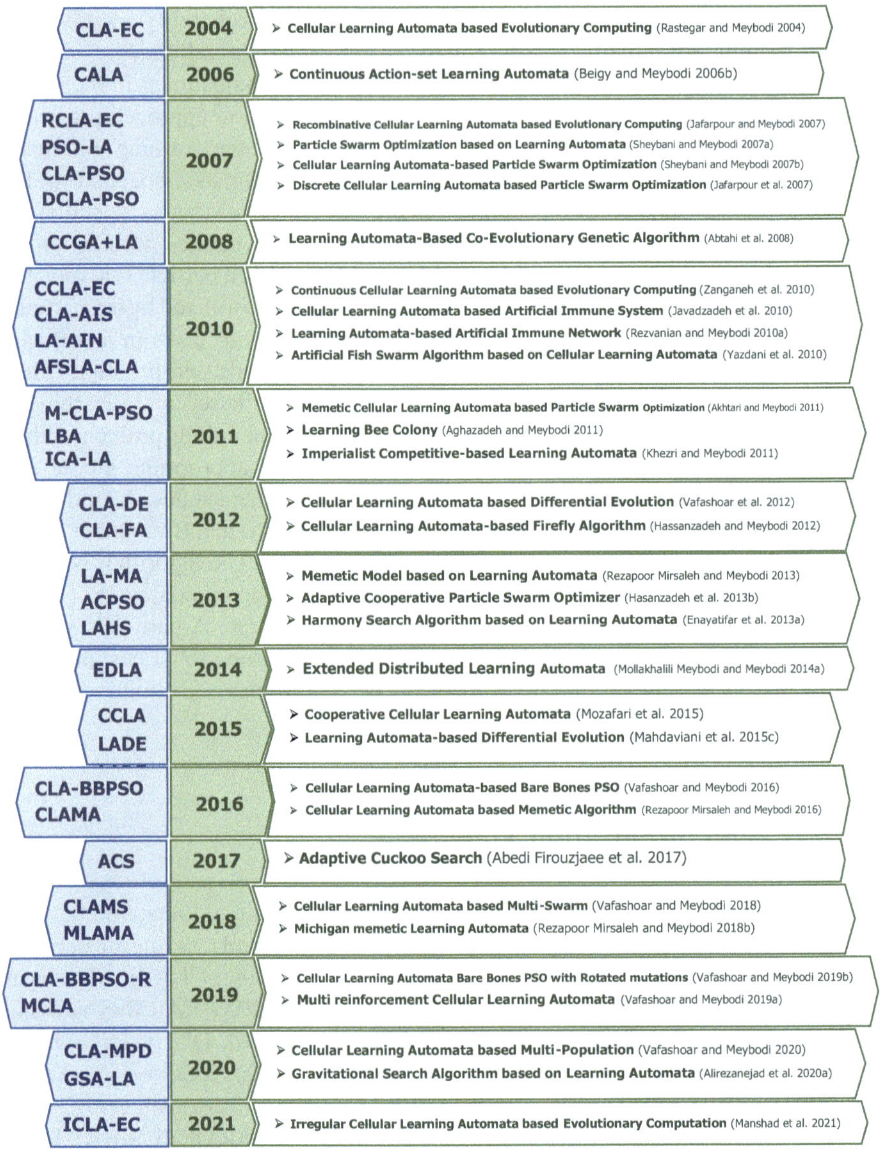

Fig. 1.10 Timeline of evolution for learning automata models in optimization

MLAMA (Rezapoor Mirsaleh and Meybodi 2018c), CLA-BBPSO-R (Vafashoar and Meybodi 2019b), MCLA (Vafashoar and Meybodi 2019a), CLA-MPD (Vafashoar and Meybodi 2020), GSA-LA (Alirezanejad et al. 2020), and ICLA-EC (Manshad et al. 2021). The timeline of appearing these models is shown in Fig. 1.10.

1.6 Chapter Map

The book begins with an introductory chapter providing background tutorials for learning automata theory and models and optimization by evolutionary algorithms, which most later chapters build upon them. Chapter 2 describes the history and research trends of the LA and optimization as bibliometric perspectives. Chapter 3 covers hybrid techniques for optimization using cellular automata, learning automata, and cellular learning automata. Chapter 4 is dedicated to learning automata for behavior control in evolutionary computation. Chapter 5 provides a tutorial on a memetic model based on fixed structure learning automata for solving NP-Hard problems. Chapter 6 is devoted to a brief description of the applications of object migration automaton (OMA)-memetic algorithm for solving NP-Hard problems. Then the next set of chapters describes several approaches for optimization in dynamic environments. Chapter 7 provides an overview of multi-population methods for dynamic environments. Chapter 8 is dedicated to introducing learning automata for online function evaluation management in evolutionary multi-population methods for dynamic optimization problems. Finally, Chapter 9 gives a brief description of function management in multi-population methods with a variable number of populations using a variable action learning automaton approach.

1.7 Conclusion

This chapter begins with the basic concepts of learning automata theory and optimization methods using evolutionary algorithms. We described some hybrid evolutionary algorithms based on learning automata definitions, and we also reported recent applications of learning automata models. Besides, recent models of learning automata models for solving optimization problems, like the CLA-EC, CLA-DE, CLA-AIS, LACS, CLA-MPD, and ICLA-EC, are introduced.

References

Abdolzadeh, M., Rashidi, H.: An approach of cellular learning automata to job shop scheduling problem. Int. J. Simul. Syst. Sci. Technol. **34**, 391–401 (2010)

Abedi Firouzjaee, H., Kazemi Kordestani, J., Meybodi, M.R.: Cuckoo search with composite flight operator for numerical optimization problems and its application in tunnelling. Eng. Opt. **49**, 597–616 (2017). https://doi.org/10.1080/0305215X.2016.1206535

Abshouri, A.A,. Meybodi, M.R., Bakhtiary, A.: New firefly algorithm based on multi swarm & learning automata in dynamic environments. In: IEEE Proceedings, pp. 989–993 (2011)

Abtahi, F., Meybodi, M.R., Ebadzadeh, M.M., Maani, R.: Learning automata-based co-evolutionary genetic algorithms for function optimization. In: Proceedings of the 6th International Symposium on Intelligent Systems and Informatics, (SISY), pp. 1–5 (2008)

Adinehvand, K., Sardari, D., Hosntalab, M., Pouladian, M.: An efficient multistage segmentation method for accurate hard exudates and lesion detection in digital retinal images. J. Intell. Fuzzy Syst. **33**, 1639–1649 (2017). https://doi.org/10.3233/JIFS-17199

Agache, M., Oommen, B.J.: Generalized pursuit learning schemes: new families of continuous and discretized learning automata. IEEE Trans. Syst. Man Cybern. Part B Cybern. **32**, 738–749 (2002). https://doi.org/10.1109/TSMCB.2002.1049608

Aghababa, A.B., Fathinavid, A., Salari, A., Zavareh, S.E.H.: A novel approach for malicious nodes detection in ad-hoc networks based on cellular learning automata. In: 2012 World Congress on Information and Communication Technologies, pp. 82–88. IEEE (2012)

Aghazadeh, F., Meybodi, M.R.: Learning bees algorithm for optimization. In: International Conference on Information and Intelligent Computing, pp. 115–122 (2011)

Ahangaran, M., Taghizadeh, N., Beigy, H.: Associative cellular learning automata and its applications. Appl. Soft Comput. **53,** 1–18 (2017). https://doi.org/10.1016/j.asoc.2016.12.006

Akbari Torkestani, J., Meybodi, M.R.: Learning automata-based algorithms for finding minimum weakly connected dominating set in stochastic graphs. Int. J. Uncertainty Fuzziness Knowl. Based Syst. **18**, 721–758 (2010). https://doi.org/10.1142/S0218488510006775

Akbari Torkestani, J., Meybodi, M.R.: A learning automata-based heuristic algorithm for solving the minimum spanning tree problem in stochastic graphs. J. Supercomputing **59**, 1035–1054 (2012). https://doi.org/10.1007/s11227-010-0484-1

Akhtari, M., Meybodi, M.R.: Memetic-CLA-PSO: a hybrid model for optimization. In: 2011 UkSim 13th International Conference on Computer Modelling and Simulation, pp. 20–25. IEEE (2011)

Aldrees, M., Ykhlef, M.: A seeding cellular learning automata approach for viral marketing in social network. In: Proceedings of the 16th International Conference on Information Integration and Web-Based Applications & Services - iiWAS 2014, pp. 59–63. ACM Press, New York (2014)

Ali, K.I., Brohi, K.: An adaptive learning automata for genetic operators allocation probabilities. In: 2013 11th International Conference on Frontiers of Information Technology, pp. 55–59. IEEE (2013)

Alipour, M.M., Razavi, S.N., Feizi Derakhshi, M.R., Balafar, M.A.: A hybrid algorithm using a genetic algorithm and multiagent reinforcement learning heuristic to solve the traveling salesman problem. Neural Comput. Appl. **30**, 2935–2951 (2018). https://doi.org/10.1007/s00521-017-2880-4

Alirezanejad, M., Enayatifar, R., Motameni, H., Nematzadeh, H.: GSA-LA: gravitational search algorithm based on learning automata. J. Exp. Theoret. Artif. Intell. 1–17 (2020). https://doi.org/10.1080/0952813X.2020.1725650

Amirazodi, N., Saghiri, A.M., Meybodi, M.: An adaptive algorithm for super-peer selection considering peer's capacity in mobile peer-to-peer networks based on learning automata. Peer-to-Peer Network. Appl. **11**, 74–89 (2018). https://doi.org/10.1007/s12083-016-0503-y

Amiri, F., Yazdani, N., Faili, H., Rezvanian, A.: A novel community detection algorithm for privacy preservation in social networks. In: Intelligent Informatics, pp. 443–450 (2013)

Arora, S., Anand, P.: Learning automata-based butterfly optimization algorithm for engineering design problems. Int. J. Comput. Mater. Sci. Eng. **07,** 1850021 (2018). https://doi.org/10.1142/S2047684118500215

Aso, H., Kimura, M.: Absolute expediency of learning automata. Inf. Sci. **17**, 91–112 (1979). https://doi.org/10.1016/0020-0255(79)90034-3

Barnard, C.J., Sibly, R.M.: Producers and scroungers: a general model and its application to captive flocks of house sparrows. Anim. Behav. **29**, 543–550 (1981)

Barto, A.G., Sutton, R.S., Anderson, C.W.: Neuronlike adaptive elements that can solve difficult learning control problems. IEEE Trans. Syst. Man Cybern. **SMC-13**, 834–846 (1983). https://doi.org/10.1109/TSMC.1983.6313077

Beheshtifard, Z., Meybodi, M.R.: An adaptive channel assignment in wireless mesh network: the learning automata approach. Comput. Electr. Eng. **72**, 79–91 (2018). https://doi.org/10.1016/j.compeleceng.2018.09.004

Beigy, H., Meybodi, M.R.: A mathematical framework for cellular learning automata. Adv. Complex Syst. **07,** 295–319 (2004). https://doi.org/10.1142/S0219525904000202

Beigy, H., Meybodi, M.R.: Utilizing distributed learning automata to solve stochastic shortest path problems. Int. J. Uncertainty Fuzziness Knowl. Based Syst. **14**, 591–615 (2006a). https://doi.org/10.1142/S0218488506004217

Beigy, H., Meybodi, M.R.: A new continuous action-set learning automaton for function optimization. J. Franklin Inst. **343**, 27–47 (2006b)

Beigy, H., Meybodi, M.R.: Open synchronous cellular learning automata. Adv. Complex Syst. **10**, 527–556 (2007)

Beigy, H., Meybodi, M.R.: Asynchronous cellular learning automata. Automatica **44**, 1350–1357 (2008)

Beigy, H., Meybodi, M.R.: Cellular learning automata with multiple learning automata in each cell and its applications. IEEE Trans. Syst. Man Cybern. Part B (Cybernetics) **40**, 54–65 (2010). https://doi.org/10.1109/TSMCB.2009.2030786

Betka, A., Terki, N., Toumi, A., Dahmani, H.: Grey wolf optimizer-based learning automata for solving block matching problem. Signal Image Video Process. **14**, 285–293 (2020). https://doi.org/10.1007/s11760-019-01554-w

Boveiri, H.R., Javidan, R., Khayami, R.: An intelligent hybrid approach for task scheduling in cluster computing environments as an infrastructure for biomedical applications. Expert Syst. (2020). https://doi.org/10.1111/exsy.12536

Bushehrian, O., Nejad, S.E.: Health-care pervasive environments: a CLA based trust management. pp. 247–257 (2017)

Chen, Y., He, H., Zhou, N.: Traffic flow modeling and simulation based on a novel cellular learning automaton. In: 2018 IEEE International Conference of Intelligent Robotic and Control Engineering (IRCE), pp. 233–237. IEEE (2018)

Dai, C., Wang, Y., Ye, M., Xue, X., Liu, H.: An orthogonal evolutionary algorithm with learning automata for multiobjective optimization. IEEE Trans. Cybern. **46,** 3306–3319 (2016). https://doi.org/10.1109/TCYB.2015.2503433

Daliri Khomami, M.M., Haeri, M.A., Meybodi, M.R., Saghiri, A.M.: An algorithm for weighted positive influence dominating set based on learning automata. In: 2017 IEEE 4th International Conference on Knowledge-Based Engineering and Innovation (KBEI), pp. 0734–0740. IEEE (2017)

Daliri Khomami, M.M., Rezvanian, A., Bagherpour, N., Meybodi, M.R.: Minimum positive influence dominating set and its application in influence maximization: a learning automata approach. Appl. Intell. **48**, 570–593 (2018). https://doi.org/10.1007/s10489-017-0987-z

Daliri Khomami, M.M., Rezvanian, A., Saghiri, A.M., Meybodi, M.R.: SIG-CLA: a significant community detection based on cellular learning automata. In: 2020 8th Iranian Joint Congress on Fuzzy and intelligent Systems (CFIS). pp. 039–044 (2020b)

Daliri Khomami, M.M., Rezvanian, A., Saghiri, A.M., Meybodi, M.R.: Utilizing cellular learning automata for finding communities in weighted networks. In: 2020 6th International Conference on Web Research (ICWR), pp. 325–329 (2020a)

Damerchilu, B., Norouzzadeh, M.S., Meybodi, M.R.: Motion estimation using learning automata. Mach. Vis. Appl. **27**, 1047–1061 (2016). https://doi.org/10.1007/s00138-016-0788-0

Deng, X., Jiang, Y., Yang, L.T., Yi, L., Chen, J., Liu, Y., Li, X.: Learning automata based confident information coverage barriers for smart ocean Internet of Things. IEEE Internet Things J. 1 (2020). https://doi.org/10.1109/JIOT.2020.2989696

Di, C., Zhang, B., Liang, Q., Li, S., Guo, Y.: Learning automata based access class barring scheme for massive random access in machine-to-machine communications. IEEE Internet Things J. 1 (2018). https://doi.org/10.1109/JIOT.2018.2867937

Di, C., Su, Y., Han, Z., Li, S.: Learning automata based SVM for intrusion detection, pp. 2067–2074 (2019)

Eiben, A.E., Smith, J.E.: Introduction to Evolutionary Computing. Springer, Heidelberg (2015)

El Hatri, C., Boumhidi, J.: Q-learning based intelligent multi-objective particle swarm optimization of light control for traffic urban congestion management. In: 2016 4th IEEE International Colloquium on Information Science and Technology (CiSt), pp. 794–799. IEEE (2016)
Enayatifar, R., Yousefi, M., Abdullah, A.H., Darus, A.N.: LAHS: a novel harmony search algorithm based on learning automata. Commun. Nonlinear Sci. Numer. Simul. **18**, 3481–3497 (2013). https://doi.org/10.1016/j.cnsns.2013.04.028
Esnaashari, M., Meybodi, M.R.: A cellular learning automata based clustering algorithm for wireless sensor networks. Sensor Lett. **6**, 723–735 (2008)
Esnaashari, M., Meybodi, M.R.M.: A cellular learning automata-based deployment strategy for mobile wireless sensor networks. J. Parallel Distrib. Comput. **71**, 988–1001 (2011)
Esnaashari, M., Meybodi, M.R.: Deployment of a mobile wireless sensor network with k-coverage constraint: a cellular learning automata approach. Wirel. Netw. **19**, 945–968 (2013). https://doi.org/10.1007/s11276-012-0511-7
Esnaashari, M., Meybodi, M.R.: Irregular cellular learning automata. IEEE Trans. Cybern. **45**, 1622–1632 (2018). https://doi.org/10.1016/j.jocs.2017.08.012
Estahbanati, M.J.: Hybrid probabilistic-harmony search algorithm methodology in generation scheduling problem. J. Exp. Theoret. Artif. Intell. **26**, 283–296 (2014)
Fahimi, M., Ghasemi, A.: A distributed learning automata scheme for spectrum management in self-organized cognitive radio network. IEEE Trans. Mob. Comput. **16**, 1490–1501 (2017). https://doi.org/10.1109/TMC.2016.2601926
FathiNavid, A., Aghababa, A.B.: Irregular cellular learning automata-based method for intrusion detection in mobile ad hoc networks. In: 51st International FITCE (Federation of Telecommunications Engineers of the European Community), pp. 1–6 (2012)
Friedman, E., Shenker, S.: Synchronous and asynchronous learning by responsive learning automata (1996)
Ge, H., Huang, J., Di, C., Li, J., Li, S.: Learning automata based approach for influence maximization problem on social networks. In: 2017 IEEE Second International Conference on Data Science in Cyberspace (DSC), pp. 108–117. IEEE (2017)
Geshlag, M.B.M., Sheykhzadeh, J.: A new particle swarm optimization model based on learning automata using deluge algorithm for dynamic environments. J. Basic Appl. Sci. Res. **3**, 394–404 (2012)
Ghamgosar, M., Khomami, M.M.D., Bagherpour, N., Meybodi, M.R.: An extended distributed learning automata based algorithm for solving the community detection problem in social networks. In: 2017 Iranian Conference on Electrical Engineering (ICEE), pp. 1520–1526. IEEE (2017)
Ghavipour, M., Meybodi, M.R.: An adaptive fuzzy recommender system based on learning automata. Electron. Commer. Res. Appl. **20**, 105–115 (2016). https://doi.org/10.1016/j.elerap.2016.10.002
Ghavipour, M., Meybodi, M.R.: Irregular cellular learning automata-based algorithm for sampling social networks. Eng. Appl. Artif. Intell. **59**, 244–259 (2017). https://doi.org/10.1016/j.engappai.2017.01.004
Ghavipour, M., Meybodi, M.R.: A dynamic algorithm for stochastic trust propagation in online social networks: learning automata approach. Comput. Commun. **123**, 11–23 (2018a). https://doi.org/10.1016/j.comcom.2018.04.004
Ghavipour, M., Meybodi, M.R.: Trust propagation algorithm based on learning automata for inferring local trust in online social networks. Knowl. Based Syst. **143**, 307–316 (2018b). https://doi.org/10.1016/j.knosys.2017.06.034
Ghavipour, M., Meybodi, M.R.: A streaming sampling algorithm for social activity networks using fixed structure learning automata. Appl. Intell. **48**, 1054–1081 (2018c). https://doi.org/10.1007/s10489-017-1005-1
Ghosh, L., Ghosh, S., Konar, D., Konar, A., Nagar, A.K.: EEG-induced error correction in path planning by a mobile robot using learning automata. In: Soft Computing for Problem Solving, pp. 273–285 (2019)

Goodwin, M., Yazidi, A.: Distributed learning automata-based scheme for classification using novel pursuit scheme. Appl. Intell. (2020). https://doi.org/10.1007/s10489-019-01627-w

Hadavi, N., Nordin, M.d.J., Shojaeipour, A.: Lung cancer diagnosis using CT-scan images based on cellular learning automata. In: 2014 International Conference on Computer and Information Sciences (ICCOINS), pp. 1–5. IEEE (2014)

Han, Z., Li, S.: Opportunistic routing algorithm based on estimator learning automata, pp. 2486–2492 (2019)

Hariri, A., Rastegar, R., Zamani, M.S., Meybodi, M.R.: Parallel hardware implementation of cellular learning automata based evolutionary computing (CLA-EC) on FPGA. In: 13th Annual IEEE Symposium on Field-Programmable Custom Computing Machines (FCCM 2005), pp. 311–314. IEEE (2005)

Farsi, H., Nasiripour, R., Mohammadzadeh, S.: Eye gaze detection based on learning automata by using SURF descriptor. J. Inf. Syst. Telecommun. (JIST) **21**, 1–10 (2018). https://doi.org/10.7508/jist.2018.21.006

Hasanzadeh, M., Meybodi, M.R.: Grid resource discovery based on distributed learning automata. Computing **96**, 909–922 (2014). https://doi.org/10.1007/s00607-013-0337-x

Hasanzadeh, M., Meybodi, M.R., Ebadzadeh, M.M.: A robust heuristic algorithm for cooperative particle swarm optimizer: a learning automata approach. In: ICEE 2012 - 20th Iranian Conference on Electrical Engineering, Tehran, Iran, pp. 656–661 (2012)

Hasanzadeh, M., Meybodi, M.R., Ebadzadeh, M.M.: Adaptive cooperative particle swarm optimizer. Appl. Intell. **39**, 397–420 (2013). https://doi.org/10.1007/s10489-012-0420-6

Hasanzadeh, M., Meybodi, M.R., Ebadzadeh, M.M.: A learning automata approach to cooperative particle swarm optimizer. J. Inf. Syst. Telecommun. **6**, 56–661 (2014). Tehran, Iran

Hasanzadeh, M., Sadeghi, S., Rezvanian, A., Meybodi, M.R.: Success rate group search optimiser. J. Exp. Theoret. Artif. Intell. **28**, 53–69 (2016)

Hasanzadeh Mofrad, M, Sadeghi, S., Rezvanian, A., Meybodi, M.R.: Cellular edge detection: combining cellular automata and cellular learning automata. AEU Int. J. Electron. Commun. **69**, 1282–1290 (2015). https://doi.org/10.1016/j.aeue.2015.05.010

Hasanzadeh-Mofrad, M., Rezvanian, A.: Learning automata clustering. J. Comput. Sci. **24**, 379–388 (2018). https://doi.org/10.1016/j.jocs.2017.09.008

Hashemi, A.B., Meybodi, M.R.: A note on the learning automata based algorithms for adaptive parameter selection in PSO. Appl. Soft Comput. J. **11**, 689–705 (2011). https://doi.org/10.1016/j.asoc.2009.12.030

Hassanzadeh, T., Meybodi, M.R.: A new hybrid algorithm based on firefly algorithm and cellular learning automata. In: 20th Iranian Conference on Electrical Engineering (ICEE 2012), pp. 628–633. IEEE (2012)

He, S., Wu, Q., Saunders, J.: Group search optimizer: an optimization algorithm inspired by animal searching behavior. IEEE Trans. Evol. Comput. **13**, 973–990 (2009)

Howell, M.N., Gordon, T.J., Brandao, F.V.: Genetic learning automata for function optimization. IEEE Trans. Syst. Man Cybern. Part B (Cybern.) **32**, 804–815 (2002). https://doi.org/10.1109/TSMCB.2002.1049614

Huang, J., Ge, H., Guo, Y., Zhang, Y., Li, S.: A learning automaton-based algorithm for influence maximization in social networks, pp. 715–722 (2018)

Iima, H., Kuroe, Y.: Swarm reinforcement learning algorithms based on Sarsa method. In: 2008 SICE Annual Conference, pp. 2045–2049. IEEE (2008)

Irandoost, M.A., Rahmani, A.M., Setayeshi, S.: A novel algorithm for handling reducer side data skew in MapReduce based on a learning automata game. Inf. Sci. **501**, 662–679 (2019a). https://doi.org/10.1016/j.ins.2018.11.007

Irandoost, M.A., Rahmani, A.M., Setayeshi, S.: Learning automata-based algorithms for MapReduce data skewness handling. J. Supercomput. **75**, 6488–6516 (2019b). https://doi.org/10.1007/s11227-019-02855-0

Jafarpour, B., Meybodi, M.R.: Recombinative CLA-EC. In: Proceedings - International Conference on Tools with Artificial Intelligence, ICTAI. IEEE, pp. 415–422 (2007)

Jafarpour, B., Meybodi, M.R., Shiry, S.: A hybrid method for optimization (Discrete PSO + CLA). In: 2007 International Conference on Intelligent and Advanced Systems, ICIAS 2007, pp. 55–60 (2007)

Jalali Moghaddam, M., Esmaeilzadeh, A., Ghavipour, M., Zadeh, A.K.: Minimizing virtual machine migration probability in cloud computing environments. Cluster Comput. (2020). https://doi.org/10.1007/s10586-020-03067-5

Javadi, M.S., Saniei, M., Rajabi Mashhadi, H.: An augmented NSGA-II technique with virtual database to solve the composite generation and transmission expansion planning problem. J. Exp. Theoret. Artif. Intell. **26**, 211–234 (2014). https://doi.org/10.1080/0952813X.2013.815280

Javadi, M., Mostafaei, H., Chowdhurry, M.U., Abawajy, J.H.: Learning automaton based topology control protocol for extending wireless sensor networks lifetime. J. Netw. Comput. Appl. **122**, 128–136 (2018). https://doi.org/10.1016/j.jnca.2018.08.012

Javadzadeh, R., Afsahi, Z., Meybodi, M.R.: Hybrid model base on artificial immune system and cellular learning automata (CLA-AIS). In: IASTED Technology Conferences/705: ARP/706: RA/707: NANA/728: CompBIO. ACTAPRESS, Calgary, AB, Canada (2010)

Jobava, A., Yazidi, A., Oommen, B.J., Begnum, K.: On achieving intelligent traffic-aware consolidation of virtual machines in a data center using Learning Automata. J. Comput. Sci. **24**, 290–312 (2018). https://doi.org/10.1016/j.jocs.2017.08.005

John Oommen, B., Agache, M.: Continuous and discretized pursuit learning schemes: various algorithms and their comparison. IEEE Trans. Syst. Man Cybern. Part B Cybern. **31**, 277–287 (2001). https://doi.org/10.1109/3477.931507

Kaelbling, L.P., Littman, M.L., Moore, A.W.: Reinforcement learning: a survey. J. Artif. Intell. Res. **4**, 237–285 (1996). https://doi.org/10.1613/jair.301

Kahani, N., Fallah, M.S.: A reactive defense against bandwidth attacks using learning automata. In: Proceedings of the 13th International Conference on Availability, Reliability and Security - ARES 2018, pp. 1–6. ACM Press, New York (2018)

Kamarian, S., Yas, M.H., Pourasghar, A., Daghagh, M.: Application of firefly algorithm and ANFIS for optimisation of functionally graded beams. J. Exp. Theoret. Artif. Intell. **26**, 197–209 (2014). https://doi.org/10.1080/0952813X.2013.813978

Kavousi-Fard, A., Kavousi-Fard, F.: A new hybrid correction method for short-term load forecasting based on ARIMA, SVR and CSA. J. Exp. Theoret. Artif. Intell. **25**, 559–574 (2013). https://doi.org/10.1080/0952813X.2013.782351

Kazemi Kordestani, J., Meybodi, M.R., Rahmani, A.M.: A two-level function evaluation management model for multi-population methods in dynamic environments: hierarchical learning automata approach. J. Exp. Theoret. Artif. Intell. 1–26 (2020). https://doi.org/10.1080/0952813X.2020.1721568

Khadangi, E., Bagheri, A., Shahmohammadi, A.: Biased sampling from facebook multilayer activity network using learning automata. Appl. Intell. **45**, 829–849 (2016). https://doi.org/10.1007/s10489-016-0784-0

Khani, M., Ahmadi, A., Hajary, H.: Distributed task allocation in multi-agent environments using cellular learning automata. Soft Comput. (2017). https://doi.org/10.1007/s00500-017-2839-5

Kheradmand, S., Meybodi, M.R.: Price and QoS competition in cloud market by using cellular learning automata. In: 2014 4th International Conference on Computer and Knowledge Engineering (ICCKE), pp. 340–345. IEEE (2014)

Khezri, S., Meybodi, M.R.: Improving imperialist competitive algorithm using learning automata. In: 16th Annual CSI Computer Conference (CSI 2011), Tehran, Iran (2011)

Khomami, M.M.D., Rezvanian, A., Meybodi, M.R.: Distributed learning automata-based algorithm for community detection in complex networks. Int. J. Mod. Phys. B **30**, 1650042 (2016b). https://doi.org/10.1142/S0217979216500429

Khomami, M.M.D., Bagherpour, N., Sajedi, H., Meybodi, M.R.: A new distributed learning automata based algorithm for maximum independent set problem. In: 2016 Artificial Intelligence and Robotics (IRANOPEN), Qazvin, Iran, Iran, pp. 12–17. IEEE (2016a)

Khomami, M.M.D., Rezvanian, A., Meybodi, M.R.: A new cellular learning automata-based algorithm for community detection in complex social networks. J. Comput. Sci. **24**, 413–426 (2018). https://doi.org/10.1016/j.jocs.2017.10.009

Khomami, M.M.D., Rezvanian, A., Saghiri, A.M., Meybodi, M.R.: Overlapping community detection in social networks using cellular learning automata. In: 2020 28th Iranian Conference on Electrical Engineering (ICEE), pp. 1–6. IEEE (2020)

Khomami, M.M.D., Rezvanian, A., Meybodi, M.R., Bagheri, A.: CFIN: a community-based algorithm for finding influential nodes in complex social networks. J. Supercomput. 2207–2236 (2021). https://doi.org/10.1007/s11227-020-03355-2

King-Sun, F.: Learning control systems–review and outlook. IEEE Trans. Autom. Control **15**, 210–221 (1970). https://doi.org/10.1109/TAC.1970.1099405

Kordestani, J.K., Rezvanian, A., Meybodi, M.R.: CDEPSO: a bi-population hybrid approach for dynamic optimization problems. Appl. Intell. **40**, 682–694 (2014a). https://doi.org/10.1007/s10489-013-0483-z

Kordestani, J.K., Ahmadi, A., Meybodi, M.R.: An improved differential evolution algorithm using learning automata and population topologies. Appl. Intell. **41**, 1150–1169 (2014b). https://doi.org/10.1007/s10489-014-0585-2

Kordestani, J.K., Firouzjaee, H.A., Meybodi, M.R.: An adaptive bi-flight cuckoo search with variable nests for continuous dynamic optimization problems. Appl. Intell. **48**, 97–117 (2018). https://doi.org/10.1007/s10489-017-0963-7

Kordestani, J.K., Ranginkaman, A.E., Meybodi, M.R., Novoa-Hernández, P.: A novel framework for improving multi-population algorithms for dynamic optimization problems: a scheduling approach. Swarm Evol. Comput. **44**, 788–805 (2019). https://doi.org/10.1016/j.swevo.2018.09.002

Krishna, P.V., Misra, S., Joshi, D., Obaidat, M.S.: Learning Automata Based Sentiment Analysis for recommender system on cloud. In: 2013 International Conference on Computer, Information and Telecommunication Systems (CITS), pp. 1–5. IEEE (2013)

Krishna, P.V., Misra, S., Joshi, D., Gupta, A., Obaidat, M.S.: Secure socket layer certificate verification: a learning automata approach. Secur. Commun. Netw. **7**, 1712–1718 (2014). https://doi.org/10.1002/sec.867

Kumar, N., Lee, J.-H., Rodrigues, J.J.: Intelligent mobile video surveillance system as a Bayesian coalition game in vehicular sensor networks: learning automata approach. IEEE Trans. Intell. Transp. Syst. **16**, 1148–1161 (2015). https://doi.org/10.1109/TITS.2014.2354372

Kumar, N., Misra, S., Obaidat, M.S.: Collaborative learning automata-based routing for rescue operations in dense urban regions using vehicular sensor networks. IEEE Syst. J. **9**, 1081–1090 (2015). https://doi.org/10.1109/JSYST.2014.2335451

Lanctot, J.K., Oommen, B.J.: Discretized estimator learning automata. IEEE Trans. Syst. Man Cybern. **22**, 1473–1483 (1992). https://doi.org/10.1109/21.199471

Li, W., Ozcan, E., John, R.: A learning automata based multiobjective hyper-heuristic. IEEE Trans. Evol. Comput. 1 (2018). https://doi.org/10.1109/TEVC.2017.2785346

Lingam, G., Rout, R.R., Somayajulu, D.: Learning automata-based trust model for user recommendations in online social networks. Comput. Electr. Eng. **66**, 174–188 (2018). https://doi.org/10.1016/j.compeleceng.2017.10.017

Mahdaviani, M., Kordestani, J.K., Rezvanian, A., Meybodi, M.R.: LADE: learning automata based differential evolution. Int. J. Artif. Intell. Tools **24**, 1550023 (2015). https://doi.org/10.1142/S0218213015500232

Mahdaviani, M., Kordestani, J.K., Rezvanian, A., Meybodi, M.R: LADE: learning automata based differential evolution. Int. J. Artif. Intell. Tools **24**, 1550023 (2015). https://doi.org/10.1142/S0218213015500232

Mahmoudi, M., Faez, K., Ghasemi, A.: Defense against primary user emulation attackers based on adaptive Bayesian learning automata in cognitive radio networks. Ad Hoc Netw. **102**, 102147 (2020). https://doi.org/10.1016/j.adhoc.2020.102147

Manshad, M.K., Meybodi, M.R., Salajegheh, A.: A new irregular cellular learning automata-based evolutionary computation for time series link prediction in social networks. Appl. Intell. **51**, 71–84 (2021)
Manurung, R., Ritchie, G., Thompson, H.: Using genetic algorithms to create meaningful poetic text. J. Exp. Theor. Artif. Intell. **24**, 43–64 (2012). https://doi.org/10.1080/0952813X.2010.539029
Meybodi, M.R., Lakshmivarahan, S.: ε-Optimality of a general class of learning algorithms. Inf. Sci. **28**, 1–20 (1982). https://doi.org/10.1016/0020-0255(82)90029-9
Misra, S., Interior, B., Kumar, N., Misra, S., Obaidat, M., Rodrigues, J., Pati, B.: Networks of learning automata for the vehicular environment: a performance analysis study. IEEE Wirel. Commun. **21**, 41–47 (2014). https://doi.org/10.1109/MWC.2014.7000970
Mollakhalili Meybodi, M.R., Meybodi, M.R.: Extended distributed learning automata: an automata-based framework for solving stochastic graph. Appl. Intell. **41**, 923–940 (2014)
Mollakhalili Meybodi, M.R., Meybodi, M.R.: Extended distributed learning automata. Appl. Intell. **41**, 923–940 (2014). https://doi.org/10.1007/s10489-014-0577-2
Montague, P.R.: Reinforcement learning: an introduction, by Sutton, R.S. and Barto, A.G. Trends Cogn. Sci. **3**, 360 (1999). https://doi.org/10.1016/S1364-6613(99)01331-5
Moradabadi, B., Meybodi, M.R.: Link prediction based on temporal similarity metrics using continuous action set learning automata. Phys. A **460**, 361–373 (2016). https://doi.org/10.1016/j.physa.2016.03.102
Moradabadi, B., Meybodi, M.R.: Link prediction in fuzzy social networks using distributed learning automata. Appl. Intell. **47**, 837–849 (2017a). https://doi.org/10.1007/s10489-017-0933-0
Moradabadi, B., Meybodi, M.R.: A novel time series link prediction method: learning automata approach. Phys. A **482**, 422–432 (2017b). https://doi.org/10.1016/j.physa.2017.04.019
Moradabadi, B., Meybodi, M.R.: Link prediction in stochastic social networks: learning automata approach. J. Comput. Sci. **24**, 313–328 (2018a). https://doi.org/10.1016/j.jocs.2017.08.007
Moradabadi, B., Meybodi, M.R.: Link prediction in weighted social networks using learning automata. Eng. Appl. Artif. Intell. **70**, 16–24 (2018b). https://doi.org/10.1016/j.engappai.2017.12.006
Moradabadi, B., Meybodi, M.R.: Wavefront cellular learning automata. Chaos **28**, 21101 (2018c). https://doi.org/10.1063/1.5017852
Morshedlou, H., Meybodi, M.R.: Decreasing impact of SLA violations:a proactive resource allocation approachfor cloud computing environments. IEEE Trans. Cloud Comput. **2**, 156–167 (2014). https://doi.org/10.1109/TCC.2014.2305151
Morshedlou, H., Meybodi, M.R.: A new local rule for convergence of ICLA to a compatible point. IEEE Trans. Syst. Man Cybern. Syst. **47**, 3233–3244 (2017). https://doi.org/10.1109/TSMC.2016.2569464
Morshedlou, H., Meybodi, M.R.: A new learning automata based approach for increasing utility of service providers. Int. J. Commun. Syst. **31**, e3459 (2018). https://doi.org/10.1002/dac.3459
Mostafaei, H.: Stochastic barrier coverage in wireless sensor networks based on distributed learning automata. Comput. Commun. **55**, 51–61 (2015)
Mostafaei, H.: Energy-efficient algorithm for reliable routing of wireless sensor networks. IEEE Trans. Ind. Electron. 1 (2018). https://doi.org/10.1109/TIE.2018.2869345
Mostafaei, H., Obaidat, M.S.: A distributed efficient algorithm for self-protection of wireless sensor networks. In: 2018 IEEE International Conference on Communications (ICC), pp. 1–6. IEEE (2018a)
Mostafaei, H., Obaidat, M.S.: Learning automaton-based self-protection algorithm for wireless sensor networks. IET Netw. **7**, 353–361 (2018b). https://doi.org/10.1049/iet-net.2018.0005
Motiee, S., Meybodi, M.R.: Identification of web communities using cellular learning automata. In: 2009 14th International CSI Computer Conference, pp. 553–563. IEEE (2009)
Mousavian, A., Rezvanian, A., Meybodi, M.R.: Solving minimum vertex cover problem using learning automata. In: 13th Iranian Conference on Fuzzy Systems (IFSC 2013), pp. 1–5 (2013)

Mousavian, A., Rezvanian, A., Meybodi, M.R.: Cellular learning automata based algorithm for solving minimum vertex cover problem. In: 2014 22nd Iranian Conference on Electrical Engineering (ICEE), pp. 996–1000. IEEE (2014)

Mozafari, M., Shiri, M.E., Beigy, H.: A cooperative learning method based on cellular learning automata and its application in optimization problems. J. Comput. Sci. **11**, 279–288 (2015). https://doi.org/10.1016/j.jocs.2015.08.002

Nabizadeh, S., Rezvanian, A., Meybodi, M.R.: Tracking extrema in dynamic environment using multi-swarm cellular PSO with local search. Int. J. Electron. Inform. **1**, 29–37 (2012)

Kumpati, S., Narendra, M.A.L.T.: Learning Automata: An Introduction. Prentice-Hall (1989)

Narendra, K.S., Thathachar, M.A.L.: Learning automata - a survey. IEEE Trans. Syst. Man. Cybern. **SMC-4**, 323–334 (1974). https://doi.org/10.1109/TSMC.1974.5408453

Nesi, L.C., da Righi, R.R.: H2-SLAN: a hyper-heuristic based on stochastic learning automata network for obtaining, storing, and retrieving heuristic knowledge. Expert Syst. Appl. **153**, 113426 (2020). https://doi.org/10.1016/j.eswa.2020.113426

Oommen, B.J., Ma, D.C.Y.: Deterministic learning automata solutions to the equipartitioning problem. IEEE Trans. Comput. **37**, 2–13 (1988)

Papadimitriou, G.I., Vasilakos, A.V., Papadimitriou, G.I., Paximadis, C.T.: A new approach to the design of reinforcement schemes for learning automata: stochastic estimator learning algorithms. In: Conference Proceedings 1991 IEEE International Conference on Systems, Man, and Cybernetics, pp. 1387–1392. IEEE (1991)

Papadimitriou, G.I., Pomportsis, A.S., Kiritsi, S., Talahoupi, E.: Absorbing stochastic estimator learning algorithms with high accuracy and rapid convergence. In: Proceedings ACS/IEEE International Conference on Computer Systems and Applications. IEEE Comput. Soc, pp. 45–51 (2002)

Parvanak, A.R., Jahanshahi, M., Dehghan, M.: A cross-layer learning automata based gateway selection method in multi-radio multi-channel wireless mesh networks. Computing (2018). https://doi.org/10.1007/s00607-018-0648-z

Qavami, H.R., Jamali, S., Akbari, M.K., Javadi, B.: A learning automata based dynamic resource provisioning in cloud computing environments. In: 2017 18th International Conference on Parallel and Distributed Computing, Applications and Technologies (PDCAT), pp. 502–509. IEEE (2017)

Qureshi, M.N., Tiwana, M.I., Haddad, M.: Distributed self optimization techniques for heterogeneous network environments using active antenna tilt systems. Telecommun. Syst. **70**, 379–389 (2019). https://doi.org/10.1007/s11235-018-0494-5

Rahmani, P., Javadi, H.H.S., Bakhshi, H., Hosseinzadeh, M.: TCLAB: a new topology control protocol in cognitive MANETs based on learning automata. J. Network Syst. Manage. **26**, 426–462 (2018). https://doi.org/10.1007/s10922-017-9422-3

Rahmanian, A.A., Ghobaei-Arani, M., Tofighy, S.: A learning automata-based ensemble resource usage prediction algorithm for cloud computing environment. Future Gener. Comput. Syst. **79,** 54–71 (2018). https://doi.org/10.1016/j.future.2017.09.049

Rasouli, N., Razavi, R., Faragardi, H.R.: EPBLA: energy-efficient consolidation of virtual machines using learning automata in cloud data centers. Cluster Comput. (2020). https://doi.org/10.1007/s10586-020-03066-6

Rastegar, R., Meybodi, M.R.: A new evolutionary computing model based on cellular learning automata. In: IEEE Conference on Cybernetics and Intelligent Systems, 2004, pp. 433–438. IEEE (2004)

Rastegar, R., Rahmati, M., Meybodi, M.R.: A clustering algorithm using cellular learning automata based evolutionary algorithm. In: Adaptive and Natural Computing Algorithms, pp. 144–150. Springer, Vienna (2005)

Ren, J., Wu, G., Su, X., Cui, G., Xia, F., Obaidat, M.S.: Learning automata-based data aggregation tree construction framework for cyber-physical systems. IEEE Syst. J. **12**, 1467–1479 (2018). https://doi.org/10.1109/JSYST.2015.2507577

Rezaee Jordehi, A., Jasni, J.: Parameter selection in particle swarm optimisation: a survey. J. Exp. Theoret. Artif. Intell. **25**, 527–542 (2013)

Rezapoor Mirsaleh, M., Meybodi, M.R.: LA-MA: a new memetic model based on learning automata. In: 18th National Conference of Computer Society of Iran, pp 1–6 (2013)
Rezapoor Mirsaleh, M., Meybodi, M.R.: A learning automata-based memetic algorithm. Genet. Program. Evol. Mach. **16**, 399–453 (2015). https://doi.org/10.1007/s10710-015-9241-9
Rezapoor Mirsaleh, M., Meybodi, M.R.: A new memetic algorithm based on cellular learning automata for solving the vertex coloring problem. Memetic Comput. **8**, 211–222 (2016). https://doi.org/10.1007/s12293-016-0183-4
Rezapoor Mirsaleh, M., Meybodi, M.R.: Assignment of cells to switches in cellular mobile network: a learning automata-based memetic algorithm. Appl. Intell. **48**, 3231–3247 (2018a). https://doi.org/10.1007/s10489-018-1136-z
Rezapoor Mirsaleh, M., Meybodi, M.R.: A Michigan memetic algorithm for solving the vertex coloring problem. J. Comput. Sci. **24**, 389–401 (2018b). https://doi.org/10.1016/j.jocs.2017.10.005
Rezapoor Mirsaleh, M., Meybodi, M.R.: Balancing exploration and exploitation in memetic algorithms: a learning automata approach. Comput. Intell. **34**, 282–309 (2018c). https://doi.org/10.1111/coin.12148
Rezvanian, A., Meybodi, M.R.: An adaptive mutation operator for artificial immune network using learning automata in dynamic environments. In: 2010 Second World Congress on Nature and Biologically Inspired Computing (NaBIC), pp. 479–483. IEEE (2010a)
Rezvanian, A., Meybodi, M.R.: Tracking extrema in dynamic environments using a learning automata-based immune algorithm. In: Communications in Computer and Information Science, pp. 216–225. Springer, Heidelberg (2010b)
Rezvanian, A., Meybodi, M.R.: LACAIS: Learning automata based cooperative artificial immune system for function optimization. In: Communications in Computer and Information Science, pp. 64–75. Springer, Heidelberg (2010c)
Rezvanian, A., Meybodi, M.R.: An adaptive mutation operator for artificial immune network using learning automata in dynamic environments. In: 2010 Second World Congress on Nature and Biologically Inspired Computing (NaBIC), pp. 479–483. IEEE (2010d)
Rezvanian, A., Meybodi, M.R.: Finding maximum clique in stochastic graphs using distributed learning automata. Int. J. Uncertainty Fuzziness Knowl. Based Syst. **23**, 1–31 (2015a). https://doi.org/10.1142/S0218488515500014
Rezvanian, A., Meybodi, M.R.: Finding minimum vertex covering in stochastic graphs: a learning automata approach. Cyber. Syst. **46**, 698–727 (2015b). https://doi.org/10.1080/01969722.2015.1082407
Rezvanian, A., Meybodi, M.R.: Stochastic Social Networks: Measures and Algorithms. LAP LAMBERT Academic Publishing (2016a)
Rezvanian, A., Meybodi, M.R.: Stochastic graph as a model for social networks. Comput. Hum. Behav. **64**, 621–640 (2016b). https://doi.org/10.1016/j.chb.2016.07.032
Rezvanian, A., Meybodi, M.R.: Sampling algorithms for stochastic graphs: a learning automata approach. Knowl. Based Syst. **127**, 126–144 (2017a). https://doi.org/10.1016/j.knosys.2017.04.012
Rezvanian, A., Meybodi, M.R.: A new learning automata-based sampling algorithm for social networks. Int. J. Commun. Syst. **30**, e3091 (2017b). https://doi.org/10.1002/dac.3091
Rezvanian, A., Rahmati, M., Meybodi, M.R.: Sampling from complex networks using distributed learning automata. Physica A **396**, 224–234 (2014). https://doi.org/10.1016/j.physa.2013.11.015
Rezvanian, A., Saghiri, A.M., Vahidipour, S.M., Esnaashari, M., Meybodi, M.R.: Learning automata theory. In: Recent Advances in Learning Automata, pp. 3–19. Springer (2018a)
Rezvanian, A., Saghiri, A.M., Vahidipour, S.M., Esnaashari, M., Meybodi, M.R.: Recent Advances in Learning Automata. Springer (2018b)
Rezvanian, A., Saghiri, A.M., Vahidipour, S.M., Esnaashari, M., Meybodi, M.R.: Cellular Learning Automata. pp 21–88 (2018c)

Rezvanian, A., Vahidipour, S.M., Esnaashari, M.: New applications of learning automata-based techniques in real-world environments. J. Comput. Sci. **24**, 287–289 (2018d). https://doi.org/10.1016/j.jocs.2017.11.012
Rezvanian, A., Saghiri, A.M., Vahidipour, S.M., Esnaashari, M., Meybodi, M.R.: Learning automata for cognitive peer-to-peer networks. In: Recent Advances in Learning Automata, pp. 221–278 (2018e)
Rezvanian, A., Saghiri, A.M., Vahidipour, S.M., Esnaashari, M., Meybodi, M.R.: Learning automata for wireless sensor networks. In: Recent Advances in Learning Automata, pp. 91–219 (2018f)
Rezvanian, A., Moradabadi, B., Ghavipour, M., Daliri Khomami, M.M., Meybodi, M.R.: Social recommender systems. In: Learning Automata Approach for Social Networks, pp. 281–313. Springer (2019a)
Rezvanian, A., Moradabadi, B., Ghavipour, M., Daliri Khomami, M.M., Meybodi, M.R.: Wavefront cellular learning automata: a new learning paradigm. In: Learning Automata Approach for Social Networks, pp. 51–74. Springer (2019b)
Rezvanian, A., Moradabadi, B., Ghavipour, M., Daliri Khomami, M.M., Meybodi, M.R.: Social networks and learning systems: a bibliometric analysis. In: Learning Automata Approach for Social Networks, pp. 75–89. Springer (2019c)
Rezvanian, A., Moradabadi, B., Ghavipour, M., Khomami, M.M.D., Meybodi, M.R.: Social link prediction. In: Learning Automata Approach for Social Networks, pp. 169–239. Springer (2019d)
Rezvanian, A., Moradabadi, B., Ghavipour, M., Daliri Khomami, M.M., Meybodi, M.R.: Social trust management. In: Learning Automata Approach for Social Networks, pp. 241–279. Springer (2019e)
Rezvanian, A., Moradabadi, B., Ghavipour, M., Daliri Khomami, M.M., Meybodi, M.R.: Learning Automata Approach for Social Networks. Springer International Publishing (2019f)
Rezvanian, A., Moradabadi, B., Ghavipour, M., Daliri Khomami, M.M., Meybodi, M.R.: Introduction to learning automata models. In: Learning Automata Approach for Social Networks, pp. 1–49. Springer (2019g)
Willianms, R.J.: Toward a Theory of Reinforcement-Learning Connectionist Systems. Northeastern University (1988)
Roohollahi, S., Bardsiri, A.K., Keynia, F.: Using an evaluator fixed structure learning automata in sampling of social networks. J AI Data Min. **8**, 127–148 (2020). https://doi.org/10.22044/JADM.2019.7145.1842
Ruan, X., Jin, Z., Tu, H., Li, Y.: Dynamic cellular learning automata for evacuation simulation. IEEE Intell. Transp. Syst. Mag. **11**, 129–142 (2019). https://doi.org/10.1109/MITS.2019.2919523
Rummery, G.A.A., Niranjan, M.: On-line Q-learning using connectionist systems. University of Cambridge, Department of Engineering (1994)
Safara, F., Souri, A., Deiman, S.F.: Super peer selection strategy in peer-to-peer networks based on learning automata. Int. J. Commun. Syst. **33,** e4296 (2020). https://doi.org/10.1002/dac.4296
Saghiri, A.M., Meybodi, M.R.: An approach for designing cognitive engines in cognitive peer-to-peer networks. J. Netw. Comput. Appl. **70**, 17–40 (2016). https://doi.org/10.1016/j.jnca.2016.05.012
Saghiri, A.M., Meybodi, M.R.: A closed asynchronous dynamic model of cellular learning automata and its application to peer-to-peer networks. Genet. Program. Evol. Mach. **18**, 313–349 (2017a). https://doi.org/10.1007/s10710-017-9299-7
Saghiri, A.M., Meybodi, M.R.: A distributed adaptive landmark clustering algorithm based on mOverlay and learning automata for topology mismatch problem in unstructured peer-to-peer networks. Int. J. Commun. Syst. **30**, e2977 (2017b). https://doi.org/10.1002/dac.2977
Saghiri, A.M., Meybodi, M.R.: An adaptive super-peer selection algorithm considering peers capacity utilizing asynchronous dynamic cellular learning automata. Appl. Intell. **48**, 271–299 (2018a). https://doi.org/10.1007/s10489-017-0946-8
Saghiri, A.M., Meybodi, M.R.: Open asynchronous dynamic cellular learning automata and its application to allocation hub location problem. Knowl. Based Syst. **139**, 149–169 (2018b). https://doi.org/10.1016/j.knosys.2017.10.021

Saleem, A., Afzal, M.K., Ateeq, M., Kim, S.W., Bin, Z.Y.: Intelligent learning automata-based objective function in RPL for IoT. Sustain. Cities Soc. **59**, 102234 (2020). https://doi.org/10.1016/j.scs.2020.102234

Samma, H., Lim, C.P., Mohamad Saleh, J.: A new reinforcement learning-based memetic particle swarm optimizer. Appl. Soft Comput. **43**, 276–297 (2016). https://doi.org/10.1016/j.asoc.2016.01.006

Santoso, J., Riyanto, B., Adiprawita, W.: Dynamic path planning for mobile robots with cellular learning automata. J. ICT Res. Appl. **10**, 1–14 (2016). https://doi.org/10.5614/itbj.ict.res.appl.2016.10.1.1

Saraeian, S., Shirazi, B., Motameni, H.: Optimal autonomous architecture for uncertain processes management. Inf. Sci. **501**, 84–99 (2019). https://doi.org/10.1016/j.ins.2019.05.095

Savargiv, M., Masoumi, B., Keyvanpour, M.R.: A new ensemble learning method based on learning automata. J. Ambient Intell. Human. Comput. (2020). https://doi.org/10.1007/s12652-020-01882-7

Schwartz, A.: A reinforcement learning method for maximizing undiscounted rewards. In: Machine Learning Proceedings 1993, pp. 298–305 (1993)

Sengupta, A., Chakraborti, T., Konar, A., Kim, E., Nagar, A.K.: An adaptive memetic algorithm using a synergy of differential evolution and learning automata. In: 2012 IEEE Congress on Evolutionary Computation, pp. 1–8. IEEE (2012)

Seyyedi, S.H., Minaei-Bidgoli, B.: Estimator learning automata for feature subset selection in high-dimensional spaces, case study: email spam detection. Int. J. Commun. Syst. **31**, e3541 (2018). https://doi.org/10.1002/dac.3541

Shen, X.-N., Minku, L.L., Marturi, N., Guo, Y.-N., Han, Y.: A Q-learning-based memetic algorithm for multi-objective dynamic software project scheduling. Inf. Sci. **428**, 1–29 (2018). https://doi.org/10.1016/j.ins.2017.10.041

Sheng, X., Xu, W.: Solving the economic dispatch problem with q-learning quantum-behaved particle swarm optimization method. In: 2015 14th International Symposium on Distributed Computing and Applications for Business Engineering and Science (DCABES), pp. 98–101. IEEE (2015)

Sheybani, M., Meybodi, M.R.: PSO-LA: a new model for optimization. In: 12th Annual International Computer Society of Iran Computer Conference CSICC2007, Iran, pp. 1162–1169 (2007a)

Sheybani, M., Meybodi, M.R.: CLA-PSO: a new model for optimization. In: Proceedings of the 15th Conference on Electrical Engineering, Volume on Computer, Telecommunication Research Center, Tehran, Iran, pp. 1–8 (2007b)

Shi, Y., Eberhart, R.: A modified particle swarm optimizer. In: 1998 IEEE International Conference on Evolutionary Computation Proceedings. IEEE World Congress on Computational Intelligence (Cat. No. 98TH8360), pp. 69–73. IEEE (1998)

Shyu, S.J., Yin, P.-Y., Lin, B.M., Haouari, M.: Ant-tree: an ant colony optimization approach to the generalized minimum spanning tree problem. J. Exp. Theoret. Artif. Intell. **15,** 103–112 (2003)

Sikeridis, D., Tsiropoulou, E.E., Devetsikiotis, M., Papavassiliou, S.: Socio-physical energy-efficient operation in the internet of multipurpose things. In: 2018 IEEE International Conference on Communications (ICC), pp. 1–7. IEEE (2018)

Simha, R., Kurose, J.F.: Relative reward strength algorithms for learning automata. IEEE Trans. Syst. Man Cybern. **19**, 388–398 (1989). https://doi.org/10.1109/21.31041

Sohrabi, M.K., Roshani, R.: Frequent itemset mining using cellular learning automata. Comput. Hum. Behav. **68,** 244–253 (2017). https://doi.org/10.1016/j.chb.2016.11.036

Soleimani-Pouri, M., Rezvanian, A., Meybodi, M.R.: Solving maximum clique problem in stochastic graphs using learning automata. In: 2012 Fourth International Conference on Computational Aspects of Social Networks (CASoN), pp. 115–119. IEEE (2012)

Soleimani-pouri, M., Rezvanian, A., Meybodi, M.R.: An ant based particle swarm optimization algorithm for maximum clique problem in social networks. In: Can, F., Özyer, T., Polat, F. (eds.) State of the Art Applications of Social Network Analysis, pp. 295–304. Springer (2014)

Stuart, R., Norvig, P.: Artificial Intelligence: A Modern Approach, 3rd edn. Printice-Hall (2002)

Su, Y., Qi, K., Di, C., Ma, Y., Li, S.: Learning automata based feature selection for network traffic intrusion detection. In: 2018 IEEE Third International Conference on Data Science in Cyberspace (DSC), pp. 622–627. IEEE (2018)

Sutton, R.S., Barto, A.G.: Reinforcement Learning: An Introduction. MIT Press, Cambridge (1998)

Thakur, D., Khatua, M.: Cellular Learning Automata-Based Virtual Network Embedding in Software-Defined Networks, pp. 173–182 (2019)

Thathachar, M.A.L., Harita, B.R.: Learning automata with changing number of actions. IEEE Trans. Syst. Man Cybern. **17**, 1095–1100 (1987). https://doi.org/10.1109/TSMC.1987.6499323

Thathachar, M.A.L., Ramachandran, K.M.: Asymptotic behaviour of a learning algorithm. Int. J. Control **39**, 827–838 (1984). https://doi.org/10.1080/00207178408933209

Thathachar, M.A.L., Sastry, P.S.: A new approach to the design of reinforcement schemes for learning automata. IEEE Trans. Syst. Man Cybern. **SMC-15**, 168–175 (1985a). https://doi.org/10.1109/TSMC.1985.6313407

Thathachar, M.A.L., Sastry, P.S.: A class of rapidly converging algorithms for learning automata. IEEE Trans. Syst. Man Cybern. **SMC-15**, 168–175 (1985b)

Thathachar, M., Sastry, P.: Estimator algorithms for learning automata. In: Proceedings of the Platinum Jubilee Conference on Systems and Signal Processing, Bengalore, India (1986)

Thathachar, M.A.L., Sastry, P.S.: Varieties of learning automata: an overview. IEEE Trans. Syst. Man Cybern. Part B Cybern. **32**, 711–722 (2002). https://doi.org/10.1109/TSMCB.2002.1049606

Thathachar, M.A.L., Sastry, P.S.: Networks of Learning Automata. Springer, Boston (2004)

Toffolo, T.A.M., Christiaens, J., Van Malderen, S., Wauters, T., Vanden Berghe, G.: Stochastic local search with learning automaton for the swap-body vehicle routing problem. Comput. Oper. Res. **89**, 68–81 (2018). https://doi.org/10.1016/j.cor.2017.08.002

Toozandehjani, H., Zare-Mirakabad, M.-R., Derhami, V.: Improvement of recommendation systems based on cellular learning automata. In: 2014 4th International Conference on Computer and Knowledge Engineering (ICCKE), pp. 592–597. IEEE (2014)

Tsetlin, M.L.: On the behavior of finite automata in random media. Autom. Remote Control **22**, 1210–1219 (1962)

Vafaee Sharbaf, F., Mosafer, S., Moattar, M.H.: A hybrid gene selection approach for microarray data classification using cellular learning automata and ant colony optimization. Genomics **107**, 231–238 (2016). https://doi.org/10.1016/j.ygeno.2016.05.001

Vafashoar, R., Meybodi, M.R.: Multi swarm bare bones particle swarm optimization with distribution adaption. Appl. Soft Comput. J. **47**, 534–552 (2016). https://doi.org/10.1016/j.asoc.2016.06.028

Vafashoar, R., Meybodi, M.R.: Multi swarm optimization algorithm with adaptive connectivity degree. Appl. Intell. **48**, 909–941 (2018). https://doi.org/10.1007/s10489-017-1039-4

Vafashoar, R., Meybodi, M.R.: Reinforcement learning in learning automata and cellular learning automata via multiple reinforcement signals. Knowl. Based Syst. **169**, 1–27 (2019a). https://doi.org/10.1016/j.knosys.2019.01.021

Vafashoar, R., Meybodi, M.R.: Cellular learning automata based bare bones PSO with maximum likelihood rotated mutations. Swarm Evol. Comput. **44**, 680–694 (2019b). https://doi.org/10.1016/j.swevo.2018.08.016

Vafashoar, R., Meybodi, M.R.: A multi-population differential evolution algorithm based on cellular learning automata and evolutionary context information for optimization in dynamic environments. Appl. Soft Comput. **88**, 106009 (2020). https://doi.org/10.1016/j.asoc.2019.106009

Vafashoar, R., Meybodi, M.R., Momeni Azandaryani, A.H.: CLA-DE: a hybrid model based on cellular learning automata for numerical optimization. Appl. Intell. **36**, 735–748 (2012). https://doi.org/10.1007/s10489-011-0292-1

Vahidipour, S.M., Meybodi, M.R., Esnaashari, M.: Finding the shortest path in stochastic graphs using learning automata and adaptive stochastic petri nets. Int. J. Uncertainty Fuzziness Knowl. Based Syst. **25**, 427–455 (2017b). https://doi.org/10.1142/S0218488517500180

Vahidipour, S.M., Meybodi, M.R., Esnaashari, M.: Adaptive Petri net based on irregular cellular learning automata with an application to vertex coloring problem. Appl. Intell. **46**, 272–284 (2017a). https://doi.org/10.1007/s10489-016-0831-x

Vahidipour, S.M., Esnaashari, M., Rezvanian, A., Meybodi, M.R.: GAPN-LA: a framework for solving graph problems using Petri nets and learning automata. Eng. Appl. Artif. Intell. **77**, 255–267 (2019). https://doi.org/10.1016/j.engappai.2018.10.013

Vasilakos, A.V., Paximadis, C.T.: Faulttolerant routing algorithms using estimator discretized learning automata for high-speed packet-switched networks. IEEE Trans. Reliab. **43**, 582–593 (1994). https://doi.org/10.1109/24.370222

Velusamy, G., Lent, R.: Dynamic cost-aware routing of web requests. Future Internet **10**, 57 (2018). https://doi.org/10.3390/fi10070057

Verbeeck, K., Nowé, A., Nowe, A.: Colonies of learning automata. IEEE Trans. Syst. Man Cybern. Part B Cybern. **32**, 772–780 (2002). https://doi.org/10.1109/TSMCB.2002.1049611

Watkins, C.C.J.H.: Learning from Delayed Rewards (1989)

Wolfram, S.: Theory and applications of cellular automata. World Scientific Publication (1986)

Wu, G., Mallipeddi, R., Suganthan, P.N.: Ensemble strategies for population-based optimization algorithms – a survey. Swarm Evol. Comput. **44**, 695–711 (2019). https://doi.org/10.1016/j.swevo.2018.08.015

Xue, L., Sun, C., Wunsch, D.C.: A game-theoretical approach for a finite-time consensus of second-order multi-agent system. Int. J. Control Autom. Syst. **17**, 1071–1083 (2019). https://doi.org/10.1007/s12555-017-0716-8

Yas, M.H., Kamarian, S., Pourasghar, A.: Application of imperialist competitive algorithm and neural networks to optimise the volume fraction of three-parameter functionally graded beams. J. Exp. Theoret. Artif. Intell. **26**, 1–12 (2014)

Yazdani, D., Golyari, S., Meybodi, M.R.: A new hybrid algorithm for optimization based on artificial fish swarm algorithm and cellular learning automata. In: In: Proceedings of 2010 5th International Symposium on Telecommunications (IST), Tehran, Iran, pp. 932–937 (2010)

Yazidi, A., Bouhmala, N., Goodwin, M.: A team of pursuit learning automata for solving deterministic optimization problems. Appl. Intell. (2020). https://doi.org/10.1007/s10489-020-01657-9

Yu, X., Gen, M.: Introduction to Evolutionary Algorithms. Springer, London (2010)

Zamani, M.S., Mehdipour, F., Meybodi, M.R.: Implementation of cellular learning automata on reconfigurable computing systems. In: CCECE 2003 - Canadian Conference on Electrical and Computer Engineering. Toward a Caring and Humane Technology (Cat. No. 03CH37436), pp. 1139–1142. IEEE (2003)

Zanganeh, S., Meybodi, M.R., Sedehi, M.H.: Continuous CLA-EC. In: 2010 Fourth International Conference on Genetic and Evolutionary Computing, pp. 186–189. IEEE (2010)

Zarei, B., Meybodi, M.R.: Improving learning ability of learning automata using chaos theory. J. Supercomputing (2020). https://doi.org/10.1007/s11227-020-03293-z

Zhang, J., Xu, L., Li, J., Kang, Q., Zhou, M.: Integrating particle swarm optimization with learning automata to solve optimization problems in noisy environment. In: 2014 IEEE International Conference on Systems, Man, and Cybernetics (SMC), pp. 1432–1437. IEEE (2014)

Zhang, J., Xu, L., Ma, J., Zhou, M.: A learning automata-based particle swarm optimization algorithm for noisy environment. In: 2015 IEEE Congress on Evolutionary Computation (CEC), pp. 141–147 (2015)

Zhang, F., Wang, X., Li, P., Zhang, L.: An energy aware cellular learning automata based routing algorithm for opportunistic networks. Int. J. Grid Distrib. Comput. **9**, 255–272 (2016). https://doi.org/10.14257/ijgdc.2016.9.2.22

Zhang, J., Zhu, X., Zhou, M.: Learning Automata-based particle swarm optimizer. In: 2018 IEEE Congress on Evolutionary Computation (CEC), pp. 1–6. IEEE (2018)

Zhao, Y., Jiang, W., Li, S., Ma, Y., Su, G., Lin, X.: A cellular learning automata based algorithm for detecting community structure in complex networks. Neurocomputing **151**, 1216–1226 (2015). https://doi.org/10.1016/j.neucom.2014.04.087

Chapter 2
Learning Automaton and Its Variants for Optimization: A Bibliometric Analysis

Javidan Kazemi Kordestani, Mehdi Razapoor Mirsaleh, Alireza Rezvanian, and Mohammad Reza Meybodi

Abstract Learning automaton (LA) is one of the powerful reinforcement learning techniques. Due to the powerful features of the LA, they have found many optimization applications in areas of evolutionary computation, cellular networks, cloud computing, computer networks, data mining, distributed systems, graph theory, internet of things, mech networks, p2p networks, social networks, to mention a few. This chapter investigates all research results and addresses the critical studies' potential learning automata and optimization trends. In this regard, first, we evaluated 245 papers indexed in Web of Science (WoS) and 506 papers indexed in Scopus until April 2021. We identified the top contributing authors, institutes, countries, journals, and key research topics based on bibliometric network analysis techniques.

2.1 Introduction

In this chapter, before describing the methodology for systematic bibliometric analysis on learning automata and optimization, we briefly introduce LA models and optimization in Sect. 2.2. The material and method are presented in Sect. 2.3, the

J. Kazemi Kordestani
Department of Computer Engineering, Science and Research Branch, Islamic Azad University, Tehran, Iran

M. Razapoor Mirsaleh
Department of Computer Engineering and Information Technology, Payame Noor University (PNU), P.O. BOX 19395-3697, Tehran, Iran
e-mail: mrezapoorm@aut.ac.ir

A. Rezvanian (✉)
Department of Computer Engineering, University of Science and Culture, Tehran, Iran
e-mail: rezvanian@usc.ac.ir

M. R. Meybodi
Computer Engineering Department, Amirkabir University of Technology (Tehran Polytechnic), Tehran, Iran
e-mail: mmeybodi@aut.ac.ir

J. Kazemi Kordestani et al. (eds.), *Advances in Learning Automata and Intelligent Optimization*, Intelligent Systems Reference Library 208, https://doi.org/10.1007/978-3-030-76291-9_2

results of the bibliometric results on the learning automata and optimization are analyzed in Sect. 2.4, and finally, Sect. 2.5 concludes the chapter.

2.2 Learning Automata Models and Optimization

Learning automaton (LA) as one of the computational intelligence techniques have been found very useful tool to solve many complex and real-world problems in networks where a large amount of uncertainty or lacking the information about the environment exists (Torkestani 2012; Mousavian et al. 2013, 2014; Mahdaviani et al. 2015; Zhao et al. 2015; Khomami et al. 2016; Mirsaleh and Meybodi 2016). A learning automaton is a stochastic model operating in reinforcement learning (Narendra and Thathachar 1989; Thathachar and Sastry 2004). This model can be classified under the reinforcement learning schemes in the temporal-difference (TD) learning methods category. *TD* learning is a combination of the Monte Carlo ideas and the dynamic programming ideas. Like Monte Carlo methods, *TD* methods can learn directly from raw experience without a model of the environment's dynamics. Like the dynamic programming, *TD* methods update estimates based on the other learned estimates without waiting for an outcome (Sutton and Barto 1998).

Cellular learning automaton (CLA) (Rezvanian et al. 2018; Vafashoar et al. 2021a), as a new reinforcement learning approach, is a combination of cellular automata (CA) and learning automata (LA). It is formed by a group of interconnected cells arranged in some regular forms, such as a grid or ring, in which each cell contains one or more LAs. A cellular learning automaton can be considered a distributed model, inheriting both the computational power of cellular automata and learning automata's learning ability. Accordingly, it is more powerful than a single learning automaton due to producing more complex behavior patterns. Also, cells' learning abilities enable cellular learning automata to produce these complex patterns by understandable behavioral rules rather than complicated mathematical functions commonly used in cellular automata. Owing to these characteristics, cellular learning automata can be successfully employed in modeling, learning, simulating, controlling, and solving challenging problems in uncertain, distributed, and complex environments.

Several studies have been conducted to investigate both the theoretical aspects and the applications of learning automata models. From a theoretical point of view, several models have been introduced, and their characteristics, such as their convergence behavior, have been studied. Also, LA has been received many attentions in a wide range of application such as channel assignment, image processing, machine vision, data mining, computer networks, scheduling, Peer-to-Peer networks, cellular networks, wireless networks, grid computing, cloud computing, Petri nets, complex network, social network analysis, and optimization, to mention a few.

Optimization problems are one of the major challenging tasks in almost every field of science. Analytical methods are not applicable in most real-world scenarios; therefore, several numerical methods such as evolutionary algorithms (Eiben and Smith 2015) and learning automata-based methods (Howell et al. 2002; Zeng and Liu 2005) have been suggested to solve these problems by estimating global optima. Most of these methods suffer from convergence to local optima and slow convergence rate. LA has been applied to a wide range of science and engineering domains (Thathachar and Sastry 2004), including different optimization problems (Sastry et al. 2009; Granmo and Oommen 2010a).

LA and CLA approaches have also been used to tackle global numerical optimization problems. Sastry et al. used a stochastic optimization method based on continuous-action-set learning automata for learning a hyperplane classifier that is robust to noise (Sastry et al. 2009). Despite its effectiveness in various domains, learning automata have been criticized for having a slow convergence rate. Zeng and Liu (Zeng and Liu 2005) have used learning automata for continuously dividing and sampling the search space. Their algorithm gradually concentrates on favorable regions and leaves out others. Beigy and Meybodi presented a new learning algorithm for continuous action learning automata (Beigy and Meybodi 2006) and proved its convergence in stationary environments. They also showed the applicability of the introduced algorithm in solving noise corrupted optimization problems. An evolutionary-inspired learning method called genetic learning automata was introduced by Howell et al. (Howell et al. 2002). In this hybrid method, genetic algorithms and learning automata are combined to compensate for both methods' drawbacks. Genetic learning automata algorithm evolves a population of probability strings for reaching the global optima. At each generation, each probability string gives rise to a one-bit string child. These children are evaluated using a fitness function, and based on this evaluation, probability strings are adjusted. Then these probability strings are further maturated using standard evolutionary operators. A somehow closer approach termed CLA-EC (Rastegar et al. 2006) is developed based on cellular learning automata and the evolutionary computing paradigm. Learning automata techniques is applied for adaptively tuning of parameters to improving the performance of the genetic algorithm (Abtahi et al. 2008), particle swarm optimization (Jafarpour et al. 2007; Hamidi and Meybodi 2008; Vafashoar and Meybodi 2016), Artificial Immune System (Rezvanian and Meybodi 2010c; Rezvanian and Meybodi 2010a, b), Improving Imperialist Competitive (Khezri and Meybodi 2011), Artificial Bee Colony (Aghazadeh and Meybodi 2011), differential evolution (Kordestani et al. 2014; Mahdaviani et al. 2015; Vafashoar and Meybodi 2020), harmony search algorithm (Enayatifar et al. 2013), Firefly Algorithm (Hassanzadeh and Meybodi 2012), Cuckoo Search (Kordestani et al. 2018), artificial fish swarm algorithm (Yazdani et al. 2010), and gravitational search algorithm (Alirezanejad et al. 2020).

2.3 Material and Method

We adopted a methodology for preparing a literature review similar to the methodology used in (Rezvanian et al. 2019; Vafashoar et al. 2021b) in five steps, including document acquisition, creating notes, structuring the literature review, writing the literature review, and building the bibliography. In this study, similar to this methodology, the research documents from Scopus and Web of Science (WoS) are collected, the results are processed and refined. The refined results are structured as key materials (e.g., top contributing authors, institutes, publications, countries, research topics, and applications), and finally, some insights into current research interests, trends, and future directions are provided. The network visualizations and bibliometric representations are made by Gephi (Bastian et al. 2009) and VOSviewer (van Eck and Waltman 2010, 2013). Gephi is an open-source network exploration, manipulation, analysis, and visualization software package written in Java on the NetBeans platform. VOSviewer provides network visualization on co-authorship, co-citation, and citation concerning authors, organizations, and countries and co-occurrence concerning keywords.

2.3.1 Data Collection and Initial Results

The data was collected from both Scopus and WoS. The WoS (previously known as Web of Knowledge) is an online subscription-based scientific citation indexing service that provides a comprehensive citation search for many different academic disciplines which covers more than 90 million records during 1900 to present from peer-reviewed journals belonging to publishing houses such as Elsevier, Springer, Wiley, IEEE, ACM, and Taylor & Francis, to name a few. It was initially produced by the Institute for Scientific Information (ISI) and is currently maintained by Clarivate Analytics (previously the Intellectual Property and Science business of Thomson Reuters) (Analytics and Clarivate 2017). Scopus is Elsevier's abstract and citation database which covers more than 69 million records consist of nearly 36,377 titles (22,794 working titles and 13,583 inactive titles) from approximately 11,678 publishers, of which 34,346 are peer-reviewed journals in top-level subject fields such as life sciences, social sciences, physical sciences, and health sciences. All journals covered in the Scopus database, regardless of whom they are published under, are reviewed each year to ensure high-quality standards are maintained.

Table 2.1 Types of documents about learning automata and optimization related research

Document type	WoS	Scopus
Conference proceedings papers	5	233
Journal papers	240	265
Total	245	498

For data collection, we searched for the ((“learning automata” OR “learning automaton”) AND (“optimization” OR “optimisation”)) as the two main keywords in the topic of papers belonging to both WoS and Scopus. The initial search resulted in 245 papers for WoS from 1990 until April 2021 and 506 papers for Scopus from 1974 until April 2021.

2.3.2 Refining the Initial Results

To refine the search results, the non-English language results such as Chinese (6), Japanese (1), and underdefined (1) for Scopus results are excluded during the data purification process. Thus, the restriction on these results produced 498 papers for Scopus results. The paper types of the search results, as presented in Table 2.1, can be categorized as conference proceedings papers and journal papers as 5 and 240 papers for WoS, and 233 and 265 papers for Scopus, respectively. The final search results consist of 498 and 245 for WoS and Scopus, respectively.

2.4 Analyzing the Results

We performed data analysis in two steps. In the first step, statistical analysis is extracted from the resulted search, networks of entities are extracted using Gephi (Bastian et al. 2009), and bibliometric analysis is performed using VOSviewer (van Eck and Waltman 2013). The network visualization and bibliometric representation, including co-authorship, co-citation, and the citation for authors, organizations, and countries, and co-occurrence to keywords. Thus, in this section, several statistics, results, and analyses are reported for the research related to “learning automata” and “optimization” based on the final resulted search.

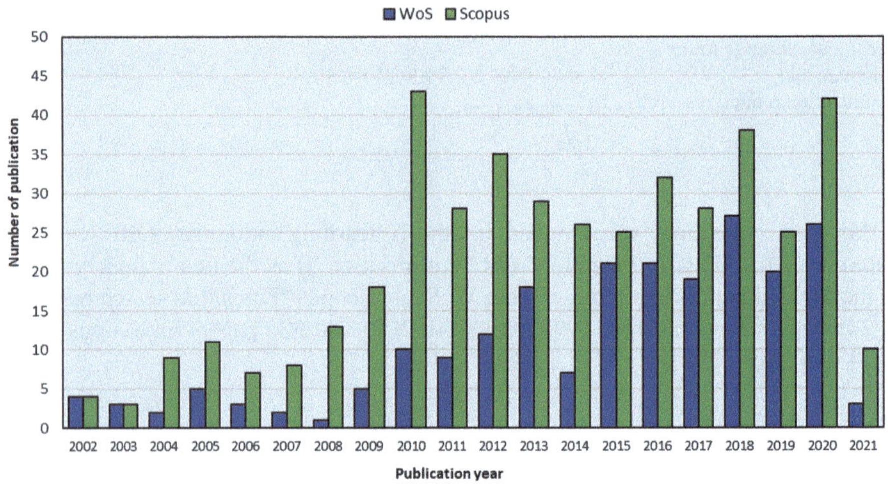

Fig. 2.1 Distribution of papers published during the time 2003–2020

2.4.1 *Initial Result Statistics*

In Fig. 2.1, the number of papers published during the time 2002–2021 is shown whose demonstrates the changing pattern of publications in the research community increasingly. It can be seen from the figure that the number of publications on the research related to the topic of "learning automata" and "optimization" was dramatically increasing from 2010. However, since then, it has been increasing slightly.

2.4.2 *Top Journals*

To understand the role of the different scientific journals and conference series, we identified the top 10 journals and conference titles appearing in the data with the most published in this field of research related to the topic of "learning automata" and "optimization." It was found that these papers have been published in 101 different journals and 136 different conferences. Table 2.2 and Table 2.3 show the distribution of the journal and conference series titles with the most publication for research related to the study topic, respectively.

Table 2.2 Distribution of the top 10 journal titles with the most publication related to the topic of "learning automata" and "optimization."

	Journal name	Publisher	No. publications
1	Applied Intelligence	Springer Nature	15
2	IEEE Transactions on Systems Man and Cybernetics Part B Cybernetics	IEEE	11
3	Applied Soft Computing Journal	Elsevier	9
4	International Journal of Systems Science	Taylor & Francis	8
5	Computer Communications	Elsevier	6
6	IEEE Transactions On Systems Man And Cybernetics	IEEE	6
7	Information Sciences	Elsevier	5
8	Journal of Computational Science	Elsevier	5
9	Neurocomputing	Elsevier	5
10	Swarm and Evolutionary Computation	Elsevier	5

The citation network for the journals with a minimum of one citation for each journal in research related to the topic of "learning automata" and "optimization" is shown in Fig. 2.2, whose node size is proportional to the number of documents published each journal.

2.4.3 *Top Researchers*

From the resulted search for the research related to the topic of "learning automata" and "optimization," the top ten contributing authors based on the number of publications in particular for this research topic of study are extracted. These results are summarized in Table 2.4, in which in the second column is the author's name, the third column reflects the number of publications on this topic, and the last column shows the highly cited article on this topic for each author. From the results reported in Table 2.4, it can be observed that Meybodi, M. R., with 43 articles, dominates the list and is followed by Oommen, B. J. each with 15 publications.

Table 2.3 Distribution of the top 10 conference series titles for the research related to the topic of "learning automata" and "optimization."

	Conference series title	Conference abbreviation	No. publications
1	IEEE International Conference on Systems Man and Cybernetics	SMC	8
2	IEEE Congress on Evolutionary Computation	CEC	8
3	International Conference on Communications	ICC	4
4	Iranian Conference on Electrical Engineering	ICEE	4
5	World Congress of the International Federation of Automatic Control	IFAC	3
6	Iranian Conference on Intelligent Systems	ICIS	3
7	International Conference on Hybrid Intelligent Systems	HIS	2
8	Iranian Joint Congress on Fuzzy and Intelligent Systems	CFIS	2
9	International Symposium on Computational Intelligence and Informatics	CINTI	2
10	International Conference on Modelling and Simulation	UKSIM	2

2.4.3.1 Co-authorship Analysis

Co-authorship analysis can be used in authors and/or publications to track and study the relationship between authors, institutions, or countries. If applied to authors, co-authorship analysis reveals the authors' social relationships (Chen et al. 2010). To conduct co-authorship analysis, we used VOSviewer to visualization the co-authorship network based on Scopus data. The network clustering (community detection) in this co-authorship network is also applied to show the authors that probability works on similar topics in each community. The resulted network of co-authorship in the topic of "learning automata" and "optimization" consists of 734 nodes (authors) and 136 significant communities with a minimum of 10 nodes. This network with its extracted communities is depicted in Fig. 2.3. In Fig. 2.3, each node's size is proportional to the number of published articles by each author, and the community structures are revealed by both color and position of which node.

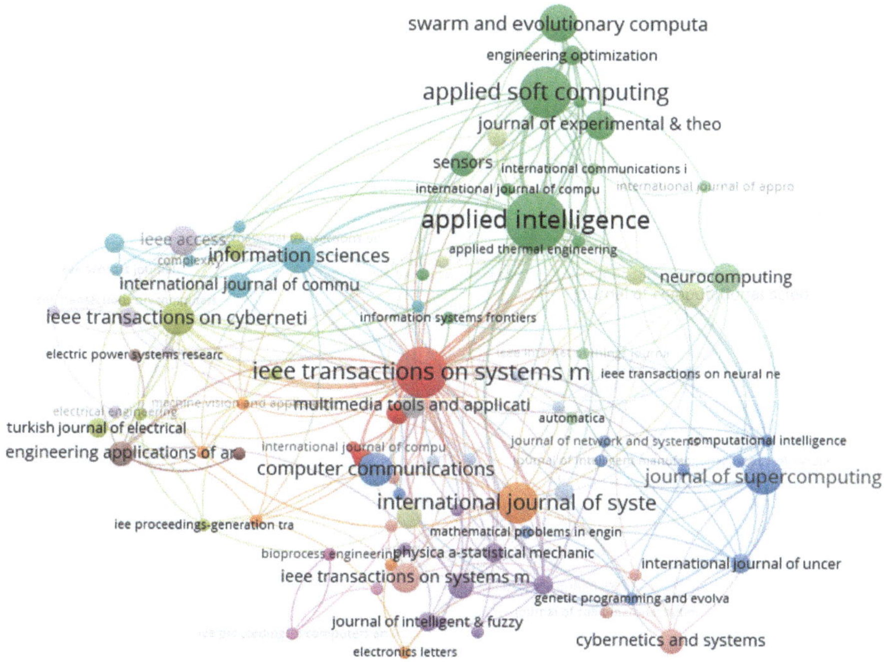

Fig. 2.2 Citation network of key journals for the research related to "learning automata" and "optimization."

This Co-authorship network, with its revealed community structures, is also colored based on the publication year as overlay visualization from 2002 through 2021 in Fig. 2.4. In Fig. 2.4, each node's size is proportional to the number of published articles by each author, and the color of each node represents the number of publications per year by each author.

The authors 'co-citation network is extracted to study the relationship between authors concerning citing each other papers. In the authors' co-citation network, a citation connection is a link between two nodes where one node cites the other. Citation links are treated as undirected by VOSviewer. We used VOSviewer to visualize the authors' co-citation network based on WoS data to conduct the co-citation analysis. The network clustering (community detection) in this co-citation network is also applied to show the authors that probability is citing on similar topics in each community. The resulted network of co-citation of the authors in the topic of "learning automata" and "optimization" consists of 623 nodes (authors) and six significant communities. This network, with its extracted communities, is depicted in Fig. 2.5. In Fig. 2.5, each node's size is proportional to the number of published articles by each author, and the community structures are revealed by both color and position of which node.

Table 2.4 Top 10 authors based on the number of publications for "learning automata" and "optimization" related research

	Author	No. publications in WoS	No. publications in Scopus	Highly cited paper
1	Meybodi, M. R	43	80	(Beigy and Meybodi 2010)
2	Oommen, B. J	15	30	(Granmo and Oommen 2010b)
3	Najim, K	11	9	(Najim and Poznyak 1996)
4	Beigy, H	10	13	(Beigy and Meybodi 2006)
5	Granmo, O. C	9	25	(Granmo et al. 2007)
6	Yazidi, A	9	15	(Yazidi et al. 2012)
7	Torkestani, J. A	8	3	(Torkestani and Meybodi 2011)
8	Kordestani, J. K	6	5	(Kordestani et al. 2018)
9	Sastry, P. S	6	4	(Thathachar and Sastry 2002)
10	Thathachar, M. A. L	6	10	(Santharam et al. 1994)

The authors' citation network with its revealed community structures is also colored based on the publication year as overlay visualization between 2002 and 2021 in Fig. 2.6. In Fig. 2.6, each node's size is proportional to the number of published articles by each author, and the color of each node represents the number of publications per year by each author.

2.4.4 *Top Papers*

The top articles for the research related to the topic of "learning automata" and "optimization" for highly cited papers (the greatest number of citations received by each paper) are reported in Table 2.5 in which the paper titles are given in the second column. The year of publication, the number of citations in WoS, the number of citations in Scopus, the average number of citations per year in WoS, and the average number of citations per year in WoS is provided in the third, fourth, fifth, sixth and the last column, respectively. From the results reported in Table 2.5, it can be observed that a review paper regarding varieties of learning automata authored by Thathachar et al. (Thathachar and Sastry 2002) with 226 citations in WoS and 292 citations in Scopus dominates the list.

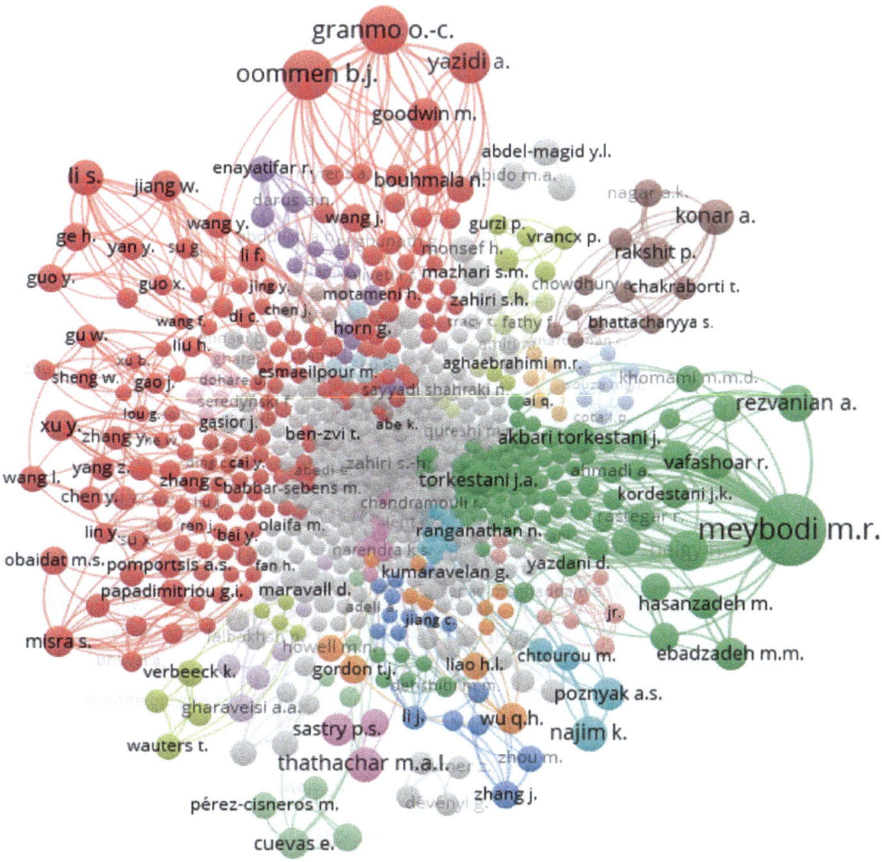

Fig. 2.3 Co-authorship network with its revealed community structures

A research study presented in the discipline of electrical power systems published by Mazhari et al. is in the second rank of the top-ten list with 8.75 and 10.37 values for this measure concerning an average number of citations received each paper in WoS and Scopus per year, respectively. Since this phenomenon has appeared in engineering applications (electrical power systems), this reason may be for the LA's practical applicability and optimization for electrical engineering application.

The network of citations for articles is demonstrated in Fig. 2.7 that the size of each node is proportional to the number of citations that each article received. In this network, nodes with dark red color represent the recent polished articles, and nodes with dark blue color represent the more past article publication.

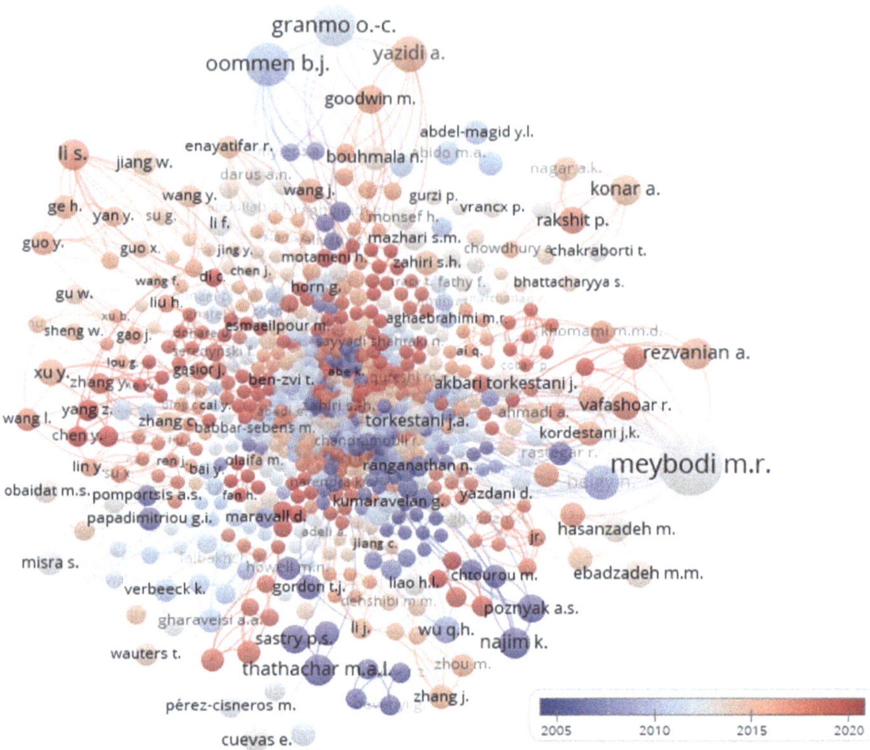

Fig. 2.4 Co-authorship network as overlay visualization during 2002–2021

2.4.5 *Top Affiliations*

The top ten institutions' plot with highly published articles is shown in Fig. 2.8, in which researchers from Amirkabir University of Technology (Tehran Polytechnic) publish most papers with 47 and 86 papers in WoS and Scopus, respectively. The list is followed by the researchers from Islamic Azad University with 43 and 42 papers in WoS and Scopus.

The corresponding to each affiliation, the country in which the institution is situated, was taken out for further analysis, and this result is shown in Fig. 2.9. As shown in Fig. 2.9, it can be seen that institutions in Iran (97 papers in WoS and 100 papers in Scopus), China (36 papers in WoS and 59 papers in Scopus), the USA (29 papers in WoS and 52 papers in Scopus), Canada (25 papers in WoS and 49 papers in Scopus), and India (23 papers in WoS and 48 papers in Scopus) are the major contributors. Researchers worldwide, particularly in Asia and North America, are attracted to learning automata and optimization.

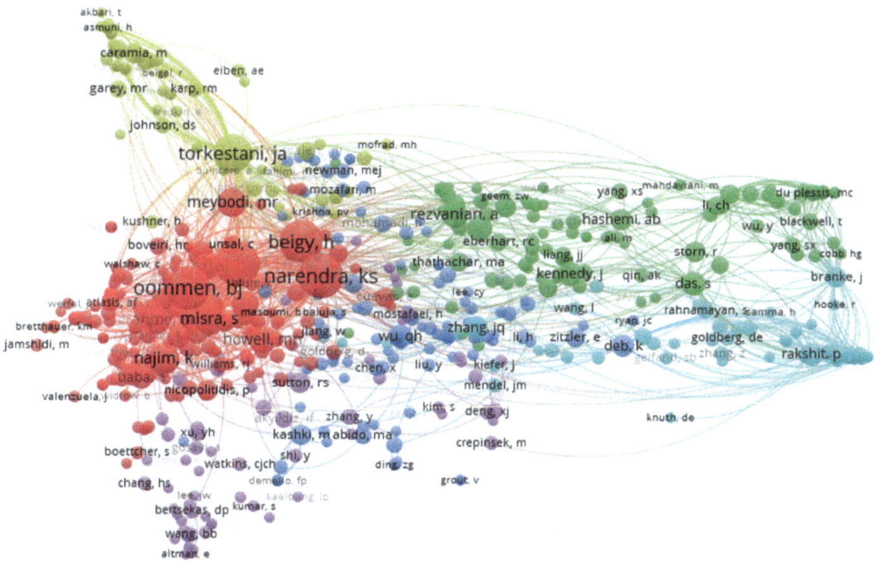

Fig. 2.5 Co-citation network of authors with its revealed community structures

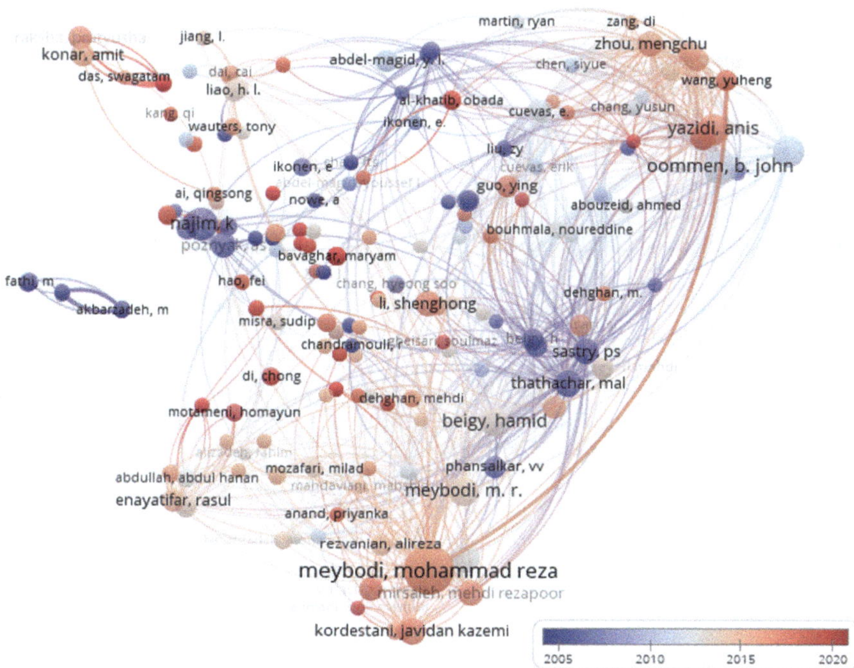

Fig. 2.6 Citation network of authors as overlay visualization during 2002–2021

Table 2.5 Highly cited articles for the research related to "learning automata" and "optimization."

	Article title (reference)	Year of the pub.	No. cit. in WoS	No. cit. in Scopus	Avg. no. cit. per year in WoS	Avg. no. cit. per year in scopus
1	Varieties of learning automata: an overview (Thathachar and Sastry 2002)	2002	226	292	11.89	15.36
2	A multi-objective PMU placement method considering measurement redundancy and observability value under contingencies (Mazhari et al. 2013)	2013	70	83	8.75	10.37
3	Closed-loop object recognition using reinforcement learning (Peng and Bhanu 1998)	1998	67	84	2.91	3.65
4	Cellular Learning Automata with Multiple Learning Automata in Each Cell and Its Applications (Beigy and Meybodi 2010)	2010	61	75	5.54	6.81
5	A note on the learning automata based algorithms for adaptive parameter selection in PSO (Hashemi and Meybodi 2011)	2011	60	78	6	7.8
6	Learning automata-based solutions to the nonlinear fractional knapsack problem with applications to optimal resource allocation (Granmo et al. 2007)	2007	59	69	4.21	4.92
7	A learning automata-based fault-tolerant routing algorithm for mobile ad hoc networks (Misra et al. 2012)	2012	55	62	6.11	6.88
8	Continuous action set learning automata for stochastic optimization (Santharam et al. 1994)	1994	55	58	2.03	2.14
9	A hybrid gene selection approach for microarray data classification using cellular learning automata and ant colony optimization (Vafaee Sharbaf et al. 2016)	2016	50	69	10	13.8
10	Reinforcement Learning for Repeated Power Control Game in Cognitive Radio Networks (Zhou et al. 2011)	2012	43	51	4.77	5.66

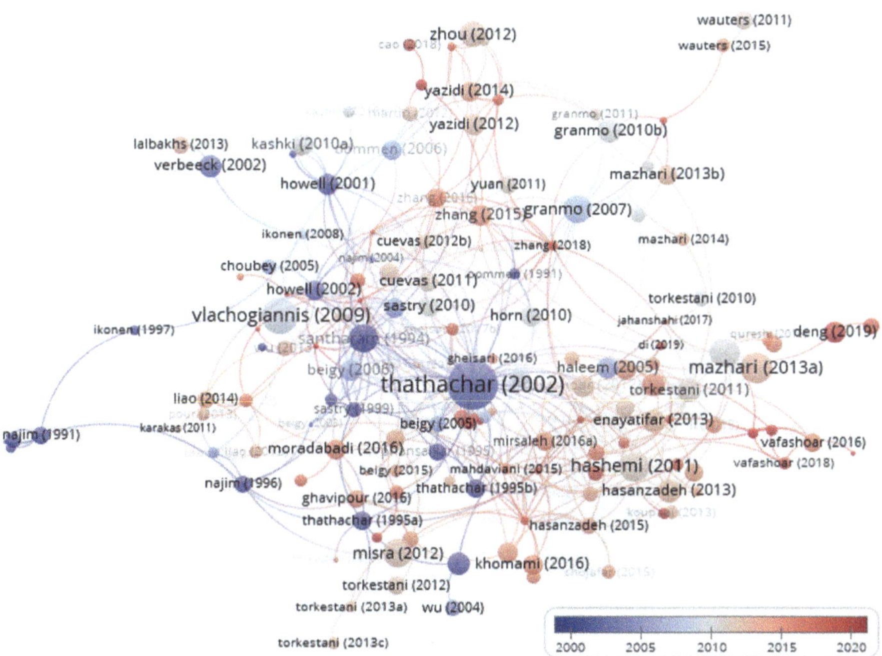

Fig. 2.7 Citation network for papers for the research related to "learning automata" and "optimization."

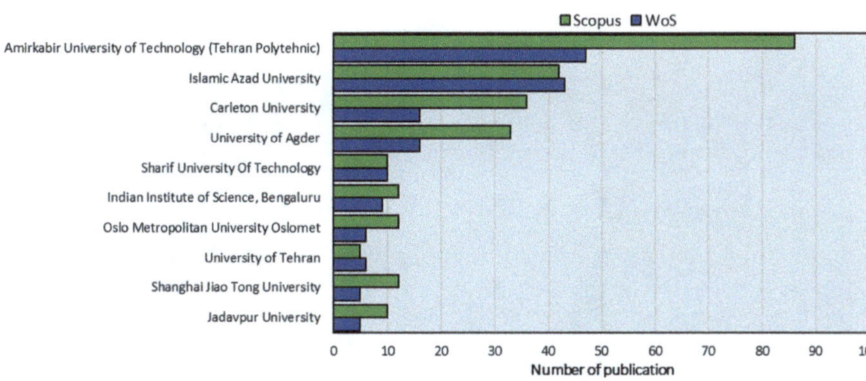

Fig. 2.8 Top 10 institutions with highly published papers for the research related to "learning automata" and "optimization."

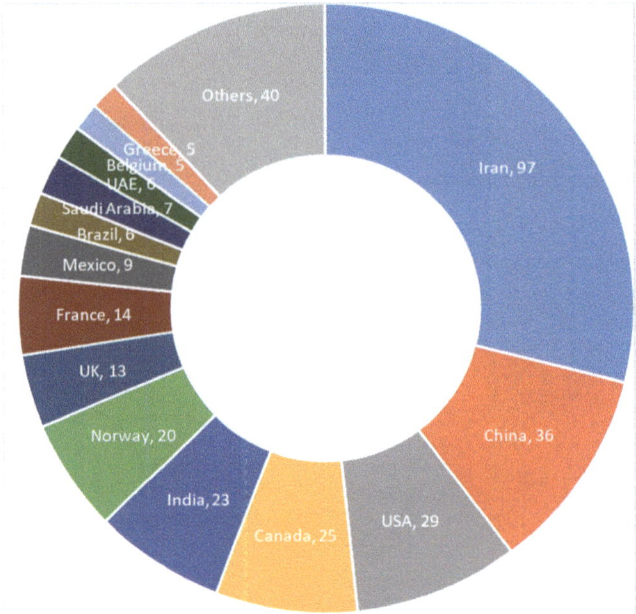

(a) WoS

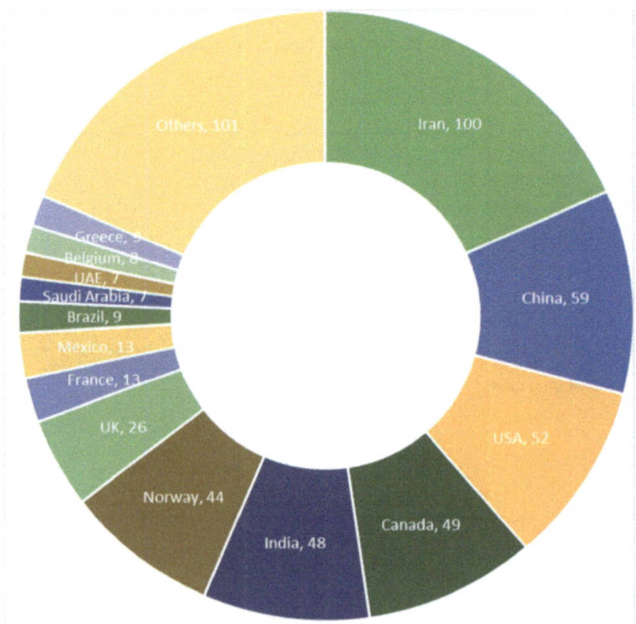

(b) Scopus

Fig. 2.9 Top 10 contributing countries for the research related to the topic of "learning automata" and "optimization."

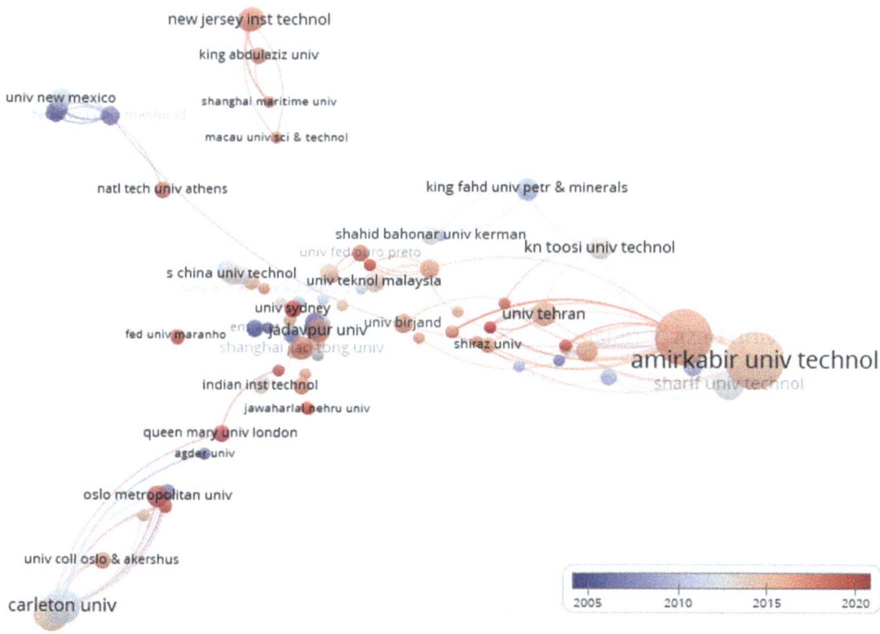

Fig. 2.10 Co-authorship network for institutions for the research related to "learning automata" and "optimization."

The network of co-authorships for contributing authors institutions is shown in Fig. 2.10, including 51 institutions in which each node size is proportional to the number of papers published by authors of those institutions. The network of co-authorships for countries of authors' institutions is shown in Fig. 2.11 that each node size is proportional to the number of citations received by each article published by authors of each countries' institutions. This network consists of 34 nodes represented as countries and ten links represented as connections between these countries' institutions.

2.4.6 *Top Keywords*

In this section, we present the results of the keyword analysis. Such a discussion helps reveal the discipline's intellectual core and identity construction by looking into keywords and using research articles and their aggregation (Sidorova et al. 2008). To do so, we adopted a similar approach to identify the top-ten keywords and the most commonly used words in both article titles and abstracts. The top 10 keywords used in the articles related to learning automata and optimization topics are listed in Table 2.6.

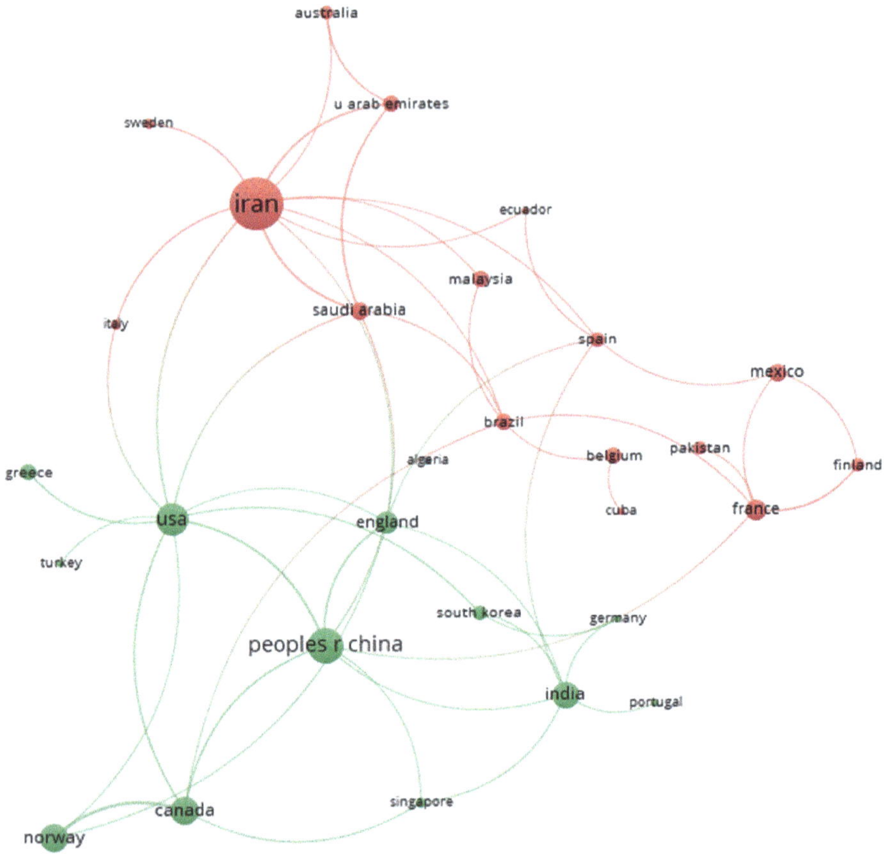

Fig. 2.11 Co-authorship network for countries

Table 2.6 Top 10 keywords of papers related to the "learning automata" and "optimization" topic

	Word	Frequency in WoS	Frequency in scopus
1	Learning automata	152	293
2	Particle swarm optimization	26	68
3	Genetic algorithm	25	63
4	Reinforcement learning	19	74
5	Cellular learning automata	18	42
6	Differential evolution	15	10
7	Local search	15	10
8	Multi-objective optimization	11	19
9	Evolutionary algorithm	9	42
10	Global optimization	8	15

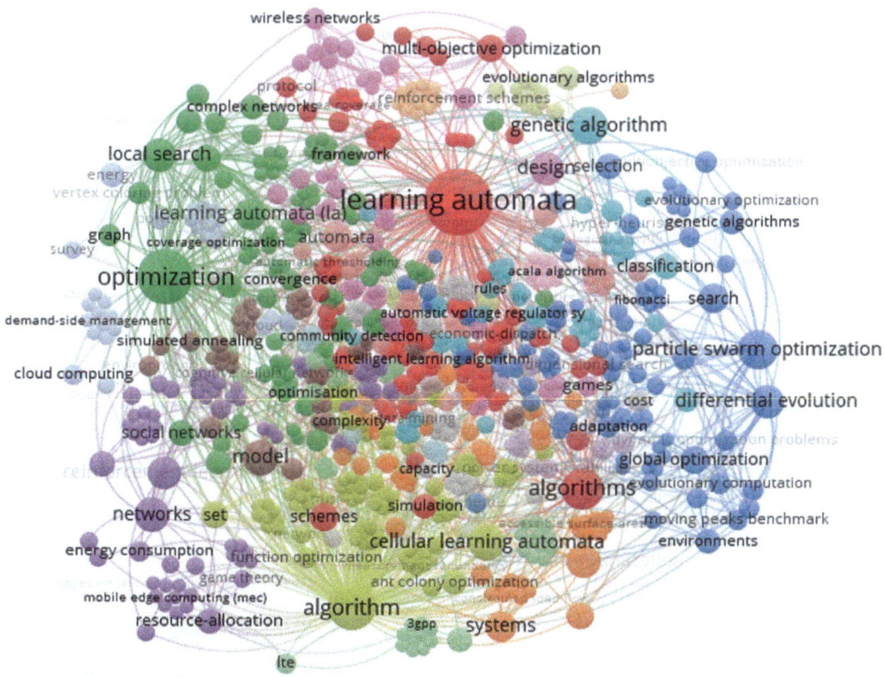

Fig. 2.12 Co-occurrence network of keywords

The network of co-occurrence keywords of papers with learning automata and optimization is depicted in Fig. 2.12. In this network, each node size is proportional to each keyword's number among all the papers with learning automata and optimization. One can see that the revealed community structures with different colors. The main keywords related to learning automata and optimization are found in the community at the network center.

The density visualization map of the co-occurrence of the most common words in both titles and abstracts of the papers with the topic of learning automata and optimization based on the Scopus is depicted in Fig. 2.13. From the result, one can observe that the main words related to the learning automata, cellular learning automata, cellular learning automata, multi-objective optimization, genetic algorithm, particle swarm optimization, and differential evolution topics are mainly positioned in the map.

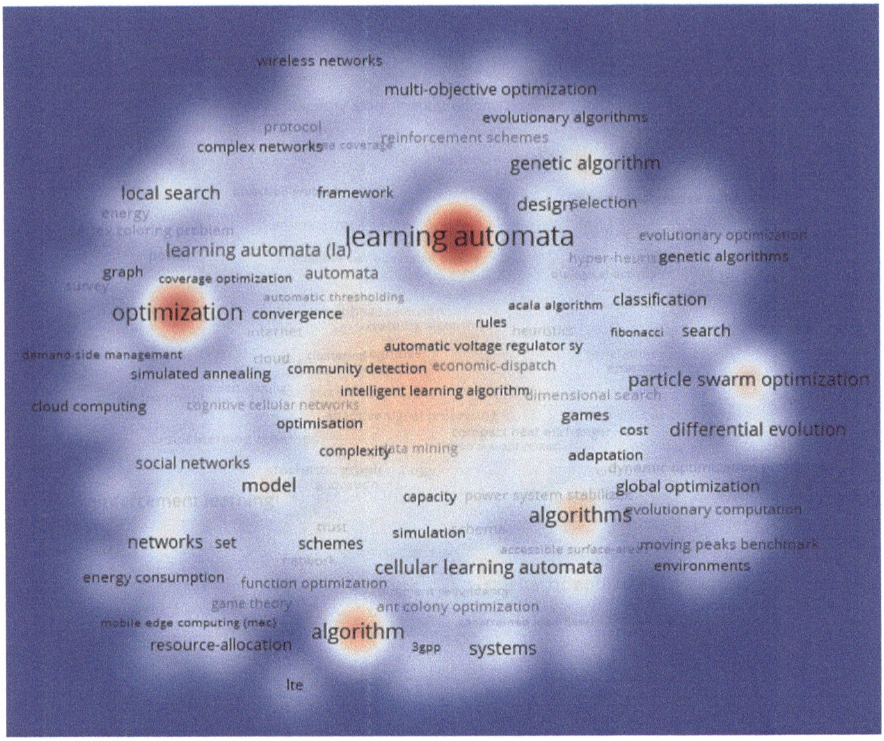

Fig. 2.13 Density visualization of the most common words belonging to the text of titles and abstracts of the papers

2.5 Conclusion

As a developing research study of reinforcement learning and optimization as an applied engineering tool, learning automata has found many applications in various engineering domains. In this chapter, based on the bibliometric network analysis, we provide a brief analysis of literature on learning automata and optimization regarding related research over 47 years (1974–2021). We also provided some insights about the contributing key scientific journals, researchers, institutes, countries, papers, keywords, and most common words in the text of title and abstracts of the papers towards advancing learning automata and optimization-related research bibliometric network analysis perspective.

References

Abtahi, F., Meybodi, M.R., Ebadzadeh, M.M., Maani, R.: Learning automata-based co-evolutionary genetic algorithms for function optimization. In: Proceedings of the 6th International Symposium on Intelligent Systems and Informatics (SISY), pp. 1–5 (2008)

Aghazadeh, F., Meybodi, M.R.: Learning bees algorithm for optimization. In: International Conference on Information and Intelligent Computing, pp. 115–122 (2011)

Alirezanejad, M., Enayatifar, R., Motameni, H., Nematzadeh, H.: GSA-LA: gravitational search algorithm based on learning automata. J. Exp. Theor. Artif. Intell. 1–17 (2020). https://doi.org/10.1080/0952813X.2020.1725650

Bastian, M., Heymann, S., Jacomy, M.: Gephi: an open source software for exploring and manipulating networks. In: Third International AAAI Conference on Weblogs and Social Media, pp. 361–362 (2009)

Beigy, H., Meybodi, M.R.: A new continuous action-set learning automaton for function optimization. J. Franklin Inst. **343**, 27–47 (2006)

Beigy, H., Meybodi, M.R.R.: Cellular learning automata with multiple learning automata in each cell and its applications . IEEE Trans. Syst. Man Cybern. Part B (cybern.) **40**, 54–65 (2010). https://doi.org/10.1109/TSMCB.2009.2030786

Chen, C., Ibekwe-SanJuan, F., Hou, J.: The structure and dynamics of cocitation clusters: a multiple-perspective cocitation analysis. J. Am. Soc. Inform. Sci. Technol. **61**, 1386–1409 (2010). https://doi.org/10.1002/asi.21309

Clarivate Analytics: Acquisition of the Thomson Reuters Intellectual Property and Science Business by Onex and Baring Asia Completed (2017). www.prnewswire.com

Eiben, A.E., Smith, J.E.: Introduction to Evolutionary Computing. Springer, Heidelberg (2015)

Enayatifar, R., Yousefi, M., Abdullah, A.H., Darus, A.N.: LAHS: a novel harmony search algorithm based on learning automata. Commun. Nonlinear Sci. Numer. Simul. **18**, 3481–3497 (2013). https://doi.org/10.1016/j.cnsns.2013.04.028

Granmo, O.-C., Oommen, B.J.: Optimal sampling for estimation with constrained resources using a learning automaton-based solution for the nonlinear fractional knapsack problem. Appl. Intell. **33**, 3–20 (2010a)

Granmo, O.-C., Oommen, B.J.: Solving stochastic nonlinear resource allocation problems using a hierarchy of twofold resource allocation automata. IEEE Trans. Comput. **59**, 545–560 (2010b)

Granmo, O.-C., Oommen, B.J., Myrer, S.A., Olsen, M.G.: Learning automata-based solutions to the nonlinear fractional knapsack problem with applications to optimal resource allocation . IEEE Trans. Syst. Man Cybern. Part B (cybern.) **37**, 166–175 (2007)

Hamidi, M., Meybodi, M.R.: New learning automata based particle swarm optimization algorithms. In: Second Iranian Data Mining Conference, pp. 1–15 (2008)

Hashemi, A.B., Meybodi, M.R.: A note on the learning automata based algorithms for adaptive parameter selection in PSO. Appl. Soft Comput. j. **11**, 689–705 (2011). https://doi.org/10.1016/j.asoc.2009.12.030

Hassanzadeh, T., Meybodi, M.R.: A new hybrid algorithm based on firefly algorithm and cellular learning automata. In: 20th Iranian Conference on Electrical Engineering (ICEE 2012), pp. 628–633. IEEE (2012)

Howell, M.N., Gordon, T.J., Brandao, F.V.: Genetic learning automata for function optimization. IEEE Trans. Syst. Man Cybern. Part B (cybern.) **32**, 804–815 (2002). https://doi.org/10.1109/TSMCB.2002.1049614

Jafarpour, B., Meybodi, M.R., Shiry, S.: A hybrid method for optimization (discrete PSO + CLA). In: 2007 International Conference on Intelligent and Advanced Systems, ICIAS 2007, pp. 55–60 (2007)

Khezri, S., Meybodi, M.R.: Improving imperialist competitive algorithm using learning automata. In: 16th Annual CSI Computer Conference (CSI 2011), Tehran, Iran (2011)

Khomami, M.M.D., Rezvanian, A., Meybodi, M.R.: Distributed learning automata-based algorithm for community detection in complex networks. Int. J. Mod. Phys. B **30**, 1650042 (2016)

Kordestani, J.K., Ahmadi, A., Meybodi, M.R.: An improved differential evolution algorithm using learning automata and population topologies. Appl. Intell. **41**, 1150–1169 (2014). https://doi.org/10.1007/s10489-014-0585-2

Kordestani, J.K., Firouzjaee, H.A., Meybodi, M.R.: An adaptive bi-flight cuckoo search with variable nests for continuous dynamic optimization problems. Appl. Intell. **48**, 97–117 (2018). https://doi.org/10.1007/s10489-017-0963-7

Mahdaviani, M., Kordestani, J.K., Rezvanian, A., Meybodi, M.R.: LADE: learning automata based differential evolution. Int. j. Artif. Intell. Tools **24**, 1550023 (2015)

Mazhari, S.M., Monsef, H., Lesani, H., Fereidunian, A.: A multi-objective PMU placement method considering measurement redundancy and observability value under contingencies. IEEE Trans. Power Syst. **28**, 2136–2146 (2013)

Mirsaleh, M.R., Meybodi, M.R.: A Michigan memetic algorithm for solving the community detection problem in complex network. Neurocomputing **214**, 535–545 (2016)

Misra, S., Krishna, P.V., Bhiwal, A., Chawla, A.S., Wolfinger, B.E., Lee, C.: A learning automata-based fault-tolerant routing algorithm for mobile ad hoc networks. J. Supercomput. **62**, 4–23 (2012)

Mousavian, A., Rezvanian, A., Meybodi, M.R.: Solving minimum vertex cover problem using learning automata. In: 13th Iranian Conference on Fuzzy Systems (IFSC 2013), pp. 1–5 (2013)

Mousavian, A., Rezvanian, A., Meybodi, M.R.: Cellular learning automata based algorithm for solving minimum vertex cover problem. In: 2014 22nd Iranian Conference on Electrical Engineering (ICEE), pp. 996–1000. IEEE (2014)

Najim, K., Poznyak, A.S.: Multimodal searching technique based on learning automata with continuous input and changing number of actions . IEEE Trans. Syst. Man Cybern. Part B (cybern.) **26**, 666–673 (1996)

Narendra, K.S., Thathachar, M.A.: Learning Automata: An Introduction. Prentice-Hall, Hoboken (1989)

Peng, J., Bhanu, B.: Closed-loop object recognition using reinforcement learning. IEEE Trans. Pattern Anal. Mach. Intell. **20**, 139–154 (1998)

Rastegar, R., Meybodi, M.R., Hariri, A.: A new fine-grained evolutionary algorithm based on cellular learning automata. Int. J. Hybrid Intell. Syst. **3**, 83–98 (2006). https://doi.org/10.3233/HIS-2006-3202

Rezvanian, A., Meybodi, M.R.: An adaptive mutation operator for artificial immune network using learning automata in dynamic environments. In: 2010 Second World Congress on Nature and Biologically Inspired Computing (NaBIC), pp. 479–483. IEEE (2010a)

Rezvanian, A., Meybodi, M.R.: LACAIS: learning automata based cooperative artificial immune system for function optimization. In: Communications in Computer and Information Science, pp. 64–75. Springer, Heidelberg (2010b)

Rezvanian, A., Meybodi, M.R.: Tracking extrema in dynamic environments using a learning automata-based immune algorithm. In: Communications in Computer and Information Science, pp. 216–225. Springer, Heidelberg (2010c)

Rezvanian, A., Saghiri, A.M., Vahidipour, S.M., Esnaashari, M., Meybodi, M.R.: Cellular learning automata. pp. 21–88 (2018)

Rezvanian, A., Moradabadi, B., Ghavipour, M., Daliri Khomami, M.M., Meybodi, M.R.: Social networks and learning systems: a bibliometric analysis. In: Learning Automata Approach for Social Networks, pp. 75–89. Springer (2019)

Santharam, G., Sastry, P.S., Thathachar, M.A.L.: Continuous action set learning automata for stochastic optimization. J. Franklin Inst. **331**, 607–628 (1994)

Sastry, P.S., Nagendra, G.D., Manwani, N.: A team of continuous-action learning automata for noise-tolerant learning of half-spaces. IEEE Trans. Syst. Man Cybern. Part B Cybern. **40**, 19–28 (2009)

Sidorova, E., Valacich, R.: Uncovering the intellectual core of the information systems discipline. MIS q. **32**, 467 (2008). https://doi.org/10.2307/25148852

Sutton, R.S., Barto, A.G.: Reinforcement Learning: An Introduction. MIT Press, Cambridge (1998)

Thathachar, M.A.L., Sastry, P.S.: Varieties of learning automata: an overview. IEEE Trans. Syst. Man Cybern. B Cybern. **32**, 711–722 (2002). https://doi.org/10.1109/TSMCB.2002.1049606
Thathachar, M.A.L., Sastry, P.S.: Networks of Learning Automata: Techniques for Online Stochastic Optimization. Springer, Boston (2004)
Torkestani, J.A.: Degree-constrained minimum spanning tree problem in stochastic graph. Cybern. Syst. **43**, 1–21 (2012)
Torkestani, J.A., Meybodi, M.R.: A cellular learning automata-based algorithm for solving the vertex coloring problem. Expert Syst. Appl. **8**, 9237–9247 (2011)
Vafaee Sharbaf, F., Mosafer, S., Moattar, M.H.: A hybrid gene selection approach for microarray data classification using cellular learning automata and ant colony optimization. Genomics **107**, 231–238 (2016). https://doi.org/10.1016/j.ygeno.2016.05.001
Vafashoar, R., Meybodi, M.R.: Multi swarm bare bones particle swarm optimization with distribution adaption. Appl. Soft Comput. J. **47**, 534–552 (2016). https://doi.org/10.1016/j.asoc.2016.06.028
Vafashoar, R., Meybodi, M.R.: A multi-population differential evolution algorithm based on cellular learning automata and evolutionary context information for optimization in dynamic environments. Appl. Soft Comput. **88**, 106009 (2020). https://doi.org/10.1016/j.asoc.2019.106009
Vafashoar, R., Morshedlou, H., Rezvanian, A., Meybodi, M.R.: Cellular Learning Automata: Theory and Applications. Springer, Cham (2021a)
Vafashoar, R., Morshedlou, H., Rezvanian, A., Meybodi, M.R.: Cellular learning automata: a bibliometric analysis. In: Cellular Learning Automata: Theory and Applications, pp. 83–109. Springer (2021b)
van Eck, N.J., Waltman, L.: Software survey: VOSviewer, a computer program for bibliometric mapping. Scientometrics **84**, 523–538 (2010). https://doi.org/10.1007/s11192-009-0146-3
van Eck, N.J., Waltman, L.: VOSviewer manual (2013)
Yazdani, D., Golyari, S., Meybodi, M.R.: A new hybrid algorithm for optimization based on artificial fish swarm algorithm and cellular learning automata. In: Proceedings of 2010 5th International Symposium on Telecommunications (IST), Tehran, Iran, pp. 932–937 (2010)
Yazidi, A., Granmo, O.C., Oommen, B.J.: Service selection in stochastic environments: a learning-automaton based solution. Appl. Intell. **36**, 617–637 (2012)
Zeng, X., Liu, Z.: A learning automata based algorithm for optimization of continuous complex functions. Inf. Sci. **174**, 165–175 (2005)
Zhao, Y., Jiang, W., Li, S., Ma, Y., Su, G., Lin, X.: A cellular learning automata based algorithm for detecting community structure in complex networks. Neurocomputing **151**, 1216–1226 (2015)
Zhou, P., Chang, Y., Copeland, J.A.: Reinforcement learning for repeated power control game in cognitive radio networks. IEEE j. Sel. Areas Commun. **30**, 54–69 (2011)

Chapter 3
Cellular Automata, Learning Automata, and Cellular Learning Automata for Optimization

Javidan Kazemi Kordestani, Mehdi Razapoor Mirsaleh, Alireza Rezvanian, and Mohammad Reza Meybodi

Abstract Since many real problems have several limitations and constraints for different environments, no standard optimization algorithms could work successfully for all kinds of problems. To enhance the abilities and improve the performance of a standard optimization algorithm for solving problems, several modifications or combinations with some techniques such as learning automata (LA), cellular automata (CA), and cellular learning automata (CLA) are presented by researchers. Thus, this chapter investigates new learning automata (LA) and cellular learning automata (CLA) models for solving optimization problems. Moreover, this chapter provides a summary of hybrid LA models for optimization problems from 2015 to 2021.

3.1 Introduction

Learning is a crucial aspect of intelligence, and machine learning has emerged as a vibrant discipline with the avowed objective of developing machines with learning capabilities. Learning has been recognized as an essential aspect of intelligent behavior. Over the last few decades, the process of learning, which was studied

J. Kazemi Kordestani
Department of Computer Engineering, Science and Research Branch, Islamic Azad University, Tehran, Iran

M. Razapoor Mirsaleh
Department of Computer Engineering and Information Technology, Payame Noor University (PNU), P.O. BOX 19395-3697, Tehran, Iran
e-mail: mrezapoorm@aut.ac.ir

A. Rezvanian (✉)
Department of Computer Engineering, University of Science and Culture, Tehran, Iran
e-mail: rezvanian@usc.ac.ir

M. R. Meybodi
Computer Engineering Department, Amirkabir University of Technology (Tehran Polytechnic), Tehran, Iran
e-mail: mmeybodi@aut.ac.ir

J. Kazemi Kordestani et al. (eds.), *Advances in Learning Automata and Intelligent Optimization*, Intelligent Systems Reference Library 208, https://doi.org/10.1007/978-3-030-76291-9_3

earlier mostly by psychologists, has become a topic of much interest to engineers as well, given its role in machine intelligence. Psychologists or biologists who conduct learning experiments on animals or human subjects try to create behavior models by analyzing experimental data. Engineers are more interested in studying learning behavior to help them to synthesize intelligent machines. While the above two goals are distinctly different, the two endeavors are nonetheless interrelated because success in one helps to improve our abilities in the other.

Learning is defined as any relatively permanent change in behavior resulting from experience, and the learning system is characterized by its ability to improve its behavior with time, in some sense tending towards an ultimate goal (Narendra 1989). The concept of learning makes it possible to design systems that can gradually improve their performance during actual operation through learning from past experiences. Every learning task consists of two parts: a learning system and environment. The learning system must learn to act in the environment. The primary learning tasks can be classified into supervised learning, semi-supervised learning, active learning, unsupervised learning, and reinforcement learning.

Cellular learning automaton (CLA) as a new reinforcement learning approach is a combination of cellular automata (CA) (Wolfram 1986) and learning automata (LA) (Rezvanian et al. 2018b; Rezvanian et al. 2019a). It is formed by a group of interconnected cells arranged in some regular forms, such as a grid or ring, in which each cell contains one or more LAs. A cellular learning automaton can be considered a distributed model, inheriting both the computational power of cellular automata and learning automata's learning ability. Accordingly, it is more powerful than a single learning automaton due to producing more complex behavior patterns. Also, cells' learning abilities enable cellular learning automata to produce these complex patterns by understandable behavioral rules rather than complicated mathematical functions commonly used in cellular automata. Owing to these characteristics, cellular learning automata can be successfully employed in modeling, learning, simulating, controlling, and solving challenging problems in uncertain, distributed, and complex environments (Rezvanian et al. 2018a).

Since many problems require that a learning automaton (LA) learn the optimal subsets of its actions, the standard LA algorithms can deal with this problem by considering all possible actions as new action sets. However, this approach is only applicable to small action spaces. Indeed, sometimes the decision-making entity receives individual feedback from the environment for each one of the actions in its chosen subset of actions (Rezvanian et al. 2018c).

This chapter aims to extend some standard learning automaton algorithms to efficiently learn the optimal subset of their actions through parallel reinforcements. These parallel reinforcements represent the favorability of each action in the performed subset of actions. The cellular learning automaton (CLA) can model several locally connected decision-making entities. All the introduced learning automaton models in this chapter can be utilized in a cellular learning automaton. Moreover, this chapter investigates the convergence properties of a CLA model in which each decision-making entity selects a subset of its actions. It is worth noting that the main part of this chapter is adopted from (Vafashoar and Meybodi 2019a). Interested readers could refer to (Vafashoar and Meybodi 2019a) for a detailed description and theoretical analysis.

3.2 Preliminaries

In this section, the required concepts and backgrounds regarding cellular automata, learning automata, and cellular learning automata are briefly described.

3.2.1 Cellular Automata

Cellular automata (CAs) are dynamical systems that exhibit complex global behavior from simple local interaction and computation (Wolfram 1982). Since the inception of cellular automaton (CA) by von Neumann in the 1950s, it has attracted the attention of many researchers over various fields for modeling different physical, natural, and real-life phenomena. CAs are (typically) spatially and temporally discrete: they are composed of a finite or denumerable set of homogenous and simple units called cells. At each time step, the cells instantiate one of a finite set of states. They evolve in parallel at discrete time steps, following state update functions or dynamical transition rules: the update of a cell state obtains by taking into account the states of cells in its local neighborhood. CAs are computational systems: they can compute functions and solve algorithmic problems. Despite functioning differently from traditional, Turing machine-like devices, CA with applicable rules can emulate a universal Turing machine (Cook 2004), and therefore compute, given Turing's thesis (Copeland 1997), anything computable.

Classically, CAs are uniform. However, non-uniformity has also been introduced in update patterns, lattice structure, neighborhood dependency, and local rule. In dynamic structure CAs, the structure of cells or local rule changes (Bhattacharjee et al. 2018). Non-uniformity also can occur in the parallelism of the updating procedure: cells can be updated asynchronously. CAs depending on their structure, can also be classified as regular CAs or irregular CAs. In Irregular CAs, the structure regularity assumption has been relaxed. The irregular CAs and CAs with structural dynamism can be generalized into models known as automata networks in the literature. An automata network is a mathematical system consisting of a network of nodes that evolves according to a set of predetermined rules.

3.2.2 Learning Automata

Learning Automata (LA) (Thathachar and Sastry 2002; Rezvanian et al. 2018b) is a stochastic optimization technique from the family of Reinforcement Learning (RL) algorithms. Having enough interaction with the unknown environment, elegance emerges, and the optimal policy will be chosen. LA is divided into two groups of fixed-structure and variable-structure automata.

VSLA can be represented by a quadruple $\langle \alpha, \beta, p, T \rangle$, where $\alpha = \{\alpha_1, \alpha_2, \ldots, \alpha_r\}$ is the set of actions, $\beta = \{\beta_1, \beta_2, \ldots, \beta_k\}$ is the set of inputs to the automaton, $p = \{p_1, p_2, \ldots, p_r\}$ is the probability vector corresponds to each action, and T is the learning algorithm. In the simplest form of VSLA, consider an automaton with r actions in a stationary environment where $\beta = \{0, 1\}$ is included in inputs. After selecting the action by the automaton, the reinforcement signal will receive from the environment. When the *favorable* response ($\beta = 0$) is received, the action probabilities are updated through Eq. (3.1).

$$p_j(t+1) = \begin{cases} p_j(t) + a\left[1 - p_j(t)\right] & j = i \\ (1-a)p_j(t) & \forall j \neq i \end{cases} \tag{3.1}$$

When the *unfavorable response* ($\beta = 1$) is received from the environment, action probabilities are updated through Eq. (3.2).

$$p_j(t+1) = \begin{cases} (1-b)p_j(t) & j = i \\ \left(\frac{b}{r-1}\right) + (1-b)p_j(t) & \forall j \neq i \end{cases} \tag{3.2}$$

where a and b are called *learning parameters,* and they are associated with the reward and penalty responses. If a and b are equal, the learning scheme is called L_{R-P} (Linear Reward-Penalty). If the learning parameter b is set to 0, then the learning scheme is named L_{R-I} (Linear Reward-Inaction). And finally, if the learning parameter b is much smaller than a, the learning scheme is called $L_{R\varepsilon P}$ (Linear Reward-epsilon-Penalty).

3.2.3 Cellular Learning Automata

Cellular learning automaton (CLA) (Rezvanian et al. 2018c; Vafashoar et al. 2021a) is a combination of cellular automaton (CA) (Packard and Wolfram 1985) and learning automaton (LA) (Narendra 1989; Rezvanian et al. 2019b). The basic idea of CLA is to use LA for adjusting the state transition probability of a stochastic CA. This model, which opens a new learning paradigm, is superior to CA because of its ability to learn and is also superior to single LA because it consists of a collection of LAs interacting with each other (Beigy and Meybodi 2004). A CLA is a CA in which some LAs are assigned to every cell. Each LA residing in a particular cell determines its action (state) based on its action probability vector. Like CA, there is a local rule that the CLA operates under. The local rule of the CLA and the actions selected by the neighboring LAs of any particular LA determine the reinforcement signal to that LA. The neighboring LAs (cells) of any particular LA (cell) constitute the local environment of that LA (cell). The local environment of an LA (cell) is non-stationary due to the fact that the action probability vectors of the neighboring LAs vary during the evolution of the CLA. The operation of a CLA could be described as the following steps (Fig. 3.1): At the first step, the internal state of every cell is determined on the basis of the action probability vectors of the LAs residing in that cell. In the second step, the local rule of the CLA determines the reinforcement signal to each LA residing in that cell. Finally, each LA updates its action probability vector based on the supplied reinforcement signal and the chosen action. This process continues until the desired result is obtained.

CLA can be either *synchronous* or *asynchronous*. In a synchronous CLA, LAs in all cells are activated simultaneously using a global clock. In contrast, in an asynchronous CLA (ACLA) (Beigy and Meybodi 2008), LAs in different cells are activated asynchronously. The LAs may be activated in either a time-driven or step-driven manner. In a time-driven ACLA, each cell is assumed to have an internal clock that wakes up the LAs associated with that cell. In a step-driven ACLA, a cell is selected for activation in either a random or a fixed sequence. From another point of view, CLA can be either *close* or *open*. In a close CLA, the action selection of any particular LA in the next iteration of its evolution only depends on the state of the LA's local environment (actions of its neighboring LAs) (Vafashoar et al. 2021e).

In contrast, in an open CLA (Beigy and Meybodi 2007), this depends on the local environment and the external environments. In (Beigy and Meybodi 2010), a new type of CLA, called CLA with multiple LAs in each cell, has been introduced. This model is suitable for applications such as channel assignment in cellular networks,

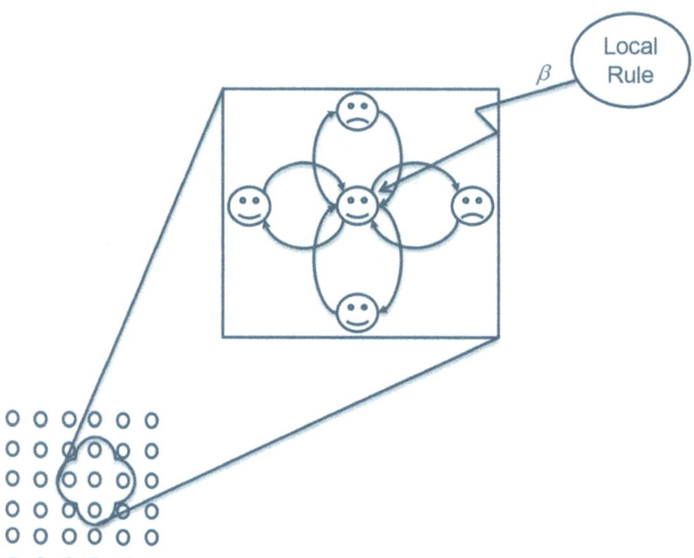

Fig. 3.1 Operation of the cellular learning automaton (CLA)

in which it is needed that each cell is equipped with multiple LAs. In (Beigy and Meybodi 2004), a mathematical framework for studying the CLA behavior has been introduced. It was shown that, for a class of local rules called commutative rules, different models of CLA converge to a globally stable state (Beigy and Meybodi 2004, 2007, 2008, 2010).

Definition 3.1 A d-dimensional cellular learning automaton is a framework (Beigy and Meybodi 2004), $\mathcal{A} = \left(Z^d,\ N\ , \Phi,\ A,\ \mathcal{F}\right)$, where.

- Z^d presents the lattice of d-tuples of integer numbers.
- $N = \{\overline{x}_1, \overline{x}_2, \ldots, \overline{x}_m\}$ is a finite subset of Z^d that is called neighborhood vector, where $\overline{x}_i \in Z^d$.
- Φ denotes the finite set of states. φ_i presents the state of the cell c_i.
- A presents the set of LAs residing in the cells of CLA.
- $F^i : \underline{\Phi_i} \to \underline{\beta}$ defines the local rule of the CLA for each cell c_i, where $\underline{\beta}$ is the set of possible values for the reinforcement signal, and it calculates the reinforcement signal for each LA using the chosen actions of neighboring LAs.

In what follows, a CLA with n cells and neighborhood function $\overline{N}(i)$ is considered. A learning automaton denoted by A_i, which has a finite action set $\underline{\alpha}_i$ is associated with cell i (for $i = 1, 2, \ldots, n$) of CLA. Let cardinality of $\underline{\alpha}_i$ be m_i and the state of CLA represented by $\underline{p} = (\underline{p}_1^{'}, \underline{p}_2^{'}, \ldots, \underline{p}_n^{'})^{'}$, where $\underline{p}_i = (p_{i1}, \ldots, p_{im_i})^{'}$ is the action probability vector of A_i. The local environment for each learning automaton is the learning automata residing in its neighboring cells. From the repeated application of simple local rules and simple learning algorithms, the global behavior of CLA can be very complicated.

The operation of CLA takes place in the following iterations. At iteration k, each learning automaton chooses an action. Let $\alpha_i \in \underline{\alpha}_i$ be the action chosen by A_i. Then all learning automata receive a reinforcement signal. Let $\beta_i \in \underline{\beta}$ be the reinforcement signal received by A_i. The application of the local rule produces this reinforcement signal $\mathcal{F}^i(\alpha_{i+\overline{x}_1}, \alpha_{i+\overline{x}_2}, \ldots, \alpha_{i+\overline{x}_{\overline{m}}}) \to \underline{\beta}$. The higher value of β_i means that the chosen action of A_i is more rewarded. Since each set $\underline{\alpha}_i$ is finite, rule $\mathcal{F}^i(\alpha_{i+\overline{x}_1}, \alpha_{i+\overline{x}_2}, \ldots, \alpha_{i+\overline{x}_{\overline{m}}}) \to \underline{\beta}$ can be represented by a hyper matrix of dimensions $m_1 \times m_2 \times \ldots \times m_{\overline{m}}$. These n hyper matrices together constitute the rule of CLA.

When all of these n hyper matrices are equal, the rule is uniform; otherwise, the rule is nonuniform. For the sake of simplicity in our presentation, the rule $\mathcal{F}^i(\alpha_{i+\overline{x}_1}, \alpha_{i+\overline{x}_2}, \ldots, \alpha_{i+\overline{x}_{\overline{m}}})$ denoted by $\mathcal{F}^i(\alpha_1, \alpha_2, \ldots, \alpha_{\overline{m}})$. Based on set $\underline{\beta}$, the CLA can be classified into three groups: P-model, Q-model, and S-model Cellular learning automata. When $\underline{\beta} = \{0, 1\}$, we refer to CLA as *P-model cellular learning automata,* when $\underline{\beta} = \{b_1, \ldots, b_l\}$, (for $l < \infty$), we refer to CLA as *Q-model cellular learning automata,* and when $\underline{\beta} = [b_1, b_2]$, we refer to CLA as *S-model cellular learning automata.* If learning automaton A_i uses learning algorithm L_i, we denote CLA by the $CLA(L_1, \ldots, L_n)$. If $L_i = L$ for all $i = 1, 2, \ldots, n$, then we denote the CLA by the *CLA(L).* In the following, some definitions and notations are presented.

Definition 3.2 A configuration of CLA is a map $\mathcal{K} : Z^d \rightarrow \underline{p}$ that associates an action probability vector with every cell. We will denote the set of all configurations of $\mathcal{A}$ by $\mathcal{K}(\mathcal{A})$ or simply $\mathcal{K}$.

The application of the local rule to every cell allows transforming a configuration into a new one.

Definition 3.3 The global behavior of a CLA is a mapping $\mathcal{G} : \mathcal{K} \rightarrow \mathcal{K}$, that describes the dynamics of CLA.

Definition 2.11 The evolution of CLA from a given initial configuration $\underline{p}(0) \in \mathcal{K}$, is a sequence of configurations $\left\{\underline{p}(k)\right\}_{k\geq 0}$, such that $\underline{p}(k+1) = \mathcal{G}\left(\underline{p}(k)\right)$.

CLA may be described as a LA network assigned to a graph's nodes (usually a finite and regular lattice). Each node is connected to a set of nodes (its neighbors), and each LA updates its action probability vector at discrete time instants, using a learning algorithm and a rule which depends on its action and that of its neighbors.

Definition 3.4 A CLA is called synchronous if all the learning automata are activated at the same time and asynchronous if at a given time only some LAs are activated.

Definition 3.5 A CLA is called uniform if the neighborhood function and the local rule are the same for all CLA cells.

Definition 3.6 A rule is called uniform if all cells' rule is the same; otherwise, it is called nonuniform.

Definition 3.7 A configuration p is called deterministic if the action probability vector of each learning automaton is a unit vector; otherwise, it is called probabilistic. Hence, the set of all deterministic configurations, $\mathcal{K}^*$, set of probabilistic configurations, $\mathcal{K}$, in CLA are

$$\mathcal{K}^* = \left\{\underline{p} | \underline{p} = (\underline{p}_1', \underline{p}_2', \ldots, \underline{p}_n')', \underline{p}_i = (p_{i1}, \ldots, p_{im_i})', p_{iy} = 0 \text{ or } 1 \quad \forall y, i, \sum\nolimits_y p_{iy} = 1 \ \forall i\right\}$$

and

$$\mathcal{K} = \left\{\underline{p} | \underline{p} = (\underline{p}_1', \underline{p}_2', \ldots, \underline{p}_n')', \underline{p}_i = (p_{i1}, \ldots, p_{im_i})', 0 \leq p_{iy} \leq 1 \quad \forall y, i, \sum\nolimits_y p_{iy} = 1 \ \forall i\right\},$$

respectively.

3.2.3.1 Cellular Learning Automata Models

Up to now, various CLA models such as open CLA (Beigy and Meybodi 2007), asynchronous CLA (Beigy and Meybodi 2008), irregular CLA (Esnaashari and

Meybodi 2018), associative CLA (Ahangaran et al. 2017), dynamic irregular CLA (Esnaashari and Meybodi 2018), asynchronous dynamic CLA (Saghiri and Meybodi 2018a), and wavefront CLA (Moradabadi and Meybodi 2018; Rezvanian et al. 2019c) are developed and successfully applied on different application domains. Table 3.1 summarizes some application areas for these models.

3.3 CA, CLA, and LA Models for Optimization

In what follows, we first provide a summary of CA, CLA, and LA models for optimization problems during 2015 to 2021 as listed in Table 3.2, and then we will briefly introduce some recent CLA and LA models for optimizations.

3.3.1 Cellular Learning Automata-Based Evolutionary Computing (CLA-EC)

CLA-EC model, presented in (Rastegar and Meybodi 2004; Rastegar et al. 2006), is a combination of *CLA* and the evolutionary computing model (Fig. 3.2). In this model, each cell of the *CLA* is equipped with an *m*-bit binary genome. Each genome has two components; model genome and string genome. The model genome is composed of *m* learning automata. Each *LA* has two actions; 0 and 1. The set of actions selected by the set of *LA*s of a particular cell concatenated to each other to form the second component of the genome, i.e., the string genome. The operation of a CLA-EC, in any particular cell c_i, takes place as follows. Each *LA* residing within the cell c_i randomly selects one of its actions according to its action probability vector. The actions selected by the set of *LA*s of the cell c_i are concatenated to form a new string genome for that cell. The fitness of this newly generated genome is then evaluated. If the newly generated genome is better than the cell's current genome, then the cell ci's current genome is replaced by the newly generated genome. Next, according to the fitness evaluation of their corresponding genomes, many of the cell c_i's neighboring cells are selected for mating. Note that the mating in this context is not reciprocal, i.e., a cell selects another cell for mating but not necessarily vice versa.

The mating process results in the cell c_i are some reinforcement signals, one for each *cell's LA*. The process of computing the reinforcement signal for each *LA* is described in Fig. 3.3. Each *LA* updates its action probability vector based on the supplied reinforcement signal and its selected action. This process continues until a termination condition is satisfied.

Table 3.1 Some applications of different CLA models

CLA model	Application
Associative CLA	Clustering (Ahangaran et al. 2017; Hasanzadeh-Mofrad and Rezvanian 2018), classification (Ahangaran et al. 2017), image segmentation (Ahangaran et al. 2017)
Asynchronous CLA	Edge detection (Bohlool and Meybodi 2007), super-peer selection (Saghiri and Meybodi 2018b), hub allocation problem (Saghiri and Meybodi 2018a), topology mismatch problem in unstructured peer-to-peer networks (Saghiri and Meybodi 2017), web recommendation (Talabeigi et al. 2010; Rezvanian et al. 2019d)
Asynchronous dynamic CLA	Topology mismatch problem in unstructured peer-to-peer networks (Saghiri and Meybodi 2017), super-peer selection (Saghiri and Meybodi 2018b), hub allocation (Saghiri and Meybodi 2018a)
Basic CLA	Image processing (Hasanzadeh Mofrad et al. 2015), Frequent itemset mining (Sohrabi and Roshani 2017), gene selection (Vafaee Sharbaf et al. 2016), text summarization (Abbasi-ghalehtaki et al. 2016), skin segmentation (Abin et al. 2008), Lung cancer diagnosis (Hadavi et al. 2014), skin detector (Abin et al. 2009), stock market (Mozafari and Alizadeh 2013), edge detection (Sinaie et al. 2009)
CLA-DE	Optimization (Vafashoar et al. 2012)
CLA-EC	Hardware design (Hariri et al. 2005a), FPGA (Hariri et al. 2005b), clustering (Rastegar et al. 2005), optimization (Rastegar et al. 2006), dynamic optimization (Manshad et al. 2011), link prediction in social networks (Khaksar Manshad et al. 2020)
CLA-PSO	Optimization (Akhtari and Meybodi 2011)
Cellular Petri-net	Vertex coloring problem (Vahidipour et al. 2017b), graph problems (Vahidipour et al. 2019)
Closed Asynchronous Dynamic CLA	Peer-to-peer networks (Saghiri and Meybodi 2017)
Cooperative CLA-EC	Optimization (Masoodifar et al. 2007; Mozafari et al. 2015)
Dynamic irregular CLA	Deployment of mobile wireless sensor networks (Esnaashari and Meybodi 2011), transportation (Ruan et al. 2019)
Irregular CLA	Channel assignment (Morshedlou and Meybodi 2017), sampling social networks (Ghavipour and Meybodi 2017), community detection in complex social networks (Zhao et al. 2015; Khomami et al. 2018), dynamic point coverage in wireless Sensor (Esnaashari and Meybodi 2010), clustering in wireless sensor networks (Esnaashari and Meybodi 2008), link prediction in social network (Khaksar Manshad et al. 2020)

(continued)

Table 3.1 (continued)

CLA model	Application
Open CLA	Hub allocation problem (Saghiri and Meybodi 2018a), edge detection (Bohlool and Meybodi 2007), optimization (Saghiri and Meybodi 2018b)
Re-combinative CLA-EC	Optimization (Jafarpour and Meybodi 2007)
Wavefront CLA	Online social network problems such as prediction and sampling (Moradabadi and Meybodi 2018)

3.3.2 Cooperative Cellular Learning Automata-Based Evolutionary Computing (CLA-EC)

The CLA-EC model's performance is highly dependent on the learning rate of the LAs that reside in its constituting cells. Large learning rate values decrease the model's performance, whereas small learning rate values decrease its convergence speed. The idea behind the cooperative CLA-EC model (Masoodifar et al. 2007) is to use two CLA-EC models with different learning rates in cooperation with each other. The operations of the CLA-ECs in the cooperative CLA-EC model are synchronized. At any iteration, each CLA-EC performs its normal operation independent of the other CLA-EC. Any m iteration, the genomes of the corresponding cells in the two CLA-ECs are exchanged, as depicted in Fig. 3.4. This way, one of the CLA-ECs (the one with the high learning rate) explores through the space of the problem, and the other CLA-EC (the one with the small learning rate) exploits and enhances the solutions found thus far.

In (Khaksarmanshad and Meybodi 2010), a different cooperative CLA-EC model has been presented in which a CLA-EC model cooperates with a self-adaptive genetic algorithm. The Cooperative CLA-EC model has been used in function optimization (Khaksarmanshad and Meybodi 2010; Masoodifar et al. 2007) and solving the 0/1 knapsack problem (Masoodifar et al. 2007).

3.3.3 Recombinative Cellular Learning Automata-Based Evolutionary Computing (RCLA-EC)

CLA-EC model suffers from a lack of explorative behavior. In (Jafarpour and Meybodi 2007), recombinative CLA-EC (RCLA-EC) has been presented to overcome this problem by adding a recombination step to the basic operation step of the CLA-EC model. The authors argue that there might be some string genomes with poor fitness, which contain good partial solutions. However, the CLA-EC model has no mechanism for partial string exchange between cells, and thus, it cannot reap the benefits of such good partial solutions. To overcome this problem, in recombinative CLA-EC, a shuffle recombination (Burkowski 1999) step has been added to the basic

Table 3.2 Summary of LA models for optimization problems during 2015 to 2021

Application/optimization problem	LA model
Bayesian network structure training	SLA (Gheisari and Meybodi 2017; Gheisari et al. 2017), VSLA (Gheisari et al. 2016)
Channel assignment problem	MCLA (Vafashoar and Meybodi 2019c), OMA-LA (Mirsaleh and Meybodi 2018), CLA (Jahanshahi et al. 2017), SLA (Salehi et al. 2018)
Community detection problem	ICLA (Daliri Khomami et al. 2020b) (Zhao et al. 2015), CLA (Khomami et al. 2018), (Daliri Khomami et al. 2020a), DLA (Khomami et al. 2016), EDLA (Ghamgosar et al. 2017), MLA-MA (Mirsaleh and Meybodi 2016a)
Coverage problem	SLA (Singh et al. 2020) (Li et al. 2015) (Ben-Zvi 2018) (Jameii et al. 2016)
Deployment problem	SLA (Zhang et al. 2018), CLA (Lin et al. 2018)
Dynamic optimization problem	HSLA (Kazemi Kordestani et al. 2021), CLA (Vafashoar and Meybodi 2020) (Vafashoar et al. 2021c), SLA (Kordestani et al. 2019) (Cao and Cai 2018), VSLA (Kordestani et al. 2018)
Energy optimization problem	SLA (Zhang et al. 2020) (Gasior et al. 2018) (Velusamy and Lent 2019) (Nouri 2019) (Hao et al. 2019), DGPA (Abbas et al. 2018), DLA (Kumar et al. 2019), CLA (Subha and Malarkkan 2017)
Graph coloring problem	VSLA (Akbari Torkestani 2016), CLA-MA (Mirsaleh and Meybodi 2016b), SLA (Vahidipour et al. 2019), MA-LA (Mirsaleh and Meybodi 2017)
Graph partitioning problem	SLA (Kazemitabar et al. 2018) (Díaz-Cortés et al. 2017), OMA (Jobava et al. 2017)
Image segmentation problem	CLA (Adinehvand et al. 2017), CALA (Anari et al. 2017)
Link prediction problem	CALA (Moradabadi and Meybodi 2016), DLA (Moradabadi and Meybodi 2017)
Load balancing problem	SLA (Mohajer et al. 2020a), DLA (Mohajer et al. 2020b)
Maximum clique problem	DLA (Soleimani-Pouri et al. 2012) (Rezvanian and Meybodi 2015)
Maximum satisfiability problem (MAX-SAT)	FLA (Bouhmala 2015) (Bouhmala et al. 2016), PLA (Yazidi et al. 2020)
Minimum vertex covering	DLA (Rezvanian and Meybodi 2015), CLA (Mousavian et al. 2014)

(continued)

Table 3.2 (continued)

Application/optimization problem	LA model
Minimum spanning tree problem	NLA (Zojaji et al. 2020), WCLA (Moradabadi and Meybodi 2018) (Rezvanian et al. 2019g)
Multi-objective optimization (MOO) problem	SLA (Sayyadi Shahraki and Zahiri 2020b) (Zhao and Zhang 2020) (Cota et al. 2019) (Jameii et al. 2016) (Li et al. 2019) (Dai et al. 2016) (Chowdhury et al. 2016) (Li et al. 2015), MOLA (Sayyadi Shahraki and Zahiri 2020a)
Neural network training	SLA (Xiao et al. 2019) CALA (Fakhrmoosavy et al. 2018) (Lindsay and Gigivi 2020)
Numeric/function optimization	DRLA (Sayyadi Shahraki and Zahiri 2021), VSLA (Alirezanejad et al. 2021), LABOA (Arora and Anand 2018), CLA (Vafashoar et al. 2021c) (Vafashoar and Meybodi 2018) (Mozafari et al. 2015) (Vafashoar and Meybodi 2016) (Vafashoar and Meybodi 2019d), CARLA (Guo et al. 2017) (Guo et al. 2015), SLA (Abedi Firouzjaee et al. 2017) (Rakshit 2020) (Rakshit et al. 2019) (Rakshit et al. 2017) (Lin et al. 2016) (Rakshit and Konar 2018), MA-LA (Rezapoor Mirsaleh and Meybodi 2018), LABSO (Xu et al. 2020), LAPSO (Zhang et al. 2015), LADE (Mahdaviani et al. 2015)
Pilot allocation problem	PLA (Raharya et al. 2020a), (Raharya et al. 2020b)
Resource allocation problem	HSLA (Zhu et al. 2017), GLA (Bayessa et al. 2018), SLA (Rauniyar et al. 2020) (Liu et al. 2020) (Zhao et al. 2019), BLA (Yang et al. 2020b) (Yang et al. 2020a)
Routing problem	SLA (Khot and Naik 2021; Saritha et al. 2017; Suma et al. 2020) (Velusamy and Lent 2019) (Velusamy and Lent 2018) (Hao et al. 2019), CLA (Jahanshahi et al. 2017), MOLA (Hashemi and He 2017), VSLA (Gheisari 2020),
Scheduling problem	SLA (Cota et al. 2019) (Su et al. 2018) (Wauters et al. 2015)
Stochastic shortest path problem	DLA (Beigy and Meybodi 2020) (Vahidipour et al. 2017a)
Stochastic point location (SPL) problem	SLA (Yazidi and Oommen 2017) (Yazidi and Oommen 2015)
Vehicle routing problem	SLA (Zhou et al. 2020) (Toffolo et al. 2018)

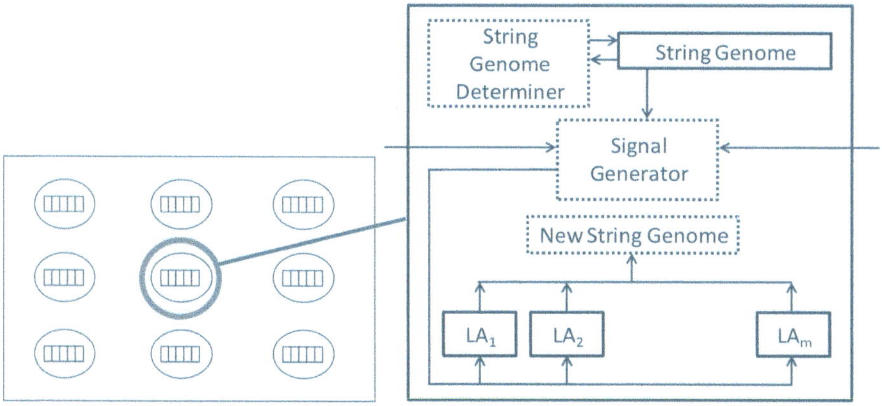

Fig. 3.2 The structure of the CLA-EC model

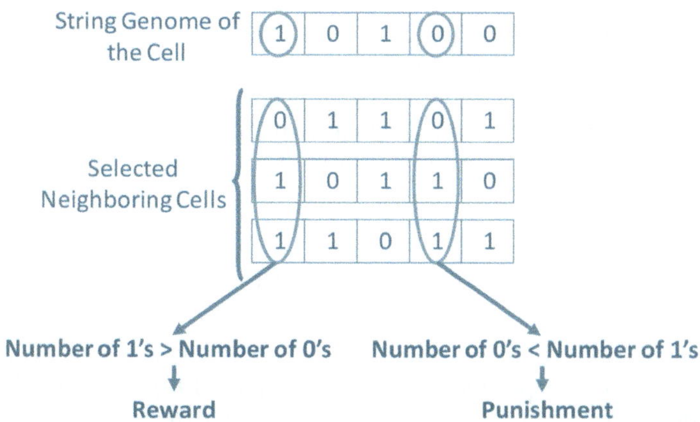

Fig. 3.3 The process of computing the reinforcement signal for each *LA*

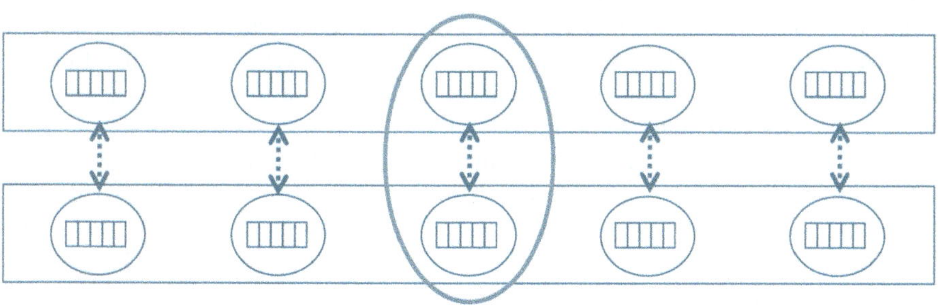

Fig. 3.4 Cooperative CLA-EC model

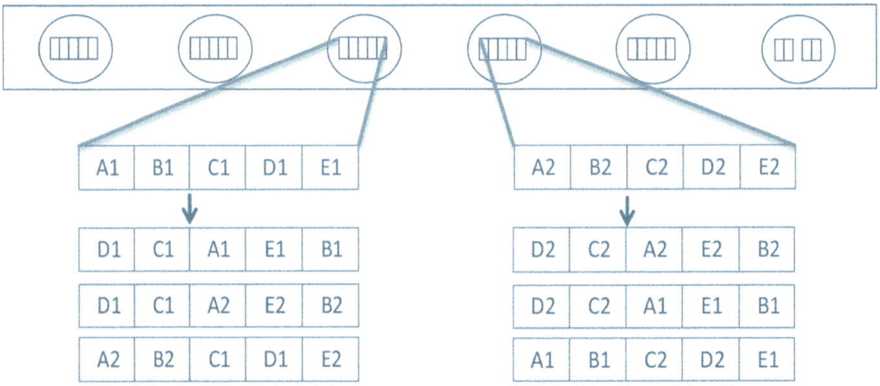

Fig. 3.5 Recombination step of the recombinative CLA-EC model

steps of the operation of the CLA-EC model. In this step, every cell c_i (for even i) recombines its string genome with that of the cell c_{i+1}. Next, the string genomes of the cells c_i and c_{i+1} are replaced by the string genomes resulted from the recombination. The recombination step of the recombinative CLA-EC model is presented in Fig. 3.5.

3.3.4 CLA-EC with Extremal Optimization (CLA-EC-EO)

Extremal Optimization (EO) (Boettcher and Percus 2002) is an optimization heuristic inspired by the Bak-Sneppen model (Bak and Sneppen 1993) of the self-organized criticality from the field of statistical physics. This heuristic is designed to address complex combinatorial optimization problems such as the traveling salesman problem and spin glasses. However, the technique has been demonstrated to function in optimization domains. Unlike genetic algorithms that work with a population of candidate solutions, EO evolves a single solution and makes local modifications to the worst components. The technique can be regarded as a fine-grained search that tries to enhance a candidate solution gradually.

The EO technique may lead to a deterministic search for some types of problems, such as the single spin-flip neighborhood for the spin Hamiltonian (Boettcher and Percus 2002). The basic EO algorithm has been slightly modified in the τ-EO algorithm (Boettcher and Percus 2002). In this algorithm, unlike the basic EO algorithm in which the worst component is modified at each step, any component of the candidate solution has a chance of being modified at each step. Another drawback of the EO algorithm is that a general definition of the individual components' fitness may prove ambiguous or even impossible. This problem has been addressed in the generalized EO (GEO) algorithm presented by Sousa and Ramos (de Sousa and Ramos 2002). In this algorithm, the candidate solution is represented as a binary string. The fitness of each component i of the candidate solution (fitness of the ith bit) is assumed to be the difference between the candidate solution's fitness and the fitness of the candidate solution in which the ith bit is flipped.

The CLA-EC-EO model, presented in (Khatamnejad and Meybodi 2008), is a combination of the CLA-EC and the GEO models. This model's main idea is that the GEO model enhances the CLA-EC model's solutions. The operation of this model takes place as follows. The CLA-EC model explores the search space until it finds a candidate solution with the fitness above a specified threshold. This candidate solution is then given to the GEO model for enhancement. The enhanced candidate solution is given back to the CLA-EC model. This process continues until a termination condition is met.

3.3.5 Cellular Learning Automata-Based Differential Evolution (CLA-DE)

Differential evolution (DE), presented by Storn and Price (Storn and Price 1997), is an evolutionary algorithm that optimizes a problem by iteratively trying to improve a population of candidate solutions concerning a given measure of quality. DE is used for multidimensional real-valued functions but does not use the gradient of the problem being optimized. This means that DE does not require the optimization problem to be differentiable, as is required by classic optimization methods such as gradient descent.

The operation of the basic DE algorithm takes place as follows. First, the best candidate solution in the current population is identified as *BCS*. Next, candidate solutions of the population are selected one by one. For each candidate solution CS_i, three other candidate solutions, namely RCS_{i1}, RCS_{i2}, and RCS_{i3}, are selected randomly. A new candidate solution NCS_i is then created by combining CS_i, RCS_{i1}, RCS_{i2}, and RCS_{i3}, and *BCS* through one of the following formulae

$$NCS_i = RCS_{i1} + F.(RCS_{i2} - RCS_{i3}),\ F \in [0, 2]. \tag{3.3}$$

$$NCS_i = BCS + F.(RCS_{i1} - RCS_{i2}),\ F \in [0, 2]. \tag{3.4}$$

$$NCS_i = CS_i + F.(BCS - CS_i) + F.(RCS_{i1} - RCS_{i2}),\ F \in [0, 2]. \tag{3.5}$$

The new candidate solution NCS_i is recombined with CSi using either binomial or exponential crossover operators. If this newly recombined candidate solution is more fitted than the candidate solution CS_i, then it is replaced with CS_i in the current population.

The CLA-DE model is presented in (Vafashoar et al. 2012) to improve the accuracy of the convergence rate and CLA model results. In this model, each cell of the CLA is equipped with m learning automata, where m is the dimension of the problem at hand. The LA assigned to the ith dimension is responsible for learning the admissible interval for the *candidate solution's ith dimens*ion. Each dimension is divided into some intervals, and the LA assigned to each dimension has to choose an interval among the intervals of that dimension. The operation of the CLA-DE model within a cell c_i takes place as follows. First, each LA selects an interval randomly according to its action probability vector. The cell's candidate solution is then created by selecting random numbers from the selected intervals in each dimension. Next, the neighboring cells' candidate solutions are evaluated, and the most fitted candidate solution (BCS) is identified. Three other neighboring cells are selected at random (RCS_{i1}, RCS_{i2}, and RCS_{i3}). Using different combinations of RCS_{i1}, RCS_{i2}, and RCS_{i3}, and using Eqs. (3.3) and (3.4), a set of candidate solutions is generated. The candidate solution of the cell recombines all newly generated candidate solutions. Finally, the most fitting candidate solution among these newly generated candidate solutions, and the cell's current candidate solution, is selected as the cell's new candidate solution. The action selected by the LA, assigned to the ith dimension, is rewarded if one of the following two conditions is met:

1. The interval selected by the LA coincides with the selection of more than half of the neighboring cells.
2. The candidate solution is better than the candidate solution of more than half of the neighboring cells.

3.3.6 Cellular Particle Swarm Optimization (Cellular PSO)

The particle swarm optimization (PSO) algorithm, introduced by Kennedy and Eberhart (Kennedy and Eberhart 1995), is based on the social behavior metaphor. In this algorithm, a group of particles flies through a D dimensional space, adjusting their positions in the search space according to their own and other particles' experiences. Each particle i is represented with a position vector $p_i = (p_{i1}, p_{i2}, \ldots, p_{iD})^T$ and a velocity vector $v_i = (v_{i1}, v_{i2}, \ldots, v_{iD})^T$. The elements of the vectors p_i and v_i are initially selected randomly. In every time instant k, each particle i updates its velocity and position vectors according to the following equations

$$v_i(k+1) = wv_i(k) + c_1r_1(pbest_i(k) - p_i(k)) + c_2r_2(gbest - p_i(k)), \quad (3.6)$$

$$p_i(k+1) = p_i(k) + v_i(k+1), \quad (3.7)$$

w is the inertial weight, and c1 and c2 are positive acceleration coefficients used to scale the contribution's cognitive and social components. r_1 and r_2 are uniform random variables in the range [0, 1]. $pbest_i$ is the most suitable position, visited by the particle i during its lifetime. *gbest* is the best position found by the swarm at any time. The standard PSO algorithm is also known as the *gbest* PSO.

Another basic PSO algorithm, which differs from the *gbest* PSO in the size of the particles' neighborhood, is *lbest* PSO. This algorithm defines a neighborhood for each particle. The social component reflects information exchanged within the neighborhood of the particle, reflecting local knowledge of the environment. Regarding the velocity equation, the social contribution to a particle's velocity is proportional to the particle's distance and the best position found by that particle's neighboring particles. The velocity is calculated as Eq. (3.8) given below

$$v_i(k+1) = wv_i(k) + c_1r_1(pbest_i(k) - p_i(k)) + c_2r_2(lbest_i(k) - p_i(k)), \quad (3.8)$$

where $lbest_i$ is the local best position in the neighborhood of particle i.

The main difference between the *gbest* PSO and the *lbest* PSO algorithms is their convergence characteristics (Kiink et al. 2002). Due to the larger particle interconnectivity in the *gbest* PSO, it converges faster than the *lbest* PSO. However, this faster convergence comes at the cost of faster-losing diversity than the *lbest* PSO. The larger diversity of the *lbest* PSO makes it less susceptible to being trapped in local minima. Generally, the neighborhood structures, such as the ring topology (Kennedy 1999; Mendes et al. 2003; Peer et al. 2003) and inertia weight strategy (Nabizadeh et al. 2010; Kordestani et al. 2014; Kordestani et al. 2016) improve the performance of the PSO in multi-modal environments.

Many real-world problems are dynamic optimization problems where the environment changes over time. Optimization algorithms in such dynamic environments must find the optimal solution in a short time and track it after each change in the environment. To design PSO algorithms for dynamic environments, two important points have to be considered: old memory and diversity loss. Old memory refers to the condition in which memory of the particles, which is the best location visited in the past and corresponding fitness, may no longer be valid after a change in the environment (Blackwell 2007). Diversity will be lost when all particles of the swarm converge on a few peaks in the landscape. This situation reduces the swarm's capability to find new peaks if an environmental change occurs (Blackwell 2007).

Cellular PSO (CPSO), recently introduced in (Hashemi and Meybodi 2009), is one of the best PSO-based algorithms used to optimize dynamic environments. In this algorithm, a cellular automaton (CA) is embedded in the search space and divides it into some partitions; each corresponds to one cell in the CA. Figure 3.6 illustrates a 2-D CA embedded into a two-dimensional search space. CA implicitly divides the swarm into several sub-swarms, each of which resides in a cell. The swarm that resides in the cell ci is responsible for finding the highest peak in ci and its neighboring cells. Initially, particles are distributed randomly among the cells of the CA. The particles may migrate from one cell to another to provide more searching power for the cells in which peaks may exist with a high probability. When a cell becomes crowded, some randomly chosen particles move through the CA to randomly select cells to search for new search space areas. When a change in the environment is detected, the particles around any peak begin a random search around that peak to track the changes. The outdated memory problem is solved in the CPSO algorithm by re-evaluating the

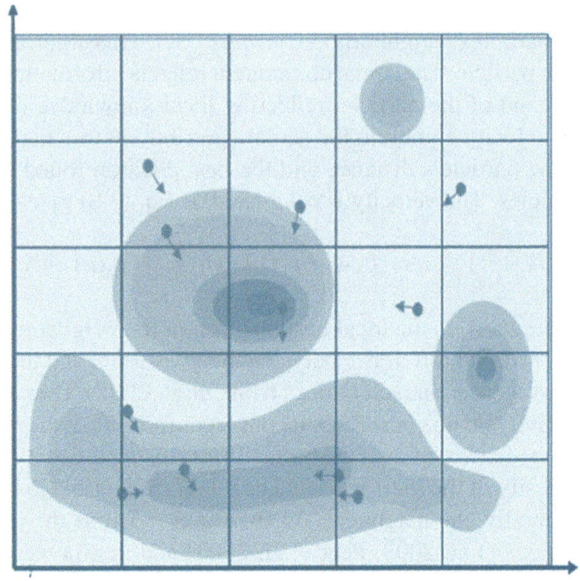

Fig. 3.6 Partitioning of 2-D search space by the CA

memory (Eberhart and Yuhui Shi 2001), and the diversity loss problem is prevented by imposing a limit on the number of particles searching in each cell.

Cell capacity has a considerable effect on the performance of the CPSO algorithm. Therefore, making this parameter adaptive to the environmental changes significantly improves the performance of the CPSO algorithm. To this end, an adaptive version of the CPSO, called CLA-PSO, is presented in (Hashemi and Meybodi 2009), in which the capacity of each cell is determined using a learning automaton. In this algorithm, instead of a CA, a CLA is embedded in the search space. The LA of each cell is responsible for adjusting the value of the capacity of that cell. Each LA has two actions: INCREASE (the cell's capacity) and DECREASE (the cell's capacity). The CLA-PSO process in each cell c_i occurs as follows: First, the particles reside within the cell ci update their velocities and positions according to the Eqs. (3.6) and (3.7). Then the LA of the cell c_i selects one of its actions. According to the selected action, the capacity of the cell ci is changed. Afterward, the LA updates its action probability vector, as described below.

If the LA's selected action is INCREASE, and currently, the cell's particle density is higher than its capacity, the LA will be rewarded. Otherwise, it will be penalized. This is because if a cell has attracted too many particles, it may contain a peak; thus, it will be a wise choice to increase its capacity. On the other hand, if the particle density in a cell is currently less than its capacity, there is no need to increase the cell capacity. Therefore, in this case, the cell's LA will be rewarded if it has selected DECREASE action. Otherwise, it will be penalized.

3.3.7 *Firefly Algorithm Based on Cellular Learning Automata (CLA-FA)*

In this section, the firefly algorithm based on cellular learning automata called CLA-FA (Hassanzadeh and Meybodi 2012) is briefly introduced. In the CLA-FA, there are swarms of one-dimensional fireflies equal to the number of search space dimensions that each swarm is placed in a cell of cellular learning automata. Each cell is equipped with a learning automaton that determines and adjusts the corresponding next movement of fireflies.

In the CLA-FA, the authors tried to balance global search and local search using the learning ability of learning automata and cellular automata's portioning ability. The CLA-FA performs: instead of using one swarm with n D-dimensional fireflies, the CLA-FA uses a D-swarm of n one-dimensional fireflies. Each swarm's objective is to optimize a component of the optimization vector that corresponds to that swarm. To achieve the desired result, all these swarms should cooperate. In order to cooperate, the best firefly of each swarm would be chosen. The vector that is being optimized for all swarms is called the "main vector." Each swarm introduces the best firefly as representative of that dimension. The main vector, the D-dimensional vector, consists of each of the D swarms' best firefly values. In each iteration of algorithm performance, for fireflies of swarm i, just the element of that i-th dimension of the main vector is changed.

Indeed, for cooperation among swarms, for calculating the fitness of a firefly in the j-th swarm, vector value for that firefly is replaced in the j-th element of the main vector to obtain a D-dimensional vector, and then its fitness is calculated in D-dimensional space. In CLA-FA, every cell is assigned to one of the main vector components. The structure of cellular learning automata is one-dimensional (linear) with a Moor neighborhood with periodic boundary (D-th cell is the first cell neighbor). Therefore, the i-th cell corresponds to the *main vector's i-th component* and includes the i-th swarm of fireflies. Available learning automata is equal in all cells, and all are updated together (cellular learning automata is homogenous). An available learning automaton in each cell includes two actions and has a variable structure with a linear learning algorithm. Learning automata's actions are as 1) Decrease in Alpha and Gamma parameters by multiplying their values by smaller value than one; 2) Performing Reset action for a defined percentage of available fireflies in the swarm.

The first action of learning automata is balancing the global search ability and local search ability of the CLA-FA algorithm. For this purpose, first, we consider the large parameter's value of Alpha and Gamma, so the algorithm's global search ability is high. Thus, the swarm moves faster toward the global optimum and passes local optimums with more ability. With performing the first action of learning automata, with fireflies movement toward global optimum, the gradually parameters value of Alpha and Gamma are decreased, so the algorithm could finally perform acceptable local search near-global optimum.

In the CLA-FA, each of the fireflies has corresponding Alpha and Gamma for itself, which is the values of these parameters, can be different in two different fireflies. The second action of learning automata is considered for avoiding premature convergence to increase escaping from local optimums. For this purpose, by doing this action, a defined percentage of fireflies of swarm randomly disperse in problem space, and their Alpha and Gamma parameters values have become equal to their initial values (Reset action). Thus, fireflies that have left the swarm return to swarm by the FA search strategy; they search remote areas to find better fitness values that it can cause to discover positions with better fitness.

In the cellular learning automata used in CLA-FA, each cell is neighbor with two previous and next cells. Dominated local rule on cellular learning automata determines whether the action learning automata has done should receive reward or penalty.

After that, learning automata in cells do an action, and each cell will evaluate the amount of its neighbor's improvement in the previous iteration. If the ratio of at least one of neighbors is higher than a defined percentage of swarm improvement corresponding to considered cell, performed action by learning automata will be fined; otherwise, it will be rewarded if the neighbor has considerable improvement to considered swarm. In the corresponding dimension of this swarm, the algorithm is trapped in the local optimum. Thus, there is a balance between global search and local searchability and the capability to avoid convergence in the CLA-FA algorithm.

3.3.8 Harmony Search Algorithm Based on Learning Automata (LAHS)

Previous studies have proved that fine-tuning the harmony search parameters (HS) algorithm can noticeably affect its accuracy and efficiency. This section presents a learning automaton mechanism for tuning parameters in the HS algorithm presented by Enayatifar et al. (Enayatifar et al. 2013). The learning-based harmony search, so-called LAHS, performs similarly to the HS algorithm, with an additional section at the end of each iteration to select HS parameter values. This supplementary section includes two phases: the learning phase and parameter selection. The parameter setting is based on an adventurous method in which the new value for a parameter could be selected from an acceptable range without considering its previous value. A parameter can change radically from one end of its range to the other in two consecutive iterations. A learning automaton with N actions is employed for the parameter, where each action corresponds to one of the N equally discretized distance values in the acceptable range of this parameter. At each iteration, one action of LA_p is selected, and the corresponding value of this action is set as the new value for parameter q. The process of learning parameter q is shown in Fig. 3.7. A Roulette-wheel selection is employed for choosing the corresponding action of each automaton in every iteration. Therefore, among all the actions, the ones with higher probability have higher chances to be selected.

Algorithm 3-1. The process of learning parameter q

Input: Parameter p

Output: Optimizing parameter q in each iteration

$\{p_1, p_2, \ldots, p_r\}$ Divide parameter p into N equal parts $\{a_1, a_2, \ldots, a_r\} \leftarrow$ Assign LA with N actions to p

Assign a $1/N$ percent probability to each action

while the stopping condition is not met **do**

$a_{active} \leftarrow$ Apply Rolette-Wheel for choosing an Active Action

$p_{active} \leftarrow$ Choosing parameter q according to a_{active}

Apply optimization algorithm with the value of p_{active}

if the result improves **then**

reward a_{active} and punish the others

else

punish $a_{active}$and reward the others

end if

end while

end algorithm

Fig. 3.7 The pseudocode for the process of learning parameter q in LAHS

In the HS algorithm, three parameters, i.e., *HMCR*, *PAR*, and *bw*, should be tuned during its evolution. Three learning automata with N actions are employed for each parameter, and the algorithm starts by randomly selecting one acceptable range for each parameter. New harmony is created based on the HS algorithm rules using the assigned parameter values. The parameter selection success is evaluated, and the three learning automata are updated accordingly. A "successful" parameter selection occurs when the new harmony has a better fitness value than the HM's worst existing harmony. In this case, the three learning automata are rewarded; otherwise, they are penalized. Figure 3.8 presents a flowchart of the LAHS.

3.3.9 Learning Automata Based Butterfly Optimization Algorithm (LABOA)

The authors (Arora and Anand 2018) presented the Learning Automata Based Butterfly Optimization Algorithm (LABOA). They claimed that to increase the potential of the butterfly optimization algorithm (BOA), they designed a hybrid algorithm in which it focuses on the exploration phase in the initial stages and on exploitation in the later stages of the optimization, learning automata have been embedded in BOA in which a learning automaton takes the role of configuring the behavior of a butterfly in order to create a proper balance between the process of global and local search. The introduction of learning automata accelerates global convergence speed to the true global optimum while preserving the basic BOA's main feature.

The butterfly optimization algorithm demonstrates satisfactory results as the local and global search is controlled by the switch probability factor. The diversity of solutions/butterflies is maintained because butterflies are allowed to explore the search

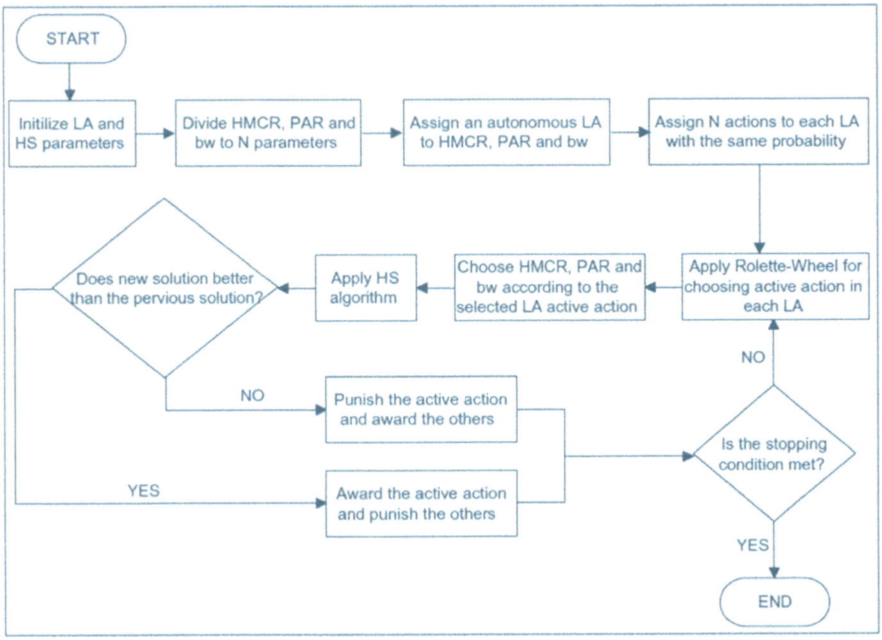

Fig. 3.8 The Flowchart of the LAHS

space during the global search process, whereas the butterflies' exploitation is encouraged when the local search is carried out. The BOA pulls off an advantage in solving complex real-world problems as these search processes occur randomly one after the other, thus maintaining the butterfly diversity. On the contrary, it is analyzed that random selection of these two phases sometimes results in BOA falling into local optima or move away from the global optimum. In the current study, learning automata are embedded in the BOA to adapt the switching probability p parameter, which critically affects the BOA's performance by altering the optimization process's search phases. In the LABOA model, an automaton is used to configure every butterfly's search behavior to create a balance between the process of global and local search. The LA has two possible actions: "Follow the best" and "Continue your way." If LA selects the "Follow the best" action, then the butterfly in the population must use the best butterfly information in the population and move towards it. If LA selects "Continue your way," then the butterfly takes random steps in the neighborhood. LABOA starts with initializing the position and fragrance of the butterflies. Then until the stopping criteria are not met, the following steps are repeated.

1) The automaton selects one of its actions based on its probability vector for each butterfly.
2) Based on the selected action, the type of search phase is selected, and then that butterfly updates its position and fragrance value.

Algorithm 3-2. LABOA

1. Objective function $f(x)$, $x = (x_1, x_2, \ldots, x_d)$
2. Generate population of n Butterflies $x_i = (i = 1, 2, \ldots n)$
3. Define c and a.
4. Calculate the fragrance of each butterfly as $f_i = cI^a$
5. **while** stopping criteria not met **do**
6. **for** each butterfly bf in the population **do**
7. Choose an action for each automaton according to its probability vector
8. Move the butterfly according to the selected action

$$x_i^{t+1} \leftarrow x_i^t + (r^2 \times g^* - x_i^t) \times f_i$$

$$x_i^{t+1} \leftarrow x_i^t + (r^2 \times x_j^t - x_k^t) \times f_i$$

9. Evaluate the new butterfly
10. **If** the new butterfly is better than the older, update it in the population, **then**
11. Update the automaton's probability vector using Eq. (3-1) and Eq. (3-2)
12. **end if**
13. **end for**
14. Find the current best butterfly g^*
15. **end while**
16. Output the best solution found

Fig. 3.9 The pseudocode of LABOA

3) According to the butterfly updating results, a reinforcement signal is generated, which will be used to update the probability vector of the automaton.

The automaton's selected action in any iteration specifies the butterfly's type of search method. Selection of the "Follow the best" action means that the best butterfly position will be used to calculate the butterfly's new position. If the "Continue your way" action is selected, the butterfly will move randomly into the search space. Evaluating the selected action is done by comparing each butterfly's new position with its old position. If the butterfly's fitness is improved, then the selected action will be evaluated as appropriate, and otherwise, the action will be evaluated as an inappropriate action. There is no doubt that the probability vector calculations an additional step, which makes the algorithm perform other steps, which w, which will consequently, however, this particular step enhances the performance of the algorithm, which justifies its addition. The pseudocode of the LABOA is shown in Fig. 3.9.

3.3.10 Grey Wolf Optimizer Based on Learning Automata (GWO-LA)

In the Grey wolf optimizer based on the learning automata (GWO-LA) algorithm (Betka et al. 2020), LA is integrated into the GWO to learn the objective function and decide whether it is a unimodal or multimodal function. Unimodal function needs good exploitation of promising areas in the search space. However, the multimodal function requires high exploration ability. The classification obtained using LA is then used to create new solutions in the appropriate areas. In the creation phase, two equations are used. The first one is based on a Gaussian distribution to enrich the exploitation for the unimodal function, and the second is based on a random distribution to support the exploration in multimodal function.

The grey wolf optimizer and every metaheuristic algorithm need to procure a good exploration and exploitation of the search space. Exploration is visiting the vast area in the search space to avoid local optimum, while exploitation concentrates the search in the promising area to find the global optimal solution. The successful algorithm procures a good balance between exploration and exploitation (Rezapoor Mirsaleh and Meybodi 2018). In the GWO-LA, because there are two main types of objective functions, including 1) Unimodal function, which has one global optimal solution and needs a high exploitation ability to find it. 2) Multimodal function has many local optimums, and a high exploration ability is required to overcome them. Thus, the authors (Betka et al. 2020) tried to develop an algorithm capable of identifying the objective function's nature. Then, it improves the exploration or exploitation of the GWO algorithm depending on the objective function nature. They have suggested integrating the LA algorithm into the GWO algorithm to learn the objective function and decide its nature.

Then, depending on the LA classification, new solutions are created in the appropriate areas within the search space to improve the exploration or exploitation. The GWO-LA algorithm's primary process is as follows: two actions, exploration and exploitation, are considered for each candidate solution. The algorithm updates the candidate solutions X_{t+1} with the conventional GWO as Eq. (3.9) and evaluates their fitness values.

$$X_{t+1} = \frac{X_1 + X_2 + X_3}{3} \tag{3.9}$$

Then for each solution, an action is selected, and as per the selected action, a new solution is created in the search space.

- Exploitation: If the selected action is exploitation, Eq. (3.10) creates a new solution in the candidate solution's neighborhood to concentrate the search in a promising area.

$$X_{new} = Gaussian\left(X_g, \alpha_g\right) + \left(rX_g - r'X_i^t\right), \tag{3.10}$$

where X_g is the global best solution and α_g is the standard deviation and computed as follows.

$$\alpha_g = \left|\frac{\log(t)}{t}\right|\left(X_i^t - X_g\right), \tag{3.11}$$

where r and r' are two random numbers taken from [0, 1].

Algorithm 3-3. GWO-LA

```
Initialize the reward and penalty of LA; a and b
Random initialization of candidate solutions; X
for t = 1 : t_max do
    Update the parameters A, a, and C
    Determine the solutions X, X_fl and X
    Update the positions of candidate solutions (2)
    Fitness evaluation of candidate solutions
    for i = 1: N do
        Select an action
        if action= exploitation then
            X_new is generated with Eq. (3-10)
        else
            X_new is generated with Eq. (3-12)
        end if
        X_new is evaluated with the fitness function
        if f(X_new) < f(X_i) then
            X_i = X_new
            The probabilities of choosing actions in LA are updated with Eq. (3-1)
        else
            The probabilities of choosing actions in LA are updated with Eq. (3-2)
        end if
    end for
end for
```

Fig. 3.10 Pseudocode of GWO-LA

- Exploration: If the selected action is exploration, Eq. (3.12) randomly creates a new solution to visit different search space areas.

$$X_{new} = (L_{max} - L_{min}) \times rand + L_{min} \tag{3.12}$$

The newly created solution replaces the candidate solution; if its fitness value is better than the current fitness value, then the probabilities are updated using Eq. (3.1). Otherwise, the probabilities are updated using Eq. (3.2). The pseudocode of the GWO-LA is depicted in Fig. 3.10.

3.3.11 Learning Automata Models with Multiple Reinforcements (MLA)

This section introduces learning automaton (LA) models (Rezvanian et al. 2018b) that can learn an optimal subset of their actions and update their internal probability vectors according to multiple environmental responses (Vafashoar and Meybodi 2019a), (Vafashoar et al. 2021d). To distinguish between a vector action and its comprising components, the terms super-action and simple-action are adopted in this chapter. It is assumed that each of the LAs has an action set like $A = \{a_1, \ldots, a_r\}$ simple-actions and aims at learning the optimal super-action, which consists of

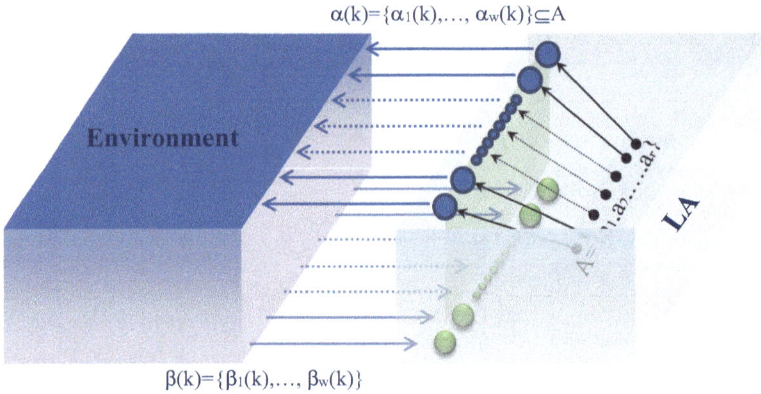

Fig. 3.11 Learning from multiple reinforcements

w simple-actions ($1 \leq w < r$). In each time step, the LA selects a super-action $\alpha = \{\alpha_1, \ldots, \alpha_w\}$ consisting of w simple-actions. Then, it performs the combination of these simple actions as a single super-action in the environment. The environment provides a w-dimensional reinforcement vector in response to the selected super-action. Each element of the reinforcement vector represents the favorability of one of the selected simple-actions in the combined super-action, as depicted in Fig. 3.11. Accordingly, each selected simple-action like $\alpha_j(k)$ receives an independent reinforcement signal like $\beta_j(k)$. The combined reinforcement for the selected super-action at step k can be considered as the sum of all elements in the reinforcement vector, i.e. $\sum_{l=1}^{w} \beta_l(k)$. The goal of the learning algorithm is to learn the super-action with the highest expected combined reinforcement value.

3.3.11.1 Multi-reinforcement Learning Automaton Type I

A probability vector ordinarily models the internal state of an LA. During the learning process, this probability vector is adapted to the environment, and the selection probability of the most reward receiving action is gradually increased. However, using this method for subset selection would be inefficient as the number of super-actions grows exponentially with some simple actions. Additionally, the reinforcements received by a typical simple action depend on the effectiveness of other simple actions in the experienced super-actions, which adversely affects the learning process. Representing the internal state as a probability vector over simple actions would also be unsuitable. As the selection probability of a particular simple-action approach to one, all other simple actions selection probabilities converge to zero.

The first multi-reinforcement learning automaton introduced in this section, called multi-reinforcement learning automaton type I (MLATI), considers an energy level for its actions (Vafashoar et al. 2021d). The internal state of an MLATI at step k is an r-dimensional vector x, where $x_i \in [0, 1]$, as presented in Fig. 3.12. Each component

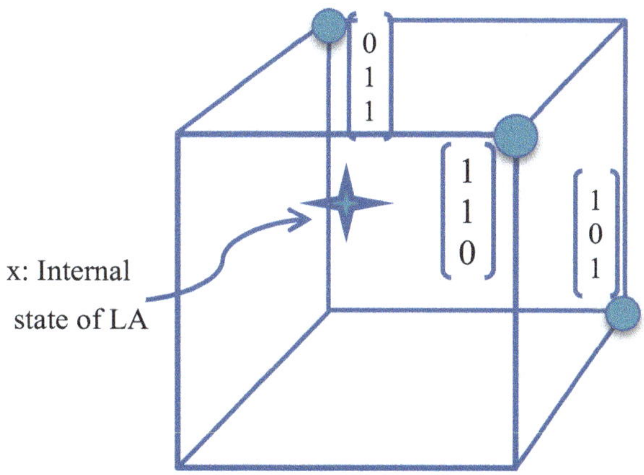

Fig. 3.12 A typical example of an MLATI aiming to find the best two actions is its three available actions. The energy vector of the MLATI changes within an r-dimensional unit hypercube

x_i of x defines the energy level of the corresponding action a_i. Actions with higher energy levels are selected with higher probability values. The automaton in Fig. 3.12 has three actions and selects a super-action consisting of two simple actions in each step. These super-actions correspond to the hypercube corners, which are represented by the black circles in the figure. The LA's objective is to move its energy vector toward the energy hypercube corner with the highest expected payoff.

The operation of MLATI follows. At each stage, a dummy super-action is selected based on the energy vector $x(k)$: each simple-action a_i chosen with probability $x_i(k)$ in the dummy super-action. This super-action corresponds to one of the LA hypercube corners and may consist of less/more than w simple actions. After this stage, the algorithm randomly selects one of the closest w-dimensional super-actions to the selected dummy super-action. Accordingly, the final selected super-action (denoted by $\alpha(k)$) consists of exactly w simple-actions. It should be noted that if we consider each element of $x(k)$ as the energy level of a simple-action, the final super-action ($\alpha(k)$) is selected based on the energy levels of the simple-actions. After action selection, the selected super-action is carried out in the environment. The LA receives a w-dimensional reinforcement signal, representing each of the selected simple actions' favorability. Based on this signal, the simple actions estimated rewards are updated according to Eq. (3.1). The algorithm maintains two additional vectors, $Z_i(k)$ and $\eta_i(k)$, in order to compute the estimated rewards of the simple actions. $Z_i(k)$ represents the total received reinforcement by the ith action, and $\eta_i(k)$ represents the number of times that the ith action is selected until step k.

$$
\begin{aligned}
Z_i(k) &= \begin{cases} Z_i(k-1) + \beta_i(k) & \text{if } a_i \in \alpha(k) \\ Z_i(k-1) & \text{otherwise} \end{cases} \\
\eta_i(k) &= \begin{cases} \eta_i(k-1) + 1 & \text{if } a_i \in \alpha(k) \\ \eta_i(k-1) & \text{otherwise} \end{cases} \\
\hat{d}_i(k) &= \frac{Z_i(k)}{\eta_i(k)}, i = 1, \ldots, r
\end{aligned}
\tag{3.13}
$$

where $\widehat{d}_i(k)$ represents the estimated reward for the ith simple-action at step k. Additionally, for simplicity, the reinforcement vector is represented by an r-dimensional vector; however, only w elements of this vector corresponding to the selected simple-actions have valid information.

The next step of the algorithm is to find w simple-actions with the highest estimated rewards. This set of simple actions corresponds to the super-action with the highest expected reward. Then, the algorithm moves the LA energy vector toward the corner of the LA hypercube associated with this super-action. Consequently, the energy levels of the promising simple-actions are increased. Figure 3.13 represents the pseudocode of the MLATI. It should be noted that the algorithm can be simply implemented without actually calculating C (in Fig. 3.13). Consequently, all of its

Algorithm 3-4. MLATI

Input

r: number of actions

$w<r$: number of actions to be selected.

λ: learning rate.

Initialization:

$Z_i(0) = c_1\ \forall i\in\{1,\ldots,r\}$, where c_1 is a constant value.

$\eta_i(0) = c_2\ \forall i\in\{1,\ldots,r\}$, where c_2 is a constant value.

$\widehat{d}_i(0) = 0\ \forall i\in\{1,\ldots,r\}$.

Set $x(0)=[0.5,\ldots,0.5]^T$, where x has r dimensions, and $x_i\in[0,1]$ represents the energy of the ith action.

Let I be an r-dimensional Boolean vector, $I_i \in\{0,1\}$.

Let $C=\{C_1,C_2,\ldots,C_o\}$ be the set of corners (vertices) of the LA hypercube with exactly w 1s and r-w 0s, where $o = |C| = \binom{r}{w}$, $C_{ij}\in\{0,1\}$, and $\sum_j C_{ij} = w\ \forall i \in \{1,\ldots,o\}$.

Set $k = 0$.

repeat

1. Set $k = k+1$ and $\alpha(k)=\{\}$.
2. Select a random set of simple actions according to the LA's energy level (i.e., according to x): each action like a_i is selected with probability x_i.
3. For each $i\in\{1,\ldots,r\}$, set $I_i=1$ if action a_i is chosen in step 2, otherwise set $I_i=0$.
4. Let C' be the subset of vertices in C, which are in the minimum distance to I. Randomly set I to one of the vertices in C'.
5. For each $i\in\{1,\ldots,r\}$, add simple-action a_i to $\alpha(k)$ if $I_i=1$.
6. Apply $\alpha(k)$ to the environment.
7. Receive an r-dimensional reinforcement signal β, where β_j is undefined if $a_j \notin \alpha(k)$.
8. Update LA information as follows:
$$Z_i(k) = \begin{cases} Z_i(k-1) + \beta_i(k) & \text{if } a_i \in \alpha(k) \\ Z_i(k-1) & \text{otherwise} \end{cases}$$
$$\eta_i(k) = \begin{cases} \eta_i(k-1) + 1 & \text{if } a_i \in \alpha(k) \\ \eta_i(k-1) & \text{otherwise} \end{cases}$$
$$\widehat{d}_i(k) = \frac{Z_i(k)}{\eta_i(k)}, i = 1,\ldots,r$$
9. Update x as follows:
10. Find w actions with the largest estimated average reinforcements represent them by I:
$$\left(I_j \in \{0,1\}\ \forall j \in \{1,\ldots,r\}\right) \text{ and } \textstyle\sum_j I_j = w \text{ and } \forall i,j\left(\left(I_i = 1 \text{ and } I_j = 0\right) \Rightarrow \widehat{d}_i \geq \widehat{d}_j\right)$$
11. Let C_l be the vertex in C corresponding to I.
12. $x(k) = x(k-1) + \lambda(C_l - x)$.

until some stopping condition is satisfied

Fig. 3.13 Pseudocode of multi-reinforcement learning automaton type I

steps can be efficiently implemented with computational complexities not exceeding o(r).

3.3.11.2 Multi-reinforcement Learning Automaton Type II

The MLATI is based on the pursuit algorithm. It has two main drawbacks, which are due to the characteristics inherent in the pursuit algorithm. In order to illustrate these drawbacks, we consider the following scenario for the pursuit algorithm. Consider a learning automaton with four actions a_1, a_2, a_3, and a_4 with the expected payoffs d_1, d_2, d_3, and d_4, respectively. Assume $d_1 > d_2 > d_3 > d_4$. Also, assume that during some early iterations of the pursuit algorithm, actions a_3 and a_4 collect larger rewards than a_1 and a_2. Accordingly, the following relations would hold for the estimated rewards and action probability values. $\widehat{d}_4(k_1) > \widehat{d}_3(k_1) > \widehat{d}_1(k_1) > \widehat{d}_2(k_1)$ and $p_4(k_1) > p_3(k_1) > p_1(k_1) > p_2(k_1)$ for some k_1. Assume that during some iterations after step k_1, a_1 receives the highest payoff from the environment, and consequently, $\widehat{d}_1(k_2) > \widehat{d}_4(k_2) > \widehat{d}_3(k_2) > \widehat{d}_2(k_2)$ with $p_1(k_2) > p_4(k_2) > p_3(k_2) > p_2(k_2)$ for some k_2. After step k_2, the received reinforcements are such that the following is always satisfied $\widehat{d}_1(k_3) = \max\left(\widehat{d}_i(k_3)\right)$ for any $k_3 > k_2$. Accordingly, after time step k_2, the selection probability of a_1 always increases, which is favorable; however, the relative order of the selection probabilities of other actions remains unaltered (i.e., $p_4(k_3) > p_3(k_3) > p_2(k_3)$ for any $k_3 > k_2$) regardless of their average received payoffs. The second issue with the pursuit algorithm is closely related to the first one. Consider again in the previous scenario that after step k_1, the average received payoff for action a_1 increases. Define k_5 as the minimum number of steps such that the following holds $\widehat{d}_1(k_1 + k_5) = \max_i\left(\widehat{d}_i(k_1 + k_5)\right)$. During these k_5 steps, the probability of action a_4 always increases regardless of the received rewards of a_4. After these k_5 steps, the probability of action a_1 starts increasing. However, it may take a long time until $p_1 > p_4$. These two issues may result in a slow convergence rate of the pursuit algorithm.

According to the first observation, the selection probabilities of actions other than those with the highest estimated reward may be unrelated to their obtained rewards. Accordingly, it would not affect the learning process if LA selects these actions with equal probabilities. The second issue can be solved if the action with the highest estimated reward is always selected with the highest probability.

Figure 3.14 summarizes the steps of the multi-reinforcement learning automaton type II (MLATII). The algorithm performs action selection in three manners: greedy, exploratory, and mutated greedy. During step k, the algorithm selects the super-action with the highest estimated reward with probability $1-\lambda(k)$. We assume that $(\lambda(k))_{k\in\mathbb{N}}$ is a decreasing sequence of numbers, which converges to zero. Hence, it is ensured that the super-action with the highest estimated reward is ultimately selected almost surely in every iteration.

Algorithm 3-5. MLATII

Input

r: number of simple-actions

$A = \{a_1, a_2, \ldots, a_r\}$: actions set

$w<r$: number of simple-actions to be selected.

$(\lambda(k))_{k\in\mathbb{N}}$: learning rate.

$(\eta(k))_{k\in\mathbb{N}}$: averaging factor.

Initialization:

Initialize the estimated rewards for the simple-actions:

$Q_i(0) = c_1 \; \forall i \in \{1, \ldots, r\}$, where c_1 is some constant value.

repeat

1. With probability $1-\lambda(k)$: // *Greedy selection*
   a. Select the *w* simple-actions with the highest expected payoffs as $\alpha(k)$.
2. Otherwise, with probability $\lambda(k)/2$: // *exploratory*
   a. Select a random super-action as $\alpha(k)$.
3. Otherwise: // *mutated greedy*
   a. Select the *w* simple-actions with the highest expected payoffs as $\alpha(k)$.
   b. Select a random action $a_l \notin \alpha(k)$ according to its probability value defined as:
      $p_l = \frac{Q_l(k)}{\sum_{j\in I} Q_j(k)} \; \forall i \in I = \{l \mid a_l \notin \alpha(k) \text{ and } a_l \in A\}$
   c. Select a random action $a_l \in \alpha(k)$ according to its probability value:
      $p_l = \frac{1-q_l}{\sum_{j\in I} 1-q_j} \; \forall l \in I$
      Where $q_h = \frac{Q_h(k)}{\sum_{j\in I} Q_j(k)} \; \forall h \in I$ and $I = \{s \mid a_s \in \alpha(k)\}$
   d. Replace a_l with a_l in $\alpha(k)$.
4. Perform $\alpha(k)$ in the environment.
5. Receive a vector reinforcement signal β.
6. For each action a_i in $\alpha(k)$ do:
   a. $Q_i(k+1) = Q_i(k) + \eta(k)(\beta_i(k) - Q_i(k))$

until some stopping condition is satisfied.

Fig. 3.14 Pseudocode of multi-reinforcement learning automaton type II

Additionally, each simple action should be selected an adequate number of times to ensure that the estimated rewards converge to their expected values (as discussed in Lemma 3.2). In this regard, with probability $\lambda(k)/2$, the introduced method randomly selects a super-action at line 2, which guarantees that each simple-action is chosen for an adequate number of times. Finally, with probability $\lambda(k)/2$, the algorithm can utilize the mutated greedy selection strategy in each iteration. The mutated greedy selection has the benefits of both exploratory and greedy selection strategies. It greedily selects a slightly perturbed version of the best-estimated super-action. Consequently, using mutated greedy selection, the algorithm can frequently select and exploit estimated better simple-actions while experiencing all simple-actions for an adequate number of times. After action selection, the selected super-action is carried out in the environment, and the estimated rewards are updated based on the received reinforcement signals.

It is easy to show that by using a decreasing sequence $(\lambda(k))_{k\in\mathbb{N}}$ which satisfies $\lambda(k) = \kappa\mu(t)$ with $\sum_k \mu(k) = \infty$ and $\kappa > 0$, each simple-action is selected infinitely often. Also, if the averaging factor satisfies the two conditions $\sum_k \eta(k) = \infty$ and $\sum_k (\eta(k))^2 < \infty$, the estimated reward of each simple action converges to the actually expected reward of the simple action. Hence as $\lambda(k) \to 0$ with $k \to \infty$, the selection probability of the most promising super-action converges to 1. Based on these observations, the algorithm is ε-optimal.

3.3.11.3 Multi-Reinforcement Learning Automata Type III

The multi-reinforcement LA introduced in the previous section is not absolutely expedient. In order to illustrate this issue, consider $\underline{a}_i$ as the optimal super-action. Also, assume that during some iteration like k, for some super-action like $\underline{a}_j$, the following holds $\sum_l I_l Q_l(k) < \sum_l J_l Q_l(k)$, where I and J are Boolean vectors representing the members of $\underline{a}_i$ and $\underline{a}_j$ respectively. Accordingly, $\underline{a}_j$ will be selected with a higher probability for some iterations after step k. During these subsequent iterations, $\sum_l J_l Q_l(k)$ starts decreasing as $\underline{a}_j$ has a lesser expected payoff in comparison to $\overline{a}_i$. However, the selection probability of $\underline{a}_j$ would increase for some iterations as $\lambda(k)$ is decreasing. This increase in the selection probability of $\underline{a}_j$ continues regardless of the received payoffs until the following condition is satisfied $\sum_l I_l Q_l(k) > \sum_l J_l Q_l(k)$. Absolute expediency is an advantageous property, which, under appropriate conditions, guarantees the convergence of a group of learning automata in a CLA structure (Beigy and Meybodi 2004, 2010; Rezvanian et al. 2018c).

One of the absolutely expedient learning automata algorithms is L_{RI} (Narendra 1989; Rezvanian et al. 2018b). In this section, another multi-reinforcement LA model is introduced, which is based on the L_{RI} algorithm. Like MLATI, each simple action has an associated energy and is selected according to its energy level in each iteration. The learning automaton's internal state is modeled as an energy vector that changes in an r-dimensional unit hypercube. However, in contrast to MLATI, the LA's total energy (energy sum of all simple-actions) is always kept constant and equal to w. Accordingly, an increase in the energy level of a particular simple action should decrease the others' energy levels. In this regard, whenever a selected simple action receives a favorable reinforcement, its energy level increases, and this increased amount is subtracted from the total energy level of the unselected simple actions.

A multi-reinforcement learning automaton type III (MLATIII) operates as follows. At each iteration k, a w-dimensional super-action is selected in the same way as MLATI. Then, this super-action is performed in the environment. The environment provides the learning system with a w-dimensional feedback signal, representing the favorability of each of the selected simple actions in the super-action. If the associated signal to a selected simple-action like a_i is favorable, i.e. $\beta_i(k) = 1$, the energy level of a_i is increased according to Eq. (3.7). This increased amount is subtracted from the energy level of unselected actions to keep the LA constant's total energy level (Eq. (3.8)).

$$x_i = x_i + \lambda\beta(k)(1 - x_i) \tag{3.14}$$

Algorithm 3-6. MLATIII

Input

r: number of simple-actions

$A = \{a_1,\ldots,a_r\}$

$w<r$: number of simple-actions to be selected.

λ: learning rate.

Initialization:

Set $x(0)=[w/r, w/r,\ldots, w/r]^T$, where $x_i \in [0,1]$ represents the energy of the ith simple-action.

Consider an r-dimensional Boolean vector I: $I_i \in \{0,1\}\ \forall i \in \{1, \ldots, r\}$.

Let $C=\{C_1,C_2,\ldots,C_o\}$ be the set of corners (vertices) of the LA hypercube with exactly w 1s and (r-w) 0s, where $o = |C| = \binom{r}{w}$, $C_{ij} \in \{0,1\}$, and $\sum_j C_{ij} = w\ \forall i \in \{1,\ldots,o\}$.

Set $k = 0$.

Repeat

1. Set $k= k+1$ and $\alpha(k)= \{\}$.
2. Select a random set of simple actions according to the LA's energy level (i.e., according to x): each action like a_i is selected with probability x_i.
3. For each $i \in \{1, \ldots, r\}$, set $I_i =1$ if action a_i is chosen in step 2, otherwise set $I_i =0$.
4. Let C' be the subset of vertices in C, which are in the minimum distance to I. Randomly set I to one of the vertices in C'.
5. For each $i \in \{1, \ldots, r\}$, add simple-action a_i to $\alpha(k)$ if $I_i =1$.
6. Apply $\alpha(k)$ to the environment.
7. Receive an r-dimensional reinforcement signal β, where β_j is undefined if $a_j \notin \alpha(k)$ (i.e. I_j=0).
8. Update LA information as follows:
   For each action a_i such that $I_i =1$ and $\beta_i(k)>0$ do:
   $x_j = x_j - \frac{\lambda\beta_i(k)(1-x_i)}{\sum_l x_l(1-I_l)}(x_j)\ \forall j \in \{h|a_h \notin \alpha(k)\ and\ a_h \in A\}$
   $x_i=x_i+\lambda\beta_i(k)(1-x_i)$

until some stopping condition

Fig. 3.15 Pseudocode of multi-reinforcement learning automata type III

$$x_j = x_j - \frac{\lambda\beta_i(k)(1-x_i)}{\sum_l x_l(1-I_l)}(x_j)\forall j \in \{h|a_h \notin \alpha(k)\ and\ a_h \in A\} \tag{3.15}$$

Here, λ is the learning rate parameter, and similar to MLATI, the selected set of simple-actions at step k is represented by $\alpha(k)$. I is an r-dimensional Boolean vector that represents the selected simple actions. Figure 3.15 illustrates the multi-reinforcement learning automaton algorithm.

3.3.12 Cellular Learning Automata Models with Multiple Reinforcements (MCLA)

The multi-reinforcement LA models can be directly integrated into the CLA model. The CLA models based on the multi-reinforcement LAs will be investigated empirically (Beigy and Meybodi 2004). This section briefly introduces a CLA model that aims to maximize the expected reward for each LA. It will be shown that the CLA model converges to a compatible point (Morshedlou and Meybodi 2017) when the local rules obey the characteristics of a potential function. Many distributed problems, such as channel assignment in mesh networks, can be modeled as potential games (Duarte et al. 2012). At each step k, the local rule of the ith cell determines the reinforcement signal to the learning automaton in the cell as follows:

$$F_i : \varphi_i \rightarrow \underline{\beta} \text{ with } \varphi_i = \prod_{j=0}^{n_i} A_{N_i(j)} \tag{3.16}$$

where N_i is the neighborhood function of the ith cell, and $N_i(j)$ returns the index of the jth neighboring cell to cell i with $N_i(0)$ defined as $N_i(0) = i$. n_i denotes the number of cells in the neighborhood of the ith cell.

Finally, $\underline{\beta}$ is the set of values that the reinforcement signal can take. We can extend the local rules of a CLA as follows:

$$T_i(\alpha_1, \ldots, \alpha_n) = F_i\left(\alpha_i, \alpha_{N_i(1)}, \ldots, \alpha_{N_i(n_i)}\right) \forall \alpha_j \in A_j \tag{3.17}$$

A local rule will be called a potential local rule if there exists a function like ψ, called a potential function, for which the condition in Eq. (3.12) is satisfied. The characteristics of local utility functions in potential games have been investigated in several recent works (Babichenko and Tamuz 2016; Ortiz 2015).

$$\begin{aligned} &T_i(\alpha_1, \ldots, \alpha_i, \ldots, \alpha_n) - T_i\left(\alpha_1, \ldots, \alpha_i', \ldots, \alpha_n\right) \\ &= \psi(\alpha_1, \ldots, \alpha_i, \ldots, \alpha_n) - \psi\left(\alpha_1, \ldots, \alpha_i', \ldots, \alpha_n\right) \forall \alpha_j \in A_j, \forall \alpha_i' \in A_i \end{aligned} \tag{3.18}$$

3.3.13 Multi-reinforcement CLA with the Maximum Expected Rewards (MCLA)

This section briefly introduces a multi-reinforcement CLA model called MCLA (Vafashoar and Meybodi 2019a). A schematic representation of multi-reinforcement CLA is presented in Fig. 3.16 (Vafashoar et al. 2021d). The model of multi-reinforcement CLA considered in this section aims at maximizing the expected reward for each LA. This CLA is closely related to the model presented in (Morshedlou and Meybodi 2017). In the model presented (Morshedlou and Meybodi 2017), each LA learns a probability distribution over its actions. This distribution leads to the maximum expected reward in response to the empirical joint distribution of actions performed by other learning automata in the neighborhood. Consequently, each LA converges to a probability distribution, which may have non-zero probabilities on several actions. However, some applications require each LA to learn a single optimal action, or super-action. In this regard, the multi reinforcement CLA estimates the payoff for each simple-action response to the neighborhood's performed

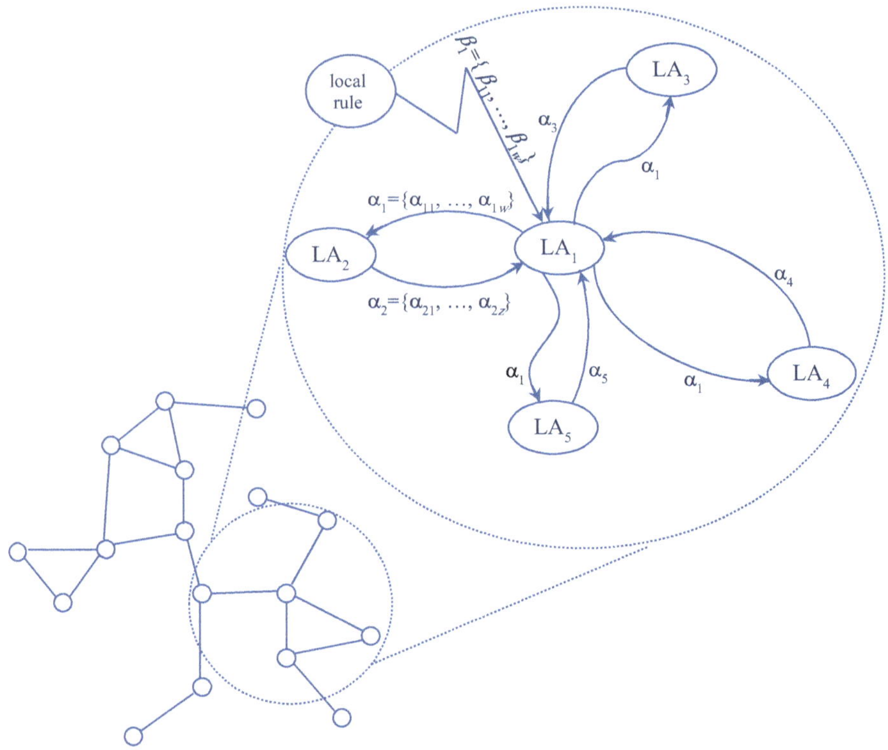

Fig. 3.16 Schematic representation of MCLA

actions. It gradually increases the most favorable super-action selection probability, consisting of simple actions with the highest expected rewards.

The three introduced LA models can be generalized into a CLA scheme, which maximizes each LA's expected reward according to the neighborhood's distribution of actions. For instance, in MLATI, each LA of the CLA selects a super-action described in Fig. 3.13. Next, it observes the selected actions in its neighborhood and updates its empirical distributions. Then, the LA receives feedback from the environment for each one of its selected simple actions. Based on the empirical probability distributions and the received feedback, each combination of actions' estimated payoff is updated. Finally, similar to Fig. 3.13, the energies of simple actions with w_i highest expected rewards are increased. Figure 3.17 gives pseudocode for the MCLA model. In this model, each learning automaton learns the empirical distribution of actions in its neighborhood. It also learns the estimated payoff of each of its simple actions combined with different super-actions performed by the neighboring learning automata.

Consequently, it can compute the expected reward for each one of its simple actions. During each iteration k, every LA-like LA_i finds its w_i simple actions with the highest expected rewards. Then, with probability $1 - \lambda(k)$, the LA performs

Algorithm 3-7. MCLA

Input

r: number of actions

w<*r*: number of actions to be selected.

$(\lambda(k))_{k\in\mathbb{N}}$: learning rate.

$(\eta(k))_{k\in\mathbb{N}}$: averaging factor.

$(\gamma(k))_{k\in\mathbb{N}}$: discount factor.

While the termination condition is not satisfied, **do**

I. **For** each cell like *i*, synchronously with other cells, **do**

1. With probability $\lambda(k)/2$ do: // exploratory selection
    1. Select a random super-action as $\alpha_i(k)$.
2. With probability $\lambda(k)/2$ do: // mutated greedy selection
    1. Select the maximum expected reward super-action. Let $\alpha_i(k)$ be this action.
    2. Define I as $I=\{l|a_{il}\in\alpha_i(k)\}$
    3. Select a random simple-action from $\alpha_i(k)$ according to its probability value defined as:
       $p_l = \frac{1-q_l(k)}{\sum_{j\in I} 1-q_j(k)}\ \forall l \in I,$
       where
       $q_h = \frac{Er(a_{ih}, \hat{p}n_i(k))}{\sum_{a_{il}\in\alpha_i(k)} Er(a_{il}, \hat{p}n_i(k))}\ \forall h \in I$
    4. Select a random simple-action not in $\alpha_i(k)$ according to the probability values defined as:
       $$p_h = \frac{Er(a_{ih}, \hat{p}n_i(k))}{\sum_{a_{is}\notin\alpha_i(k)} Er(a_{is}, \hat{p}n_i(k))}\ \forall h \in \{l|a_{il} \in A_i \& a_{il} \notin \alpha_i(k)\}$$
    5. Replace the action obtained in step 2.3 with the one obtained in step 2.4 in $\alpha_i(k)$.
3. With probability 1- $\lambda(k)$ do:
    1. Select the maximum expected reward super-action as $\alpha_i(k)$.
4. Perform the selected super-action, and receive the reinforcement vector $\beta_i(k)$.
5. $\hat{p}_{ij}(k+1) = \begin{cases}(1-\gamma(k))\hat{p}_{ij}(k) + \gamma(k) & \text{if } \alpha_i(k) = a_{ij} \\ (1-\gamma(k))\hat{p}_{ij}(k) & \text{otherwise}\end{cases}\ \forall j=1, \ldots, z_i$
6. $Q_i(k+1,:)=Q_i(k,:)$
7. $Q_i(k+1, a_{ij}, \alpha_{N_i(1)}(k), \ldots, \alpha_{N_i(n_i)}(k)) = Q_i(k, a_{ij}, \alpha_{N_i(1)}(k), \ldots, \alpha_{N_i(n_i)}(k)) + \eta(k)\big(\beta_{ij}(k) - Q_i(k, a_{ij}, \alpha_{N_i(1)}(k), \ldots, \alpha_{N_i(n_i)}(k))\big)\ \forall j\in\{l|a_{il}\in\alpha_i(k)\}$

End for

II. Set $k \leftarrow k+1$.

End while

Fig. 3.17 The pseudocode of MCLA

these actions in the environment, or with probability λ(*k*), it experiments two other action selection strategies similar to Fig. 3.14. After performing these selected simple actions, the LA updates the empirical distribution of neighboring learning automata's actions and the estimated payoff of its performed actions. It should be noted that λ(*k*) is a decreasing sequence of numbers; accordingly, the selection probability of the most promising super-actions approaches to one during the learning procedure. This model can easily be applied to channel assignment problems (Vafashoar et al. 2021d).

3.3.13.1 Extensions of MCLA Models for n-Tuple Selection

This section discusses how the multi-reinforcement technique can be used for optimal n-tuple selection problems. We can easily extend the previous methods so that each LA can select an n-tuple of its actions. As the approaches introduced in this section are very similar to the previously discussed ones, we briefly discuss the required LA extensions.

3.3.13.1.1. Reinforcement Learning Using Multiple Reinforcements for Optimal n-Tuple Selection

Multi- reinforcement models previously introduced try to tackle the subset selection problem. Using the same ideas, this section introduces a new learning automaton, which will be called multi-reinforcement n-tuple LA (MNLA). MNLA aims at finding an n-tuple of its actions under some restrictions. Assume that an MNLA has an action set $A = \{a_1, a_2, \ldots a_r\}$. During each iteration k, the MLA selects an n-tuple of its actions under the restriction that the number of distinct actions that appear in the chosen n-tuple is at most w. Then, the MNLA performs the chosen n-tuple, which is denoted by $\alpha(k) = [\alpha_1(k), \ldots, \alpha_n(k)]^T$, in the environment. In response to the performed n-tuple, the environment triggers an n-dimensional feedback signal $\beta(k)$. Each component $\beta_i(k)$ of the feedback signal represents a separate reinforcement from the environment to the associate action $\alpha_i(k)$. The objective of the MNLA is to find an n-tuple $\left(a_{j1}, a_{j2}, \ldots, a_{j1}\right)$ with the maximum expected total reinforcement signals, i.e., $E\left[\sum_{j=1}^{n} \beta_j(k)\right]$.

We represent the internal state of the MNLA by an $(n \times r)$-dimensional probability matrix M. Each row i of this probability matrix represents the selection probabilities of different actions in A for the ith position of the n-tuple. For each iteration k of the algorithm, first, a dummy n-tuple $\dot{\alpha}(k)$ is selected. Each component i of the dummy n-tuple is selected according to the probability vector M_i. Consequently, the generated dummy n-tuple may contain more than w distinct actions. To decrease the number of distinct actions in a selected n-tuple, the following procedure will be employed. Two distinct actions $a_i \in \alpha(k)$ and $a_j \in \alpha(k)$ are selected randomly. Then, the energy level of each of these actions is obtained as follows:

$$e_i = \prod\nolimits_{l \in I} M_{li} \text{ with } I = \left\{l | \dot{\alpha}_l(k) = a_i \vee \dot{\alpha}_l(k) = a_j\right\} \tag{3.19}$$

$$e_j = \prod\nolimits_{l \in I} M_{lj} \text{ with } I = \left\{l | \dot{\alpha}_l(k) = a_i \vee \dot{\alpha}_l(k) = a_j\right\} \tag{3.20}$$

After obtaining the energy levels, random action is chosen among a_i and a_j according to their probability values defined respectively as $q_i = e_i / \left(e_i + e_j\right)$ and $q_j = e_j / \left(e_i + e_j\right)$. Assume that a_l is selected in this stage. Next, all components of $\dot{\alpha}(k)$ consisting of $a_{i,}$ or a_j are replaced with a_l. The above-described procedure continues until the acquisition of a feasible n-tuple $\alpha(k)$. The chosen feasible n-tuple $\alpha(k)$ is carried out in the environment, and n-dimensional feedback signal $\beta(k)$ is received from the environment. Each component of the feedback signal is a reinforcement for its corresponding component in $\alpha(k)$. Based on the received reinforcements, the probability matrix M is updated. To update M, the following three matrices are maintained by the MNLA η, Z, and d. $\eta_{ij}(k)$ represents the number of times that action a_j is selected for the ith component of the n-tuple till iteration k, and $Z_{ij}(k)$

denotes the total reinforcement obtained by a_j when it was used as the jth component of the selected n-tuples. Finally, $d_{ij}(k)$ is the estimated feedback for action a_j when used as the ith component of the chosen n-tuples.

Based on the obtained feedback vector, the LA first updatesη, d, and Z as follows:

$$Z_{ij}(k+1) = \begin{cases} Z_{ij}(k) + \beta_i(k) \text{ if } & \alpha_i(k) = a_j \\ Z_{ij}(k) & \text{otherwise} \end{cases} \tag{3.21}$$

$$\mu_{ij}(k+1) = \begin{cases} \mu_{ij}(k) + 1 \text{ if } & \alpha_i(k) = a_j \\ \mu_{ij}(k) & \text{otherwise} \end{cases} \tag{3.22}$$

$$d_{ij}(k+1) = \mu_{ij}(k+1)/Z_{ij}(k+1) \tag{3.23}$$

Based on the obtained estimated feedbacks, the LA finds the feasible n-tuple with the highest (or lowest for minimization) sum of estimated feedbacks. The problem of finding this n-tuple is NP-complete. A greedy method for this problem will be described herein. Assume that the set of all feasible n-tuples is represented by B. the n-tuple $\delta \in B$ has the highest estimated feedbacks if the following condition is satisfied.

$$\sum\nolimits_{s=1}^{n} d_{sI(\delta_s)} \geq \sum\nolimits_{s=1}^{n} d_{sI(\rho_s)} \forall \rho \in B, \tag{3.24}$$

I is an index function, and $I(\sigma_s)$ return the index of the action in the sth component of σ. After obtaining the n-tuple δ with the minimum estimated feedbacks, the probability matrix M is updated as follows:

$$M_{ij}(k+1) = \begin{cases} M_{ij}(k) - \lambda M_{ij}(k) + \lambda \text{ if } & \delta_i = a_j \\ M_{ij}(k) - \lambda M_{ij}(k) & \text{otherwise} \end{cases} \tag{3.25}$$

3.3.13.1.2. A Greedy Method for Finding a Feasible n-Tuple

This section describes a greedy algorithm for finding a feasible n-tuple with the highest sum of estimated feedbacks. Assume that we wish to find a tuple of size n, which consists of most w distinct actions. The greedy algorithm first chooses the n-tuple $\delta(0)$, composed of actions with the highest estimated feedback. If this n-tuple consists of more than w distinct actions, the algorithm iteratively will reduce the number of its distinct actions. During each iteration t, the algorithm finds an action $a_i \in \delta(t)$, which can be replaced by other actions in $\delta(t)$ with the smallest decrease in the total feedback sum, and then replaces this action. This procedure continues until the number of distinct actions equals w. The pseudocode for the described algorithm is given in Fig. 3.18.

Algorithm 3-8. finding a feasible n-tuple

Input:

w // maximum number of distinct actions that can be selected

$A = \{a_1, a_2, \ldots, a_r\}$ *// action set*

$d_{n \times r}$ *// estimated feedback of each action for each component of an n-tuple.*

1. $\delta_i(0) \leftarrow \text{argmax}_j(d_{ij})\ \forall i \in 1 \ldots n$.
2. $t \leftarrow 0$.
3. **for** j=1 to r **do**
   a. if $j \notin \delta(0)$, then $d_{ij} \leftarrow -\infty\ \forall i \in 1 \ldots n$.
   end for
4. **while** the number of distinct actions in $\delta(t)$ is greater than w **do**
   a. $\delta(t+1) \leftarrow \varnothing$.
   b. $c \leftarrow -\infty$.
   c. **for** each distinct $j \in \delta(t)$ **do**
      i. $e \leftarrow d$.
      ii. $e_{ij} \leftarrow -\infty\ \forall i \in 1 \ldots n$.
      iii. $\delta' \leftarrow \text{argmax}_l(e_{il})\ \forall i \in 1 \ldots n$. *// a new tuple which doesn't include a_j*
      iv. $c' \leftarrow \sum_{l=1}^{n} d_{l\delta'(l)}$. *// the expected feedback sum of new tuple*
      v. **if** $c' > c$
         - $c \leftarrow c'$.
         - $\delta(t+1) \leftarrow \delta'$.
      end for
   d. $t \leftarrow t+1$.
   e. **for** j=1 to r **do**
      i. **if** $j \notin \delta(t)$, then $d_{ij} \leftarrow -\infty\ \forall i \in 1 .. n$.
      end if
   end for
   end while

Fig. 3.18 Estimating the feasible n-tuple with the highest sum of expected feedbacks

3.3.13.1.3. Modified Multi-reinforcement LA with Weak Memory for n-Tuple Selection

Like the previous section, we represent the internal state of modified multi-reinforcement LA (MMLA) by an $(n \times r)$-dimensional probability matrix M. Each row i of this probability matrix represents the selection probabilities of different actions in A for the ith position of the n-tuple. During each iteration k of the algorithm, a feasible n-tuple is selected, similar to the previous section. The chosen feasible n-tuple α(k) is carried out in the environment, and an n-dimensional feedback signal $\beta(k)$ is received from the environment. Each component of the feedback signal is a reinforcement for its corresponding component in $\alpha(k)$. Based on the received reinforcements, the probability matrix M is updated. To update the probability matrix, MMLA maintains the last received reinforcements of each selected action in an $n \times r$-dimensional matrix η. During each iteration k, η is updated based on the received reinforcement vector as follows:

$$\eta_{ij}(k) = \begin{cases} \tau\beta_i(k) + (1-\tau)\eta_{ij}(k) & \text{if} \quad \alpha_i(k) = a_j \\ \eta_{ij}(k) & \text{otherwise} \end{cases} \tag{3.26}$$

Here $\alpha_i(k)$ represents the ith component of the selected action, $\alpha(k)$. τ is the averaging factor, which puts more weight on the recently received reinforcements. Based on the updated last received reinforcements, the MMLA finds the feasible n-tuple with the highest sum of the estimated feedbacks, and the algorithm in Fig. 3.18 will be used for this purpose. After obtaining the n-tuple δ with the maximum estimated feedbacks, the probability matrix M is updated as follows:

$$M_{ij}(k+1) = \begin{cases} M_{ij}(k) - \lambda M_{ij}(k) + \lambda & \text{if } \delta_i = a_j \text{ and } \alpha_i(k) = a_j \text{ if } \delta_i = a_j \text{ and } \alpha_i(k) = a_j \\ M_{ij}(k) - \lambda M_{ij}(k) & \text{if } \delta_i = a_l \text{ and } \alpha_i(k) = a_l \text{ and } l \neq j \\ M_{ij}(k) & \text{otherwise} \end{cases}, \tag{3.27}$$

where λ is the learning rate of the MLA.

3.3.13.1.4. Modified Learning Automaton for n-Tuple Selection with Free Actions

In the previous section, a LA is introduced, which can choose an n-tuple of actions under the restriction that the number of distinct selected actions should not be higher than a constant like w. This section relaxes this constraint and introduces a new MMLA called FMLA. Assume that the action set can be partitioned into two sets A^r and A^f, $A = A^r \cup A^f$ and $A^r \cap A^f = \emptyset$. The FMLA chooses an n-tuple from A during each iteration k with the restriction that the number of distinct actions chosen from A^r is at most w; however, there is no restriction on the number of distinct actions chosen from the set A^f.

In the action selection step of MMLA, described in the previous section, a dummy n-tuple $\dot{\alpha}(k)$ is selected by the MMLA. During each step k, each action aj's selection probability for the ith component of the dummy n-tuple is $M_{ij}(k)$. Accordingly, the probability that a_j is not chosen for the ith component is $1 - M_{ij}(k)$. Hence, the probability that each action a_j is not chosen in the dummy n-tuple is $\prod_{i=1}^{n}\left(1 - M_{ij}\right)$. We define the elimination probability of each action a_j in the kth step as follows:

$$p_j(k) = \prod\nolimits_{i=1}^{n}\left(1 - M_{ij}(k)\right) \tag{3.28}$$

In the action selection step of FMLA, first, $|A^r| - w$ actions are selected from A^r according to their normalized elimination probability values. Let this set of actions be denoted by $A^e(k)$. Then, an n-tuple is selected from the set $A^f \cup (A^r - A^e)$ according to the probability matrix of the FMLA. This chosen feasible n-tuple $\alpha(k)$ is carried out in the environment, and n-dimensional feedback signal $\beta(k)$ is received from the environment. The updating procedure of FMLA is similar to that of MMLA, and Fig. 3.19 describes a greedy procedure for finding the feasible n-tuple with the highest sum of estimated feedback in FMLA.

Algorithm 3-9. finding a feasible n-tuple with the highest sum of estimated feedbacks in FMLA

```
Input:
w // maximum number of distinct actions that can be selected form Aʳ
A = {a1,a2,...,ar} // action set
A = Aʳ∪Aᶠ
η_{n×r} // estimated feedback of each action in each dimension of the selected n-tuples
δ(0)←argmax_j(η_ij) ∀i∈1..n
t←0
η_ij←-∞ ∀i∈1..n ∀j(j∉δ(0) and a_j∈Aʳ)
while number of distinct actions selected form Aʳ in δ(t) is greater than w do
        δ(t+1)←∅
        c←-∞
        for each distinct action a_j∈δ(t) such that a_j∈Aʳ do
                e←η
                e_ij←-∞ ∀i∈1..n
                δ'←argmax_j(e_ij) ∀i∈1..n
                c' ← Σ_{l=1}^{n} η_{lδ'(l)}
                If c'>c
                        c←c'
                        δ(t+1)←δ'
                End if
        End for
        t←t+1
        η_ij←-∞ ∀i∈1..n ∀j(j∉δ(t) and a_j∈Aʳ)
end while
```

Fig. 3.19 Estimating the feasible n-tuple with the highest sum of expected feedbacks in FMLA

3.3.14 *Gravitational Search Algorithm Based on Learning Automata (GSA-LA)*

Recent studies have indicated that a perfect way to make evolutionary algorithms (EAs) more accurate and efficient is to fine-tune their parameters (Alirezanejad et al. 2020). This section presents an LA mechanism for tuning gravitational constant $G(t)$ in the gravitational search algorithm (GSA). Basically, the GSA-LA performs like the original GSA with a difference at the end of each iteration where $G(t)$ should be valued. The GSA-LA removes the approach to valuing $G(T) = G_0 e^{\frac{at}{T}}$, where a is the descending constant, and T shows the maximum number of iterations. Instead, the $G(t)$ employs an LA with N actions, where each action corresponds to one of the N equally discretized distance values in the permitted range of the G(t). At each iteration, the most probable action within a set of LA actions is selected, and the value that corresponds to this action is set as the new value for $G(t)$. The learning process of the $G(t)$ is drawn in Fig. 3.20.

A new position for mass i in dimension d as $x_i^d(t+1)$ and iteration t is generated based on the GSA rules. The parameter selection success is evaluated, and the LA is updated accordingly. A 'successful' parameter selection occurs when $x_i^d(t+1)$

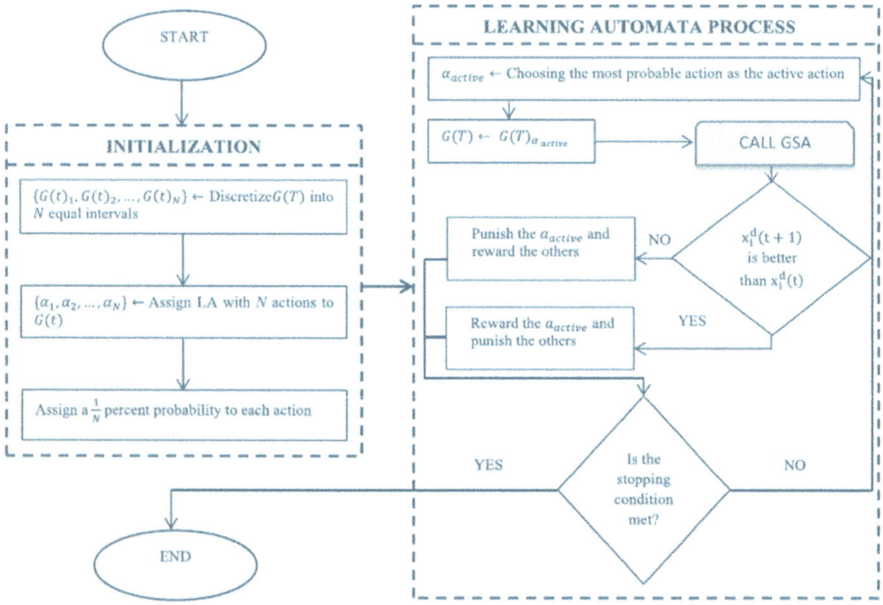

Fig. 3.20 The flowchart of the role of the learning automata process in GSA-LA

```
Algorithm 3-10. GSA-LA
Generate initial masses in the search space
Initialize a probability vector of the LA
while convergence criteria are not satisfied do
   Evaluate the fitness for each mass
   Update the G(T) based in the LA
   Update the best and worst of the population
   Calculate M for each mass
   Update the velocity and position of the masses
end while
```

Fig. 3.21 Pseudo-code of the GSA-LA

has a better fitness value than $x_i^d(t)$. In this case, the LA is rewarded; otherwise, it is punished. Figure 3.21 presents the pseudo-code for GSA-LA.

3.4 Conclusion

This chapter investigated new learning automata (LA) and cellular learning automata (CLA) models for solving optimization problems. In one of the recent models, each LA aims to learn the optimal subset or n-tuple of its actions using multiple feedback from its environment. Using a separate reinforcement for each chosen action enables the LA to learn the effectiveness of its different actions in the chosen super-actions in parallel. This decision-making is useful in scenarios like multi-agent environments where each agent has limited resources and actuators and performs several tasks simultaneously via its actuators.

References

Abbas, Z., Li, J., Yadav, N., Tariq, I.: Computational task offloading in mobile edge computing using learning automata. In: 2018 IEEE/CIC International Conference on Communications in China (ICCC), pp. 57–61. IEEE (2018)

Abbasi-ghalehtaki, R., Khotanlou, H., Esmaeilpour, M.: Fuzzy evolutionary cellular learning automata model for text summarization. Swarm Evol. Comput. **30**, 11–26 (2016). https://doi.org/10.1016/j.swevo.2016.03.004

Abedi Firouzjaee, H., Kordestani, J.K., Meybodi, M.R.: Cuckoo search with composite flight operator for numerical optimization problems and its application in tunnelling. Eng. Optim. **49**, 597–616 (2017)

Abin, A.A., Fotouhi, M., Kasaei, S.: Skin segmentation based on cellular learning automata. In: Proceedings of the 6th International Conference on Advances in Mobile Computing and Multimedia - MoMM 2008, Austria, p. 254. ACM (2008)

Abin, A.A., Fotouhi, M., Kasaei, S.: A new dynamic cellular learning automata-based skin detector. Multimed. Syst. **15**, 309–323 (2009). https://doi.org/10.1007/s00530-009-0165-1

Adinehvand, K., Sardari, D., Hosntalab, M., Pouladian, M.: An efficient multistage segmentation method for accurate hard exudates and lesion detection in digital retinal images. J. Intell. Fuzzy Syst. **33**, 1639–1649 (2017)

Ahangaran, M., Taghizadeh, N., Beigy, H.: Associative cellular learning automata and its applications. Appl. Soft Comput. **53**, 1–18 (2017). https://doi.org/10.1016/j.asoc.2016.12.006

Akbari Torkestani, J.: A learning approach to the bandwidth multicolouring problem. J. Exp. Theor. Artif. Intell. **28**, 499–527 (2016)

Akhtari, M., Meybodi, M.R.: Memetic-CLA-PSO: a hybrid model for optimization. In: 2011 UkSim 13th International Conference on Computer Modelling and Simulation, pp. 20–25. IEEE (2011)

Alirezanejad, M., Enayatifar, R., Motameni, H., Nematzadeh, H.: GSA-LA: gravitational search algorithm based on learning automata. J. Exp. Theor. Artif. Intell. **33**, 109–125 (2021)

Alirezanejad, M., Enayatifar, R., Motameni, H., Nematzadeh, H.: GSA-LA: gravitational search algorithm based on learning automata. J. Exp. Theor. Artif. Intell. 1–17 (2020). https://doi.org/10.1080/0952813X.2020.1725650

Anari, B., Torkestani, J.A., Rahmani, A.M.: Automatic data clustering using continuous action-set learning automata and its application in segmentation of images. Appl. Soft Comput. **51**, 253–265 (2017)

Arora, S., Anand, P.: Learning automata-based butterfly optimization algorithm for engineering design problems. Int. J. Comput. Mater. Sci. Eng. **7**, 1850021 (2018)

Babichenko, Y., Tamuz, O.: Graphical potential games. J. Econ. Theory **163**, 889–899 (2016). https://doi.org/10.1016/j.jet.2016.03.010

Bak, P., Sneppen, K.: Punctuated equilibrium and criticality in a simple model of evolution. Phys. Rev. Lett. **71**, 4083–4086 (1993). https://doi.org/10.1103/PhysRevLett.71.4083

Bayessa, G.A., Sah Tyagi, S.K., Parashar, V., Gao, M., Shi, J.: Novel protected sub-frame selection based interference mitigation and resource assignment in heterogeneous multi-cloud radio access networks. Sustain. Comput. Inform. Syst. **20**, 165–173 (2018). https://doi.org/10.1016/j.suscom.2018.02.009

Beigy, H., Meybodi, M.R.: Asynchronous cellular learning automata. Automatica **44**, 1350–1357 (2008)

Beigy, H., Meybodi, M.R.: Open synchronous cellular learning automata. Adv. Complex Syst. **10**, 527–556 (2007)
Beigy, H., Meybodi, M.R.: A mathematical framework for cellular learning automata. Adv. Complex Syst. **07**, 295–319 (2004). https://doi.org/10.1142/S0219525904000202
Beigy, H., Meybodi, M.R.: An iterative stochastic algorithm based on distributed learning automata for finding the stochastic shortest path in stochastic graphs. J. Supercomput. **76**, 5540–5562 (2020). https://doi.org/10.1007/s11227-019-03085-0
Beigy, H., Meybodi, M.R.R.: Cellular learning automata with multiple learning automata in each cell and its applications. IEEE Trans. Syst. Man Cybern. Part B (cybern.) **40**, 54–65 (2010). https://doi.org/10.1109/TSMCB.2009.2030786
Ben-Zvi, T.: Learning automata decision analysis for sensor placement. J. Oper. Res. Soc. **69**, 1396–1405 (2018)
Betka, A., Terki, N., Toumi, A., Dahmani, H.: Grey wolf optimizer-based learning automata for solving block matching problem. Sig. Image Video Process. **14**, 285–293 (2020)
Bhattacharjee, K., Naskar, N., Roy, S., Das, S.: A survey of cellular automata: types, dynamics, non-uniformity and applications. Nat. Comput. (2018) https://doi.org/10.1007/s11047-018-9696-8
Blackwell, T.: Particle swarm optimization in dynamic environments, pp. 29–49 (2007)
Boettcher, S., Percus, A.G.: Optimization with extremal dynamics. Complexity **8**, 57–62 (2002). https://doi.org/10.1002/cplx.10072
Bohlool, M., Meybodi, M.R.: Edge detection using open and asynchronous cellular learning automata. In: 4th Iranian Conference on Machine Vision and Image Processing, pp. 1–6 (2007)
Bouhmala, N.: A multilevel learning automata for MAX-SAT. Int. J. Mach. Learn. Cyber. **6**, 911–921 (2015). https://doi.org/10.1007/s13042-015-0355-4
Bouhmala, N., Oseland, M., Brådland, Ø.: WalkSAT based-learning automata for MAX-SAT. In: International Conference on Soft Computing-MENDEL, pp. 98–110. Springer (2016)
Burkowski, F.J.: Shuffle crossover and mutual information. In: Proceedings of the 1999 Congress on Evolutionary Computation-CEC 1999 (Cat. No. 99TH8406), pp. 1574–1580. IEEE (1999)
Cao, H., Cai, J.: Distributed opportunistic spectrum access in an unknown and dynamic environment: a stochastic learning approach. IEEE Trans. Veh. Technol. **67**, 4454–4465 (2018)
Chowdhury, A., Rakshit, P., Konar, A., Nagar, A.K.: A meta-heuristic approach to predict protein-protein interaction network. In: 2016 IEEE Congress on Evolutionary Computation (CEC), pp. 2137–2144. IEEE (2016)
Cook, M.: Universality in elementary cellular automata. Complex Syst. **15**, 1–40 (2004)
Copeland, B.: The church-turing thesis (1997)
Cota, L.P., Guimarães, F.G., Ribeiro, R.G., Meneghini, I.R., de Oliveira, F.B., Souza, M.J.F., Siarry, P.: An adaptive multi-objective algorithm based on decomposition and large neighborhood search for a green machine scheduling problem. Swarm Evol. Comput. **51**, 100601 (2019). https://doi.org/10.1016/j.swevo.2019.100601
Dai, C., Wang, Y., Ye, M., Xue, X., Liu, H.: An orthogonal evolutionary algorithm with learning automata for multiobjective optimization. IEEE Trans. Cybern. **46**, 3306–3319 (2016). https://doi.org/10.1109/TCYB.2015.2503433
Daliri Khomami, M.M., Rezvanian, A., Saghiri, A.M., Meybodi, M.R.: Utilizing cellular learning automata for finding communities in weighted networks. In: 2020 6th International Conference on Web Research (ICWR), pp. 325–329 (2020a)
Daliri Khomami, M.M., Rezvanian, A., Saghiri, A.M., Meybodi, M.R.: SIG-CLA: a significant community detection based on cellular learning automata. In: 2020 8th Iranian Joint Congress on Fuzzy and intelligent Systems (CFIS), pp. 039–044 (2020b)
de Sousa, F.L., Ramos, F.M.: Function optimization using extremal dynamics. In: ICIPE 2002, Brazil, pp. 115–119 (2002)
Díaz-Cortés, M.-A., Cuevas, E., Rojas, R.: Multi-threshold segmentation using learning automata. In: Engineering Applications of Soft Computing, pp. 101–127. Springer (2017)

Duarte, P.B.F., Fadlullah, ZMd., Vasilakos, A.V., Kato, N.: On the partially overlapped channel assignment on wireless mesh network backbone: a game theoretic approach. IEEE J. Sel. Areas Commun. **30**, 119–127 (2012). https://doi.org/10.1109/JSAC.2012.120111

Eberhart, R.C., Shi, Y.: Tracking and optimizing dynamic systems with particle swarms. In: Proceedings of the 2001 Congress on Evolutionary Computation (IEEE Cat. No. 01TH8546), pp. 94–100 (2001)

Enayatifar, R., Yousefi, M., Abdullah, A.H., Darus, A.N.: LAHS: a novel harmony search algorithm based on learning automata. Commun. Nonlinear Sci. Numer. Simul. **18**, 3481–3497 (2013). https://doi.org/10.1016/j.cnsns.2013.04.028

Esnaashari, M., Meybodi, M.R.: Dynamic point coverage problem in wireless sensor networks: a cellular learning automata approach. Ad Hoc Sens. Wirel. Netw. **10**, 193–234 (2010)

Esnaashari, M., Meybodi, M.R.: A cellular learning automata based clustering algorithm for wireless sensor networks. Sens. Lett. **6**, 723–735 (2008)

Esnaashari, M., Meybodi, M.R.: Irregular cellular learning automata. IEEE Trans. Cybern. **45**, 1622–1632 (2018). https://doi.org/10.1016/j.jocs.2017.08.012

Esnaashari, M., Meybodi, M.R.M.: A cellular learning automata-based deployment strategy for mobile wireless sensor networks. J. Parallel Distrib. Comput. **71**, 988–1001 (2011)

Fakhrmoosavy, S.H., Setayeshi, S., Sharifi, A.: A modified brain emotional learning model for earthquake magnitude and fear prediction. Eng. Comput. **34**, 261–276 (2018). https://doi.org/10.1007/s00366-017-0538-6

Gasior, J., Seredynski, F., Hoffmann, R.: Towards self-organizing sensor networks: game-theoretic e-learning automata-based approach. In: Cellular Automata: Proceedings of the 13th International Conference on Cellular Automata for Research and Industry, ACRI 2018, Como, Italy, 17–21 September 2018, p. 125. Springer (2018)

Ghamgosar, M., Khomami, M.M.D., Bagherpour, N., Meybodi, M.R.: An extended distributed learning automata based algorithm for solving the community detection problem in social networks. In: 2017 Iranian Conference on Electrical Engineering (ICEE), pp. 1520–1526. IEEE (2017)

Ghavipour, M., Meybodi, M.R.: Irregular cellular learning automata-based algorithm for sampling social networks. Eng. Appl. Artif. Intell. **59**, 244–259 (2017). https://doi.org/10.1016/j.engappai.2017.01.004

Gheisari, S.: VLA-CR: a variable action-set learning automata-based cognitive routing protocol for IoT. Comput. Commun. **164**, 162–176 (2020). https://doi.org/10.1016/j.comcom.2020.10.015

Gheisari, S., Meybodi, M.R.: A new reasoning and learning model for Cognitive Wireless Sensor Networks based on Bayesian networks and learning automata cooperation. Comput. Netw. **124**, 11–26 (2017)

Gheisari, S., Meybodi, M.R., Dehghan, M., Ebadzadeh, M.M.: Bayesian network structure training based on a game of learning automata. Int. J. Mach. Learn. Cybern. **8**, 1093–1105 (2017)

Gheisari, S., Meybodi, M.R., Dehghan, M., Ebadzadeh, M.M.: BNC-VLA: Bayesian network structure learning using a team of variable-action set learning automata. Appl. Intell. **45**, 135–151 (2016)

Guo, Y., Ge, H., Wang, F., Huang, Y., Li, S.: Function optimization via a continuous action-set reinforcement learning automata model. In: Liang, Q., Mu, J., Wang, W., Zhang, B. (eds.) Proceedings of the 2015 International Conference on Communications, Signal Processing, and Systems, pp. 981–989. Springer, Heidelberg (2016)

Guo, Y., Hao, G., Shenghong, L.: A set of novel continuous action-set reinforcement learning automata models to optimize continuous functions. Appl. Intell. **46**, 845–864 (2017)

Hadavi, N., Nordin, Md.J., Shojaeipour, A.: Lung cancer diagnosis using CT-scan images based on cellular learning automata. In: 2014 International Conference on Computer and Information Sciences (ICCOINS), pp. 1–5. IEEE (2014)

Hao, S., Zhang, H., Wang, J.: A learning automata based stable and energy-efficient routing algorithm for discrete energy harvesting mobile wireless sensor network. Wirel. Pers. Commun. **107**, 437–469 (2019)

Hariri, A., Rastegar, R., Navi, K., Zamani, M.S., Meybodi, M.R.: Cellular learning automata based evolutionary computing (CLA-EC) for intrinsic hardware evolution. In: 2005 NASA/DoD Conference on Evolvable Hardware (EH 2005), pp. 294–297. IEEE (2005a)

Hariri, A., Rastegar, R., Zamani, M.S., Meybodi, M.R.: Parallel hardware implementation of cellular learning automata based evolutionary computing (CLA-EC) on FPGA. In: 13th Annual IEEE Symposium on Field-Programmable Custom Computing Machines (FCCM 2005), pp. 311–314. IEEE (2005b)

Hasanzadeh Mofrad, M., Sadeghi, S., Rezvanian, A., Meybodi, M.R.: Cellular edge detection: combining cellular automata and cellular learning automata. AEU – Int. J. Electron. Commun. **69**, 1282–1290 (2015). https://doi.org/10.1016/j.aeue.2015.05.010

Hasanzadeh-Mofrad, M., Rezvanian, A.: Learning automata clustering. J. Comput. Sci. **24**, 379–388 (2018). https://doi.org/10.1016/j.jocs.2017.09.008

Hashemi, A.B., Meybodi, M.R.: Cellular PSO: a PSO for dynamic environments. In: Cai, Z. (ed.) Lecture Notes in Computer Science (Including Subseries Lecture Notes in Artificial Intelligence and Lecture Notes in Bioinformatics), pp. 422–433. Springer, Heidelberg (2009)

Hashemi, S.M., He, J.: LA-based approach for IoT security. J. Robot. Network. Artif. Life **3**, 240–248 (2017). https://doi.org/10.2991/jrnal.2017.3.4.7

Hassanzadeh, T., Meybodi, M.R.: A new hybrid algorithm based on firefly algorithm and cellular learning automata. In: 20th Iranian Conference on Electrical Engineering (ICEE 2012), pp. 628–633. IEEE (2012)

Jafarpour, B., Meybodi, M.R.: Recombinative CLA-EC. In: Proceedings of the International Conference on Tools with Artificial Intelligence, ICTAI, pp. 415–422. IEEE (2007)

Jahanshahi, M., Dehghan, M., Meybodi, M.R.: A cross-layer optimization framework for joint channel assignment and multicast routing in multi-channel multi-radio wireless mesh networks. Int. J. Comput. Math. **94**, 1624–1652 (2017)

Jameii, S.M., Faez, K., Dehghan, M.: AMOF: adaptive multi-objective optimization framework for coverage and topology control in heterogeneous wireless sensor networks. Telecommun. Syst. **61**, 515–530 (2016)

Jobava, A., Yazidi, A., Oommen, B.J., Begnum, K.: On achieving intelligent traffic-aware consolidation of virtual machines in a data center using Learning Automata. J. Comput. Sci. (2017). https://doi.org/10.1016/j.jocs.2017.08.005

Kazemi Kordestani, J., Meybodi, M.R., Rahmani, A.M.: A two-level function evaluation management model for multi-population methods in dynamic environments: hierarchical learning automata approach. J. Exp. Theor. Artif. Intell. **33**, 1–26 (2021)

Kazemitabar, S.J., Taghizadeh, N., Beigy, H.: A graph-theoretic approach toward autonomous skill acquisition in reinforcement learning. Evol. Syst. **9**, 227–244 (2018)

Kennedy, J.: Small worlds and mega-minds: effects of neighborhood topology on particle swarm performance. In: Proceedings of the 1999 Congress on Evolutionary Computation-CEC 1999 (Cat. No. 99TH8406), pp. 1931–1938. IEEE (1999)

Kennedy, J., Eberhart, R.: Particle swarm optimization. In: Proceedings of ICNN 1995 - International Conference on Neural Networks, pp. 1942–1948. IEEE (1995)

Khaksar Manshad, M., Meybodi, M.R., Salajegheh, A.: A new irregular cellular learning automata-based evolutionary computation for time series link prediction in social networks. Appl. Intell. (2020). https://doi.org/10.1007/s10489-020-01685-5

Khaksarmanshad, M., Meybodi, M.R.: Designing optimization algorithms based on CLA-EC for dynamic environments. In: The 4th Iran Data Mining Conference (IDMC 2010), pp. 1–6 (2010)

Khatamnejad, A., Meybodi, M.R.: A hybrid method for optimization (CLA-EC + extremal optimization). In: 13th Annual CSI Computer Conference of Iran, pp. 1–6 (2008)

Khomami, M.M.D., Rezvanian, A., Meybodi, M.R.: A New cellular learning automata-based algorithm for community detection in complex social networks. J. Comput. Sci. **24**, 413–426 (2018)

Khomami, M.M.D., Rezvanian, A., Meybodi, M.R.: Distributed learning automata-based algorithm for community detection in complex networks. Int. J. Mod. Phys. B **30**, 1650042 (2016)

Khot, P.S., Naik, U.L.: Cellular automata-based optimised routing for secure data transmission in wireless sensor networks. J. Exp. Theor. Artif. Intell. 1–19 (2021). https://doi.org/10.1080/0952813X.2021.1882002

Kiink, T., Vesterstroem, J.S., Riget, J.: Particle swarm optimization with spatial particle extension. In: IEEE Congress on Evolutionary Computation, pp. 1474–1479 (2002)

Kordestani, J.K., Firouzjaee, H.A., Meybodi, M.R.: An adaptive bi-flight cuckoo search with variable nests for continuous dynamic optimization problems. Appl. Intell. **48**, 97–117 (2018)

Kordestani, J.K., Ranginkaman, A.E., Meybodi, M.R., Novoa-Hernández, P.: A novel framework for improving multi-population algorithms for dynamic optimization problems: a scheduling approach. Swarm Evol. Comput. **44**, 788–805 (2019)

Kordestani, J.K., Rezvanian, A., Meybodi, M.R.: An efficient oscillating inertia weight of particle swarm optimisation for tracking optima in dynamic environments. J. Exp. Theor. Artif. Intell. **28**, 137–149 (2016). https://doi.org/10.1080/0952813X.2015.1020521

Kordestani, J.K., Rezvanian, A., Meybodi, M.R.: CDEPSO: a bi-population hybrid approach for dynamic optimization problems. Appl. Intell. **40**, 682–694 (2014). https://doi.org/10.1007/s10489-013-0483-z

Kumar, S., Kumar, V., Kaiwartya, O., Dohare, U., Kumar, N., Lloret, J.: Towards green communication in wireless sensor network: GA enabled distributed zone approach. Ad Hoc Netw. **93**, 101903 (2019)

Narendra, K.S., Thathachar, M.A.L.: Learning Automata: An Introduction. Prentice-Hall, Englewood Cliffs (1989)

Li, M., Miao, C., Leung, C.: A coral reef algorithm based on learning automata for the coverage control problem of heterogeneous directional sensor networks. Sensors **15**, 30617–30635 (2015)

Li, W., Özcan, E., John, R.: A learning automata-based multiobjective hyper-heuristic. IEEE Trans. Evol. Comput. **23**, 59–73 (2019). https://doi.org/10.1109/TEVC.2017.2785346

Lin, Y., Wang, L., Zhong, Y., Zhang, C.: Control scaling factor of cuckoo search algorithm using learning automata. Int. J. Comput. Sci. Math. **7**, 476–484 (2016). https://doi.org/10.1504/IJCSM.2016.080088

Lin, Y., Wang, X., Hao, F., Wang, L., Zhang, L., Zhao, R.: An on-demand coverage based self-deployment algorithm for big data perception in mobile sensing networks. Future Gener. Comput. Syst. **82**, 220–234 (2018)

Lindsay, J., Gigivi, S.: A novel way of training a neural network with reinforcement learning and without back propagation. In: 2020 International Joint Conference on Neural Networks (IJCNN), pp. 1–6 (2020)

Liu, W., Xu, Y., Qi, N., Yao, K., Zhang, Y., He, W.: Joint computation offloading and resource allocation in UAV swarms with multi-access edge computing. In: 2020 International Conference on Wireless Communications and Signal Processing (WCSP), pp. 280–285 (2020)

Mahdaviani, M., Kordestani, J.K., Rezvanian, A., Meybodi, M.R.: LADE: learning automata based differential evolution. Int. J. Artif. Intell. Tools **24**, 1550023 (2015)

Manshad, M.K., Manshad, A.K., Meybodi, M.R.: Memory/search RCLA-EC: a CLA-EC for moving parabola problem. In: 2011 6th International Conference on Computer Sciences and Convergence Information Technology (ICCIT), pp. 1–6 (2011)

Masoodifar, B., Meybodi, M.R., Hashemi, M.: Cooperative CLA-EC. In: 12th Annual CSI Computer Conference of Iran, pp. 558–559 (2007)

Mendes, R., Kennedy, J., Neves, J.: Watch thy neighbor or how the swarm can learn from its environment. In: Proceedings of the 2003 IEEE Swarm Intelligence Symposium, SIS 2003 (Cat. No. 03EX706), pp. 88–94. IEEE (2003)

Mirsaleh, M.R., Meybodi, M.R.: A Michigan memetic algorithm for solving the vertex coloring problem. J. Comput. Sci. (2017). https://doi.org/10.1016/j.jocs.2017.10.005

Mirsaleh, M.R., Meybodi, M.R.: Assignment of cells to switches in cellular mobile network: a learning automata-based memetic algorithm. Appl. Intell. **48**, 3231–3247 (2018)

Mirsaleh, M.R., Meybodi, M.R.: A Michigan memetic algorithm for solving the community detection problem in complex network. Neurocomputing **214**, 535–545 (2016a)

Mirsaleh, M.R., Meybodi, M.R.: A new memetic algorithm based on cellular learning automata for solving the vertex coloring problem. Memet. Comput. **8**, 211–222 (2016b)
Mohajer, A., Bavaghar, M., Farrokhi, H.: Reliability and mobility load balancing in next generation self-organized networks: using stochastic learning automata. Wirel. Pers. Commun. **114**, 2389–2415 (2020a)
Mohajer, A., Bavaghar, M., Farrokhi, H.: Mobility-aware load balancing for reliable self-organization networks: multi-agent deep reinforcement learning. Reliab. Eng. Syst. Saf. **202**, 107056 (2020b)
Moradabadi, B., Meybodi, M.R.: Link prediction based on temporal similarity metrics using continuous action set learning automata. Phys. A **460**, 361–373 (2016)
Moradabadi, B., Meybodi, M.R.: Wavefront cellular learning automata. Chaos **28**, 021101 (2018). https://doi.org/10.1063/1.5017852
Moradabadi, B., Meybodi, M.R.: Link prediction in fuzzy social networks using distributed learning automata. Appl. Intell. **47**, 837–849 (2017)
Morshedlou, H., Meybodi, M.R.: A new local rule for convergence of ICLA to a compatible point. IEEE Trans. Syst. Man Cybern. Syst. **47**, 3233–3244 (2017). https://doi.org/10.1109/TSMC.2016.2569464
Mousavian, A., Rezvanian, A., Meybodi, M.R.: Cellular learning automata based algorithm for solving minimum vertex cover problem. In: 2014 22nd Iranian Conference on Electrical Engineering (ICEE), pp. 996–1000. IEEE (2014)
Mozafari, M., Alizadeh, R.: A cellular learning automata model of investment behavior in the stock market. Neurocomputing **122**, 470–479 (2013). https://doi.org/10.1016/j.neucom.2013.06.002
Mozafari, M., Shiri, M.E., Beigy, H.: A cooperative learning method based on cellular learning automata and its application in optimization problems. J. Comput. Sci. **11**, 279–288 (2015). https://doi.org/10.1016/j.jocs.2015.08.002
Nabizadeh, S., Faez, K., Tavassoli, S., Rezvanian, A.: A novel method for multi-level image thresholding using Particle Swarm Optimization algorithms. In: ICCET 2010 - 2010 Proceedings of the International Conference on Computer Engineering and Technology, pp. V4-271–V4-275 (2010)
Nouri, E.: An unequal clustering-based topology control algorithm in wireless sensor networks using learning automata. In: Fundamental Research in Electrical Engineering, pp. 55–68. Springer (2019)
Ortiz, L.E.: Graphical potential games (2015)
Packard, N.H., Wolfram, S.: Two-dimensional cellular automata. J. Stat. Phys. **38**, 901–946 (1985). https://doi.org/10.1007/BF01010423
Peer, E.S., van den Bergh, F., Engelbrecht, A.P.: Using neighbourhoods with the guaranteed convergence PSO. In: Proceedings of the 2003 IEEE Swarm Intelligence Symposium, SIS 2003 (Cat. No. 03EX706), pp. 235–242. IEEE (2003)
Raharya, N., Hardjawana, W., Al-Khatib, O., Vucetic, B.: Multi-BS association and pilot allocation via pursuit learning. In: 2020 IEEE Wireless Communications and Networking Conference (WCNC), pp. 1–6 (2020a)
Raharya, N., Hardjawana, W., Al-Khatib, O., Vucetic, B.: Pursuit learning-based joint pilot allocation and multi-base station association in a distributed massive MIMO network. IEEE Access **8**, 58898–58911 (2020b). https://doi.org/10.1109/ACCESS.2020.2982974
Rakshit, P.: Improved differential evolution for noisy optimization. Swarm Evol. Comput. **52**, 100628 (2020). https://doi.org/10.1016/j.swevo.2019.100628
Rakshit, P., Konar, A.: Realization of learning induced self-adaptive sampling in noisy optimization. Appl. Soft Comput. **69**, 288–315 (2018). https://doi.org/10.1016/j.asoc.2018.04.052
Rakshit, P., Konar, A., Nagar, A.K.: Modified selection and search in learning automata based artificial bee colony in noisy environment. In: 2019 IEEE Congress on Evolutionary Computation (CEC), pp. 3173–3180 (2019)
Rakshit, P., Konar, A., Nagar, A.K.: Learning automata induced artificial bee colony for noisy optimization. In: 2017 IEEE Congress on Evolutionary Computation (CEC), pp. 984–991 (2017)

Rastegar, R., Meybodi, M.R.: A new evolutionary computing model based on cellular learning automata. In: 2004 IEEE Conference on Cybernetics and Intelligent Systems, pp. 433–438. IEEE (2004)

Rastegar, R., Meybodi, M.R., Hariri, A.: A new fine-grained evolutionary algorithm based on cellular learning automata. Int. J. Hybrid Intell. Syst. **3**, 83–98 (2006). https://doi.org/10.3233/HIS-2006-3202

Rastegar, R., Rahmati, M., Meybodi, M.R.: A clustering algorithm using cellular learning automata based evolutionary algorithm. In: Adaptive and Natural Computing Algorithms, pp. 144–150. Springer, Vienna (2005)

Rauniyar, A., Yazidi, A., Engelstad, P., Østerbo, O.N.: A reinforcement learning based game theoretic approach for distributed power control in downlink NOMA. In: 2020 IEEE 19th International Symposium on Network Computing and Applications (NCA), pp. 1–10 (2020)

Rezapoor Mirsaleh, M., Meybodi, M.R.: Balancing exploration and exploitation in memetic algorithms: a learning automata approach. Comput. Intell. **34**, 282–309 (2018)

Rezvanian, A., Meybodi, M.R.: Finding minimum vertex covering in stochastic graphs: a learning automata approach. Cybern. Syst. **46**, 698–727 (2015)

Rezvanian, A., Moradabadi, B., Ghavipour, M., Daliri Khomami, M.M., Meybodi, M.R.: Learning Automata Approach for Social Networks. Springer, Cham (2019a)

Rezvanian, A., Moradabadi, B., Ghavipour, M., Daliri Khomami, M.M., Meybodi, M.R.: Introduction to learning automata models. In: Learning Automata Approach for Social Networks, pp. 1–49. Springer (2019b)

Rezvanian, A., Moradabadi, B., Ghavipour, M., Daliri Khomami, M.M., Meybodi, M.R.: Wavefront cellular learning automata: a new learning paradigm. In: Learning Automata Approach for Social Networks, pp. 51–74. Springer (2019c)

Rezvanian, A., Moradabadi, B., Ghavipour, M., Daliri Khomami, M.M., Meybodi, M.R.: Wavefront cellular learning automata: a new learning paradigm. In: Rezvanian, A., Moradabadi, B., Ghavipour, M., Daliri Khomami, M.M., Meybodi, M.R. (eds.) Learning Automata Approach for Social Networks, pp. 51–74. Springer, Cham (2019d)

Rezvanian, A., Saghiri, A.M., Vahidipour, S.M., Esnaashari, M., Meybodi, M.R.: Recent Advances in Learning Automata. Springer, Cham (2018a)

Rezvanian, A., Saghiri, A.M., Vahidipour, S.M., Esnaashari, M., Meybodi, M.R.: Learning automata for complex social networks. In: Recent Advances in Learning Automata, pp. 279–334 (2018b)

Rezvanian, A., Saghiri, A.M., Vahidipour, S.M., Esnaashari, M., Meybodi, M.R.: Cellular learning automata, pp. 21–88 (2018c)

Rezvanian, A., Saghiri, A.M., Vahidipour, S.M., Esnaashari, M., Meybodi, M.R.: Learning automata theory. In: Recent Advances in Learning Automata, pp. 3–19. Springer (2018c)

Rezvanian, A., Moradabadi, B., Ghavipour, M., Daliri Khomami, M.M., Meybodi, M.R.: Social recommender systems. In: Learning Automata Approach for Social Networks, pp. 281–313. Springer (2019d)

Ruan, X., Jin, Z., Tu, H., Li, Y.: Dynamic cellular learning automata for evacuation simulation. IEEE Intell. Transp. Syst. Mag. **11**, 129–142 (2019). https://doi.org/10.1109/MITS.2019.2919523

Saghiri, A.M., Meybodi, M.R.: Open asynchronous dynamic cellular learning automata and its application to allocation hub location problem. Know.-Based Syst. **139**, 149–169 (2018a). https://doi.org/10.1016/j.knosys.2017.10.021

Saghiri, A.M., Meybodi, M.R.: A closed asynchronous dynamic model of cellular learning automata and its application to peer-to-peer networks. Genet. Program Evolvable Mach. **18**, 313–349 (2017). https://doi.org/10.1007/s10710-017-9299-7

Saghiri, A.M., Meybodi, M.R.: An adaptive super-peer selection algorithm considering peers capacity utilizing asynchronous dynamic cellular learning automata. Appl. Intell. **48**, 271–299 (2018b). https://doi.org/10.1007/s10489-017-0946-8

Salehi, F., Majidi, M.-H., Neda, N.: Channel estimation based on learning automata for OFDM systems. Int. J. Commun. Syst. **31**, e3707 (2018)

Saritha, V., Krishna, P.V., Misra, S., Obaidat, M.S.: Learning automata based optimized multipath routingusing leapfrog algorithm for VANETs. In: 2017 IEEE International Conference on Communications (ICC), pp. 1–5 (2017)
Sayyadi Shahraki, N., Zahiri, S.H.: Multi-objective learning automata for design and optimization a two-stage CMOS operational amplifier. Iran. J. Electr. Electron. Eng. **16**, 201–214 (2020a). https://doi.org/10.22068/IJEEE.16.2.201
Sayyadi Shahraki, N., Zahiri, S.H.: An improved multi-objective learning automata and its application in VLSI circuit design. Memet. Comput. **12**, 115–128 (2020b). https://doi.org/10.1007/s12293-020-00303-8
Sayyadi Shahraki, N., Zahiri, S.H.: DRLA: dimensionality ranking in learning automata and its application on designing analog active filters. Knowl.-Based Syst. **219**, 106886 (2021). https://doi.org/10.1016/j.knosys.2021.106886
Sinaie, S., Ghanizadeh, A., Majd, E.M., Shamsuddin, S.M.: A hybrid edge detection method based on fuzzy set theory and cellular learning automata. In: 2009 International Conference on Computational Science and Its Applications, pp. 208–214. IEEE (2009)
Singh, S., Dwivedi, A.K., Sharma, A.K., Mehra, P.S.: Learning automata based heuristics for target Q-coverage. In: 2020 8th International Conference on Reliability, Infocom Technologies and Optimization (Trends and Future Directions) (ICRITO), pp. 170–173. IEEE (2020)
Sohrabi, M.K., Roshani, R.: Frequent itemset mining using cellular learning automata. Comput. Hum. Behav. **68**, 244–253 (2017). https://doi.org/10.1016/j.chb.2016.11.036
Soleimani-Pouri, M., Rezvanian, A., Meybodi, M.R.: Solving maximum clique problem in stochastic graphs using learning automata. In: 2012 Fourth International Conference on Computational Aspects of Social Networks (CASoN), pp. 115–119 (2012)
Storn, R.M., Price, K.V.: Differential evolution – a simple and efficient heuristic for global optimization over continuous spaces. J. Glob. Optim. **11**, 341–359 (1997). https://doi.org/10.1023/A:1008202821328
Su, X., Han, W., Wu, Y., Zhang, Y., Liu, J.: A proactive robust scheduling method for aircraft carrier flight deck operations with stochastic durations. Complexity **2018**, e6932985 (2018). https://doi.org/10.1155/2018/6932985
Subha, C.P., Malarkkan, S.: Optimisation of energy efficient cellular learning automata algorithm for heterogeneous wireless sensor networks. Int. J. Network. Virtual Organ. **17**, 170–183 (2017)
Suma, R., Premasudha, B.G., Ram, V.R.: A novel machine learning-based attacker detection system to secure location aided routing in MANETs. Int. J. Network. Virtual Organ. **22**, 17–41 (2020). https://doi.org/10.1504/IJNVO.2020.104968
Talabeigi, M., Forsati, R., Meybodi, M.R.: A hybrid web recommender system based on cellular learning automata. In: 2010 IEEE International Conference on Granular Computing, pp. 453–458. IEEE (2010)
Thathachar, M.A.L., Sastry, P.S.: Varieties of learning automata: an overview. IEEE Trans. Syst. Man Cybern. B Cybern. **32**, 711–722 (2002). https://doi.org/10.1109/TSMCB.2002.1049606
Toffolo, T.A.M., Christiaens, J., Van Malderen, S., Wauters, T., Vanden Berghe, G.: Stochastic local search with learning automaton for the swap-body vehicle routing problem. Comput. Oper. Res. **89**, 68–81 (2018). https://doi.org/10.1016/j.cor.2017.08.002
Vafaee Sharbaf, F., Mosafer, S., Moattar, M.H.: A hybrid gene selection approach for microarray data classification using cellular learning automata and ant colony optimization. Genomics **107**, 231–238 (2016). https://doi.org/10.1016/j.ygeno.2016.05.001
Vafashoar, R., Meybodi, M.R.: Reinforcement learning in learning automata and cellular learning automata via multiple reinforcement signals. Knowl.-Based Syst. **169**, 1–27 (2019a). https://doi.org/10.1016/j.knosys.2019.01.021
Vafashoar, R., Meybodi, M.R.: A multi-population differential evolution algorithm based on cellular learning automata and evolutionary context information for optimization in dynamic environments. Appl. Soft Comput. **88**, 106009 (2020)
Vafashoar, R., Meybodi, M.R.: Multi swarm optimization algorithm with adaptive connectivity degree. Appl. Intell. **48**, 909–941 (2018). https://doi.org/10.1007/s10489-017-1039-4

Vafashoar, R., Meybodi, M.R.: Multi swarm bare bones particle swarm optimization with distribution adaption. Appl. Soft Comput. **47**, 534–552 (2016)

Vafashoar, R., Meybodi, M.R.: Cellular learning automata based bare bones PSO with maximum likelihood rotated mutations. Swarm Evol. Comput. **44**, 680–694 (2019c). https://doi.org/10.1016/j.swevo.2018.08.016

Vafashoar, R., Meybodi, M.R., Momeni Azandaryani, A.H.: CLA-DE: a hybrid model based on cellular learning automata for numerical optimization. Appl. Intell. **36**, 735–748 (2012). https://doi.org/10.1007/s10489-011-0292-1

Vafashoar, R., Morshedlou, H., Rezvanian, A., Meybodi, M.R.: Cellular Learning Automata: Theory and Applications. Springer, Cham (2021a)

Vafashoar, R., Morshedlou, H., Rezvanian, A., Meybodi, M.R.: Applications of cellular learning automata and reinforcement learning in global optimization. In: Cellular Learning Automata: Theory and Applications, pp. 157–224. Springer (2021b)

Vafashoar, R., Morshedlou, H., Rezvanian, A., Meybodi, M.R.: Learning from multiple reinforcements in cellular learning automata. In: Cellular Learning Automata: Theory and Applications, pp. 111–156. Springer (2021c)

Vafashoar, R., Morshedlou, H., Rezvanian, A., Meybodi, M.R.: Applications of multi-reinforcement cellular learning automata in channel assignment. In: Cellular Learning Automata: Theory and Applications, pp. 225–254. Springer (2021d)

Vafashoar, R., Morshedlou, H., Rezvanian, A., Meybodi, M.R.: Cellular learning automata: a bibliometric analysis. In: Cellular Learning Automata: Theory and Applications, pp. 83–109. Springer (2021e)

Vahidipour, S.M., Esnaashari, M., Rezvanian, A., Meybodi, M.R.: GAPN-LA: a framework for solving graph problems using Petri nets and learning automata. Eng. Appl. Artif. Intell. **77**, 255–267 (2019). https://doi.org/10.1016/j.engappai.2018.10.013

Vahidipour, S.M., Meybodi, M.R., Esnaashari, M.: Finding the shortest path in stochastic graphs using learning automata and adaptive stochastic Petri Nets. Int. J. Uncertainty Fuzziness Knowl.-Based Syst. **25**, 427–455 (2017a)

Vahidipour, S.M., Meybodi, M.R., Esnaashari, M.: Cellular adaptive Petri net based on learning automata and its application to the vertex coloring problem. Discrete Event Dyn. Syst. **27**, 609–640 (2017b). https://doi.org/10.1007/s10626-017-0251-z

Velusamy, G., Lent, R.: Evaluating an adaptive web traffic routing method for the cloud. In: 2019 IEEE ComSoc International Communications Quality and Reliability Workshop (CQR), pp. 1–6. IEEE (2019)

Velusamy, G., Lent, R.: Dynamic cost-aware routing of web requests. Future Internet **10**, 57 (2018). https://doi.org/10.3390/fi10070057

Wauters, T., Verbeeck, K., De Causmaecker, P., Vanden Berghe, G.: A learning-based optimization approach to multi-project scheduling. J. Sched. **18**, 61–74 (2015). https://doi.org/10.1007/s10951-014-0401-1

Wolfram, S.: Theory and applications of cellular automata. World Scientific Publication (1986)

Wolfram, S.: Cellular automata as simple self-organizing systems. Caltech preprint CALT-68-9385 (1982)

Xiao, G., Liu, H., Guo, W., Wang, L.: A hybrid training method of convolution neural networks using adaptive cooperative particle swarm optimiser. Int. J. Wirel. Mob. Comput. **16**, 18–26 (2019). https://doi.org/10.1504/IJWMC.2019.097418

Xu, Y., Ma, L., Shi, M.: Adaptive brain storm optimization based on learning automata. In: Pan, L., Liang, J., Qu, B. (eds.) Bio-Inspired Computing: Theories and Applications, pp. 98–108. Springer, Singapore (2020)

Yang, Z., Liu, Y., Chen, Y.: Distributed reinforcement learning for NOMA-enabled mobile edge computing. In: 2020 IEEE International Conference on Communications Workshops (ICC Workshops), pp. 1–6 (2020a)

Yang, Z., Liu, Y., Chen, Y., Al-Dhahir, N.: Cache-aided NOMA mobile edge computing: a reinforcement learning approach. IEEE Trans. Wirel. Commun. **19**, 6899–6915 (2020b). https://doi.org/10.1109/TWC.2020.3006922

Yazidi, A., Bouhmala, N., Goodwin, M.: A team of pursuit learning automata for solving deterministic optimization problems. Appl. Intell. **50**, 2916–2931 (2020)

Yazidi, A., Oommen, B.J.: The theory and applications of the stochastic point location problem. In: 2017 International Conference on New Trends in Computing Sciences (ICTCS). pp. 333–341 (2017)

Yazidi, A., Oommen, B.J.: Solving stochastic root-finding with adaptive d-ary search. In: 2015 IEEE International Conference on Evolving and Adaptive Intelligent Systems (EAIS), pp. 1–8 (2015)

Zhang, D., Du, J., Zhang, T., Yang, P., Fan, H.: New algorithm of QoS constrained routing for node energy optimization of edge computing. In: 2020 IEEE 17th International Conference on Mobile Ad Hoc and Sensor Systems (MASS), pp. 482–490. IEEE (2020)

Zhang, J., Xu, L., Ma, J., Zhou, M.: A learning automata-based particle swarm optimization algorithm for noisy environment. In: 2015 IEEE Congress on Evolutionary Computation (CEC), pp. 141–147 (2015)

Zhang, Y., Liu, S., Han, L: Optimization of node deployment in wireless sensor networks based on learning automata. In: International Conference on Human Centered Computing, pp. 79–85. Springer (2018)

Zhao, H., Wang, J., Wang, Q., Liu, F.: Queue-based and learning-based dynamic resources allocation for virtual streaming media server cluster of multi-version VoD system. Multimed. Tools Appl. **78**, 21827–21852 (2019). https://doi.org/10.1007/s11042-019-7457-z

Zhao, H., Zhang, C.: An online-learning-based evolutionary many-objective algorithm. Inf. Sci. **509**, 1–21 (2020). https://doi.org/10.1016/j.ins.2019.08.069

Zhao, Y., Jiang, W., Li, S., Ma, Y., Su, G., Lin, X.: A cellular learning automata based algorithm for detecting community structure in complex networks. Neurocomputing **151**, 1216–1226 (2015). https://doi.org/10.1016/j.neucom.2014.04.087

Zhou, B., Song, Q., Zhao, Z., Liu, T.: A reinforcement learning scheme for the equilibrium of the in-vehicle route choice problem based on congestion game. Appl. Math. Comput. **371**, 124895 (2020). https://doi.org/10.1016/j.amc.2019.124895

Zhu, J., Gu, W., Lou, G., Wang, L., Xu, B., Wu, M., Sheng, W.: Learning automata-based methodology for optimal allocation of renewable distributed generation considering network reconfiguration. IEEE Access **5**, 14275–14288 (2017). https://doi.org/10.1109/ACCESS.2017.2730850

Zojaji, M., Meybodi, M.R.M., Mirzaie, K.: A rapid learning automata-based approach for generalized minimum spanning tree problem. J. Comb. Optim. **40**, 636–659 (2020). https://doi.org/10.1007/s10878-020-00605-0

Chapter 4
Learning Automata for Behavior Control in Evolutionary Computation

Javidan Kazemi Kordestani, Mehdi Razapoor Mirsaleh, Alireza Rezvanian, and Mohammad Reza Meybodi

Abstract Global optimization plays a crucial role in different branches of science and engineering. The rapid development of technology gives rise to complex problems where several objectives may need to be satisfied. Therefore, there is a constant need to provide new optimization techniques and improving the existing methods. One of the most popular research topics in the literature on optimization is function optimization. The main reason is that many real-world optimization problems can be modeled as function optimization. Therefore, function optimization problems can be used as a benchmark for comparing the performance of different algorithms. Analytical methods encounter many difficulties when applying to complex optimization problems; thus, it is impossible to apply them for many practical cases. Evolutionary computation (EC) techniques, however, have been proven to work better than analytical methods. Although being effective, EC methods' performance often depends on the choice of their parameters, and finding the good parameter values for EC methods is a conventional approach for improving their efficiency for numerical optimization. This chapter investigates the usefulness of learning automaton (LA) to control EC methods' behavior by adjusting their parameters during the run. To this end, we are running the LA and the EC side by side, and the LA controls the EC on-the-fly parameters.

J. Kazemi Kordestani
Department of Computer Engineering, Science and Research Branch, Islamic Azad University, Tehran, Iran
e-mail: javidan.kazemi@gmail.com

M. Razapoor Mirsaleh
Department of Computer Engineering and Information Technology, Payame Noor University (PNU), P.O. BOX 19395-3697, Tehran, Iran
e-mail: mrezapoorm@aut.ac.ir

A. Rezvanian (✉)
Department of Computer Engineering, University of Science and Culture, Tehran, Iran
e-mail: rezvanian@usc.ac.ir

M. R. Meybodi
Computer Engineering Department, Amirkabir University of Technology (Tehran Polytechnic), Tehran, Iran
e-mail: mmeybodi@aut.ac.ir

J. Kazemi Kordestani et al. (eds.), *Advances in Learning Automata and Intelligent Optimization*, Intelligent Systems Reference Library 208, https://doi.org/10.1007/978-3-030-76291-9_4

4.1 Introduction

Global optimization plays a crucial role in different branches of science and engineering. The rapid development of technology gives rise to complex problems where several objectives may need to be satisfied. Therefore, there is a constant need to provide new optimization techniques and improve existing methods (Vafashoar et al. 2021). One of the most popular research topics in the literature on optimization is function optimization. The main reason is that many real-world optimization problems can be modeled as function optimization. Therefore, function optimization problems can be used as a benchmark for comparing the performance of different algorithms. Analytical methods encounter many difficulties when applying to complex optimization problems; thus, it is impossible to apply them for many practical cases. Evolutionary computation (EC) techniques, however, have been proven to work better than analytical methods.

Although being effective, EC methods' performance often depends on the choice of their parameters, and finding the good parameter values for EC methods is a conventional approach for improving their efficiency for numerical optimization.

This chapter investigates the usefulness of learning automaton (LA) (Rezvanian et al. 2018a, b, 2019) to control EC methods' behavior by adjusting their parameters during the run. To this end, we are running the LA and the EC side by side, and the LA controls the EC on-the-fly parameters. The proposed approach is evaluated through numerical experiments on different fitness landscapes. The results show the superiority of the EC calibrated by the LA-based methods.

4.2 Types of Parameter Adjustment in EC Community

EC parameter adjustment schemes introduced in the literature can be classified into three major categories: (a) EC with constant parameters, (b) EC with time-varying parameters, and (c) EC with adaptive parameters. These categories can be summarized as follows:

4.2.1 EC with Constant Parameters

The first group of methods tries to determine an exact value for the parameters of EC methods empirically. This class of studies contains strategies in which EC parameters' value is set before the actual run and remains constant during the optimization process. This strategy is by far the most basic and straightforward approach for parameter adjustment in the EC community. For example, Clerc (Clerc 1999) introduced a constriction model for particle swarm optimization (PSO) (Shi and Eberhart 1998) in which w, c_1 and c_2 has been chosen in a way to guarantee the convergence of the population. One of such constriction models is:

$$\vec{v}_i(t+1) = \chi\left[\vec{v}_i(t) + c_1 r_1(\vec{p}_i(t) - \vec{x}_i(t)) + c_2 r_2\left(\vec{p}_g(t) - \vec{x}_i(t)\right)\right] \tag{4.1}$$

where χ is computed as follows:

$$\chi = \frac{2}{\left|2 - \phi - \sqrt{\phi^2 - 4\phi}\right|}, \phi = c_1 + c_2, \phi > 4 \tag{4.2}$$

Storn and Price (Storn and Price 1997) suggested a constant range of NP, F, and CR values in differential evolution (DE). According to their experiments, a reasonable value for *NP* is in the range of $5 \times D$ to $10 \times D$, where *D* is the dimensionality of the problem. *F* should be chosen from [0.5, 1], and a good first choice for *CR* is either 0.9 or 1.

4.2.1.1 Strengths and Weaknesses

This approach is the most straightforward method for parameter adjustment in EC methods, leading to a single problem's suboptimal results. However, methods falling in this category suffer from various disadvantages:

- The process of finding an exact value for a parameter is time-consuming, especially if the parameter in hand is continuous in nature.
- For a given problem, the selected value for a parameter is not necessarily optimal.

A good value for the parameter depends on the characteristics of the problem at hand. Therefore, it would not be helpful as a general approach to solve optimization problems.

4.2.2 EC with Time-Varying Parameters

The second category contains those strategies that adjust EC parameters' value based on random values, chaotic sequence, of time-dependent functions.

4.2.2.1 EC with Random Parameters

Methods falling in this category use a range of random values for the parameters of EC methods. For example, Eberhart and Shi (Eberhart and Shi 2001) proposed a random inertia weight to enhance the PSO's ability to track the moving optima in a dynamic environment. In their method, the value of inertia weight is determined using the following equation:

$$w = 0.5 + rand() \tag{4.3}$$

where $rand()$ is a random number in [0, 1]. Therefore, inertia weight w would be a uniform random variable in the range [0.5, 1].

Das et al. (Das et al. 2005) proposed a scheme for adjusting the scaling factor F in DE, in which the value of F varies during the search process in a random manner. In their approach, the value of F is chosen randomly within the range [0.5, 1].

Brest et al. (Brest et al. 2006) introduced an algorithm called jDE, which adjusts CR and F values for each individual separately. They used a random mechanism to generate new F and CR values according to a uniform distribution in the range of [0.1, 1.0] and [0.0, 1.0], respectively.

4.2.2.2 EC with Chaotic Parameters

In this class, the value of the EC parameters is generated according to a chaotic sequence. For example, Feng et al. proposed two chaotic models for inertia weight in PSO based on Logistic mapping. Yousri et al. (Yousri et al. 2019) proposed a chaotic whale optimization variant in which some parameters of the standard whale optimization algorithm are tuned with different chaos maps.

4.2.2.3 Linearly Increasing

Another option to change a parameter's value during the search is to use a linearly increasing scheme. For example, Zheng et al. (Zheng et al. 2003a, b) proposed a linearly increasing inertia weight for PSO, which is increased from 0.4 to 0.9 throughout the optimization.

4.2.2.4 Non-linearly Increasing

Apart from linearly increasing methods, some researchers used a non-linearly increasing strategy to change EC parameters' value during the run. For instance, Jiao et al. (Zheng et al. 2003a, b) proposed an inertia weight for PSO in which the value of the w is increased according to the following equation:

$$w_{iter} = w_{initial} \times u^{iter} \tag{4.4}$$

where $w_{initial}$ is the initial inertia weight value selected in the range [0, 1], and u is a constant value in the range [1.0001, 1.005].

4.2.2.5 Linearly Decreasing

In these methods, the value of a parameter at hand is linearly decreased from an initial value to a final value. The most well-known example of this category is the linearly decreasing inertia weight in PSO (Shi and Eberhart 1999). In this method, the value of inertia weight is modified over the run according to the following equation:

$$w_{iter} = (w_{max} - w_{min}) \times \frac{(iter_{max} - iter)}{iter_{max}} + w_{min} \tag{4.5}$$

where $iter$ is the current iteration of the algorithm and $iter_{max}$ is the maximum number of iterations the PSO is allowed to continue. w_{max} and w_{min} are also the upper bound and lower bound of the inertia weight, respectively. This strategy is among the most common strategies to adjust the value of inertia weight in PSO.

Das et al. (Das et al. 2005) introduced a linearly decreasing scaling factor. In their method, the value of F is reduced from an initial value ($\mathcal{F}_{max}$) to a final value ($\mathcal{F}_{min}$) according to the following scheme:

$$\mathcal{F}_{iter} = (\mathcal{F}_{max} - \mathcal{F}_{min}) \times \frac{(iter_{max} - iter)}{iter_{max}} \tag{4.6}$$

where $\mathcal{F}_{iter}$ is the value of F in the current iteration, $\mathcal{F}_{max}$ and $\mathcal{F}_{min}$ are the upper and lower value of F, respectively, and $iter_{max}$ is the maximum number of iterations. The higher value of F enables the genomes of the population to explore wide areas of the search space during the early stages of the optimization. Moreover, F's decreasing scheme allows trial solutions in a relatively small region of the search space around the suspected global optimum at the final stages of the search process. A linearly decreasing step for Krill herd optimization is also proposed in Li et al. (2014).

4.2.2.6 Non-linearly Decreasing

These methods change the value of the parameters according to a non-linearly decreasing time-dependent function. For example, a non-linearly inertia weight for PSO was proposed by Fan and Chiu (Fan and Chiu 2007) in which at each iteration of the algorithm, the value of the w is determined according to the following equation:

$$w_{iter} = \left(\frac{2}{iter}\right)^{0.3} \tag{4.7}$$

4.2.2.7 Oscillating

This strategy's main idea is to trigger a wave of global search followed by a local search wave, forming a repeated cycle during the optimization process. This way, the EC population periodically switches from exploratory to exploitatory states of search. For example, Kazemi Kordestani et al. (Kordestani et al. 2016) proposed an oscillating triangular inertia weight for PSO to enhance the ability of PSO for tracking the moving optima. Their method uses a symmetric saw tooth function that hybridizes linearly, decreasing inertia weight and linearly increases inertia weight into a single inertia weight called oscillating triangular inertia weight. The following equation computes the oscillating triangular inertia weight:

$$w_{i,t} = w_{max} - |round(\alpha t) - \alpha t| \tag{4.8}$$

where $w_{i,t}$ is the value of inertia weight at the tth iteration after the ith change; t is the iteration counter.

which starts from zero and increases by one at the end of each iteration until the environment changes where t is set to zero again. w_{max} is the maximum value of the inertia weight. $\alpha \in [0.0, 0.5]$ is used to generate different shaped waveforms for oscillating triangular inertia weight. Finally, $round()$ is the function that rounds off real number x to its nearest integer.

4.2.2.8 Strengths and Weaknesses

Methods of this category can provide high diversity at the initial stages of the search process, which is necessary to allow the full range of the search space. Then, as the optimization reaches its final stages, the methods can fine-tune the solutions to exploit the global optima further efficiently. However, methods falling in this category also suffer from various disadvantages:

- The process of finding a suitable lower bound and upper bound for the values of a parameter is difficult.
- A good value for the upper bound and lower bound of the parameter still depends on the problem's characteristics at hand.

4.2.3 EC with Adaptive Parameters

Finally, algorithms in the third category adaptively change EC parameters by monitoring the search process's state. These methods often control the search process via one or more feedbacks received from the EC method. One way to do this is by using fuzzy logic. For example, Liu and Lampinen (Liu and Lampinen 2005) proposed a fuzzy adaptive variant of DE, named FDE, for the adaptive selection of DE parameters' value. They used a fuzzy system consisting of "9 $\times$ 2" rules for dynamic adjustment of F and CR. Each rule of the system has two inputs and one output. Parameter vector change magnitude (PC) and function value change (FC) are input variables of the system, and the value of F or CR is the system's output variable. Each fuzzy variable has three fuzzy sets: *SMALL*, *MIDDLE*, and *BIG*. Different combinations of input variables are used to determine the value of F and CR. For instance, a typical rule of their system is IF (PC is *small*) and (FC is *big*) Then (CR is *big*). Juang et al. (Juang and Chang 2011) applied fuzzy set theory to control the acceleration coefficients c_1 and c_2 in PSO during the run. They have used the amount of fitness improvement of the swarm for two consecutive iterations as feedback to monitor the progress of the algorithm as follows:

$$df^t = f\left(gbest^{t-1}\right) - f\left(gbest^t\right) \tag{4.9}$$

The fuzzy variables df, c_1 and c_2 have three fuzzy sets: SMALL, MEDIUM, and BIG. Their fuzzy system has three rules:

- IF (df is SMALL) THEN (c_1 is BIG) and (c_2 is SMALL)
- IF (df is MEDIUM) THEN (c_1 is MEDIUM) and (c_2 is MEDIUM)
- IF (df is BIG THEN) (c_1 is SMALL) and (c_2 is BIG)

Both above studies reported improved results compared to the base algorithms.

Apart from the above fuzzy methods, Qin et al. (Qin et al. 2009) proposed a self-adaptive DE, SaDE, in which the control parameters and trial vector generation strategies are adaptively adjusted based on their previous performance in generating promising solutions. Zhang and Sanderson (Zhang and Sanderson 2009) proposed JADE where F and CR values are sampled from a normal distribution and a Cauchy distribution at the individual level, respectively. In the JADE, information from the most recent successful F and CR are utilized to set the new F and CR. In this work, two different feedbacks are utilized to monitor the search progress of DE, one at population level and another at genome level, and adjust the parameter CR, accordingly.

There exist various methods in the literature that have used strategy adaptation for improving the performance of DE. For instance, Gong et al. (Gong et al. 2010) employed the *probability matching* technique for strategy adaptation in DE. In their approach, at each generation, a mutation strategy is selected for each parent from a strategy pool of four mutation schemes, i.e., "DE/rand/1", "DE/rand/2", "DE/rand-to-best/2" and "DE/current-to-rand/1", based on its probability. Afterward, the *relative fitness improvement*, which is calculated as the difference of the offspring's fitness with its parent, is gathered to update each mutation strategy's probability. Mallipeddi et al. (Mallipeddi et al. 2011) introduced an ensemble of mutation strategies and control parameters of DE (EPSDE). EPSDE contains two separate pools: a pool of distinct trial vector generation strategies (with "DE/rand/1", "DE/best/2" and "DE/current-to-rand/1") and a pool of values for the control parameters $F \in \{0.4, 0.5, 0.6, 0.7, 0.8, 0.9\}$and $CR \in \{0.1, 0.2, 0.3, 0.4, 0.5, 0.6, 0.7, 0.8, 0.9\}$. In EPSDE, successful combinations of mutation strategies and parameter values are utilized to increase the probability of generating promising offspring. Wang et al. (Wang et al. 2011) proposed a Composite DE (CoDE), which combines different trial vector generation strategies with some control parameter settings. In CoDE, they constituted a mutation strategy pool (with "DE/rand/1", "DE/rand/2" and "DE/current-to-rand/1") and a parameter candidate pool (with "$F = 1.0$, $CR = 0.1$", "$F = 1.0$, $CR = 0.9$" and "$F = 0.8$, $CR = 0.2$"). Three offspring are generated for each generation's target vector with randomly chosen parameter settings from the parameter pool. Then, the best generated offspring is transferred to the next generation, if it is fitter than its parent.

Many researchers have used LA as an adaptive tool for parameter adjustment in the literature. For example, Hashemi and Meybodi (Hashemi and Meybodi 2011) used LA to adjust the w, c_1 and c_2 in PSO. A similar idea was also used in (Kazemi Kordestani et al. 2014; Mahdaviani et al. 2015) for adjusting the parameters of DE or for adapting the parameters of AIS (Rezvanian and Meybodi 2010a,2010b). Abedi Firouzjaee et al. (Abedi Firouzjaee et al. 2017) proposed an adaptive cuckoo search algorithm that aims to automate the flight operator's selection for each individual in the cuckoo search algorithm concerning its search progress. Recently, Kazemi Kordestani et al. (Kazemi Kordestani et al. 2018) proposed a bi-flight multi-population cuckoo search algorithm that uses LA to adjust the search strategy of each

Table 4.1 Summary of learning automata-based methods for behavior control in evolutionary computation

Method	Optimizer	Description	Adjustment level	Learning algorithm
GLADE (Mahdaviani et al. 2015)	DE	The algorithm adjusts mutation strategy and CR	Population	L_{RI}, L_{RP}, $L_{R\varepsilon P}$
ILADE (Mahdaviani et al. 2015)	DE	The algorithm adjusts mutation strategy and CR	Individual	L_{RI}, L_{RP}, $L_{R\varepsilon P}$
ADE-Grid (Kazemi Kordestani et al. 2014)	DE	The algorithm adjusts mutation strategy, F, and CR	Individual	$L_{R\varepsilon P}$
UAPSO (Hashemi and Meybodi 2011)	PSO	The algorithm adjusts w, c_1 and c_2	Population	$L_{R\varepsilon P}$
IAPSO (Hashemi and Meybodi 2011)	PSO	The algorithm adjusts w, c_1 and c_2	Individual	$L_{R\varepsilon P}$
LAHS (Enayatifar et al. 2013)	HSA	The algorithm adjusts $HMCR$, PAR and bw	Population	A linear reinforcement scheme
ACS (Abedi Firouzjaee et al. 2017)	CS	The algorithm adjusts the distribution of the flight for each cuckoo	Individual	Pursuit algorithm
BFCS (Kazemi Kordestani et al. 2018)	CS	The algorithm adjusts the parameter β in Levy distribution	Population	L_{RI}
LABOA (Arora and Anand 2018)	BOA	The algorithm adjusts the switch probability	Individual	A linear reinforcement scheme
Firefly (Abshouri et al. 2011)	FF	The algorithm adjusts absorption Coefficient, maximum attractiveness value, and randomization coefficient	Individual	A linear reinforcement scheme
Lopt-aiNet (Rezvanian and Meybodi 2010c)	AIS	The algorithm adjusts scale coefficient α in hypermutation	Population	L_{RI}
GSA-LA (Alirezanejad et al. 2020)	GSA	The algorithm adjusts the parameter $G(t)$	Population	A linear reinforcement scheme

sub-population by adapting the parameter β in Levy distribution. Table 4.1 summarizes the application of LA for controlling the behavior of EC methods during the run.

4.2.3.1 Strengths and Weaknesses

This approach is the most promising method for parameter control in EC methods that can effectively solve most problems. However, extracting the rules governing the value of the parameter and the choice of feedback parameter(s) is challenging.

4.3 Differential Evolution

This section describes how to incorporate a learning automata-based parameter control mechanism in an existing EC method. Although several options exist, we have considered DE. DE has been extensively studied in the past, showing very good results, especially for complex optimization problems. DE is one of the most powerful stochastic real-parameter optimization algorithms originally proposed (Storn and Price 1995; Storn and Price 1997). Over the past decade, DE becomes very popular because of its simplicity, effectiveness, and robustness.

The main idea of DE is to use spatial differences among the population of vectors to guide the search process toward the optimum solution. In short, almost all DE variants work according to the following steps:

i. *Initialization*: A number of *NP* points are randomly sampled from the *D*-dimensional search space to form the initial population.
ii. *Repeat* steps (iii), (iv) and (v) for each vector $\vec{x}_i$ ($i \in \{1, 2, \ldots, NP\}$) of the current population.
iii. *Mutation*: A mutant vector $\vec{v}_i$ is generated for $\vec{x}_i$ according to a specified mutation strategy.
iv. *Repair*: if the mutant vector $\vec{v}_i$ is out of the feasible region, a repair operator is utilized to make it feasible.
v. *Crossover*: A crossover operator is applied to combine the information from $\vec{x}_i$ and $\vec{v}_i$, and form a trial vector $\vec{u}_i$.
vi. *Selection*: The vector with the best fitness among $\vec{x}_i$ and $\vec{u}_i$ is transferred to the next generation.

If the termination condition is not met, go to step (ii).

Different extensions of DE can be specified using the general convention *DE/x/y/z*, where DE stands for "Differential Evolution," x represents a string denoting the base vector to be perturbed, y is the number of difference vectors considered for perturbation of x, and z stands for the type of crossover being used, i.e. *exponential* or *binomial* (Das and Suganthan 2011).

DE has several advantages that make it a powerful tool for optimization tasks. Specifically, (1) DE has a simple structure and is easy to implement; (2) despite its simplicity, DE exhibits a high performance; (3) the number of control parameters in the canonical DE (Storn and Price 1997) is very few (i.e., *NP*, *F*, and *CR*); (4) due to its low space complexity, DE is suitable for handling large scale problems.

4.3.1 Initialization

DE starts with a population of *NP* randomly generated vectors in a *D*-dimensional search space. Each vector *i*, also known as *genome* or *chromosome*, is a potential solution to an optimization problem which is represented by $\vec{x}_i = (x_{i1}, x_{i2}, \ldots, x_{id})$. The initial population of vectors is simply randomized into the boundary of the search space according to a uniform distribution as follows:

$$\vec{x}_i = lb_j + rand_j[0, 1] \times \left(ub_j - lb_j\right) \tag{4.10}$$

where $i\epsilon[1, 2, \ldots, NP]$ is the index of the ith vector of the population, $j\epsilon[1, 2, \ldots, D]$ represents the jth dimension of the search space, $rand_j[0, 1]$ is a uniformly distributed random number corresponding to the jth dimension. Finally, lb_j and ub_j are the search space's lower and upper bounds corresponding to the jth dimension of the search space.

4.3.2 Difference-Vector Based Mutation

After initialization of the vectors in the search space, a mutation is performed on each genome *i* of the population to generate a donor vector $\vec{v}_i = (v_{1i}, v_{2i}, \ldots, v_{1n})$ corresponding to the target vector $\vec{x}_i$. Several strategies have been proposed for generating donor vector $\vec{v}_i$. In this chapter, we use the following mutation strategy called DE/rand/bin/1 to create a donor vector:

$$\vec{v}_{\rm i} = \vec{x}_1 + \mathcal{F}.(\vec{x}_2 - \vec{x}_3) \tag{4.11}$$

where $\vec{v}_i$ is the donor vector corresponding to the ith genome. $\vec{x}_1 \neq \vec{x}_2 \neq \vec{x}_3$ are randomly selected vectors from the population. $\mathcal{F}$ is the scaling factor used to control the amplification of the difference vector. The effect of different mutation strategies on the performance of DE has been studied in (Zheng et al. 2003a, b). If the generated mutant vector is out of the search boundary, a repair operator is used to make $\vec{v}_{\rm i}$ back to the feasible region.

4.3.3 Repair Operator

If the generated mutant vector is out of the search boundary, a repair operator is used to make $\vec{v}_i$ back to the feasible region. Different strategies have been proposed to repair out-of-bound individuals. In this chapter, if the jth element of the ith mutant vector, i.e., v_{ij}, is out of the search region $[lb_j, ub_j]$, then it is repaired as follows:

$$v_{ij} = \begin{cases} \frac{x_{ij}+lb_j}{2} \; if \, v_{ij} < lb_j \\ \frac{x_{ij}+ub_j}{2} \; if \, v_{ij} > ub_j \end{cases} \tag{4.12}$$

where x_{ij} is the jth element of the ith target vector.

4.3.4 Crossover

To introduce diversity to the population of genomes, DE utilizes a crossover operation to combine the components of the target vector $\overrightarrow{x}_i$ and donor vector $\overrightarrow{v}_i$, to form the trial vector $\overrightarrow{u}_i$. Two crossover types are commonly used in the DE community, called *binomial crossover* and *exponential crossover*. In this chapter, we use a *binomial crossover*, which is defined as follows:

$$u_{ij} = \begin{cases} v_{ij} \; if \; rand_{ij}[0, 1] \le CR \, or \; j = jrand \\ x_{ij} \; otherwise \end{cases} \tag{4.13}$$

where $rand_{ij}[0, 1]$ is a random number drawn from a uniform distribution between 0 and 1, CR is the *crossover rate* used to control the approximate number of components transferred to trial vector from donor vector. $jrand$ is a random index in the range $[1, D]$, which ensures the transmission of at least one component of the donor vector into the trial vector.

Considering Eq. (4.13), the parameter CR has a great influence on the offspring population's diversity. A large value of *CR* can maintain the individuals' diversity, which is favorable for multi-modal functions. A small value of CR, in turn, can make the trial vector a little different from the target vector. This feature is desirable for optimizing separable problems.

4.3.5 Selection

A *selection* approach is performed on vectors to determine which vector ($\overrightarrow{x}_i$ or $\overrightarrow{u}_i$) should be survived in the next generation. The most fitted vector is chosen to be the member of the next generation.

$$\vec{x}_{i,G+1} = \begin{cases} \vec{u}_{i,G} \; if \; f(\vec{x}_{i,G}) \leq f(\vec{u}_{i,G}) \\ \vec{x}_{i,G} \; otherwise \end{cases} \tag{4.14}$$

4.4 Learning Automata for Adaptive Control of Behavior in Differential Evolution

This section provides a complete description of the behavior control in DE using LA for solving the following continuous global optimization problem:

$$\begin{aligned} &\text{Minimize}: \; f(\vec{x}),\; \vec{x} = (x_1, x_2, \ldots, x_D) \in s \\ &\text{Subject to}: \; x_i \in [lb_i, ub_i] \end{aligned} \tag{4.15}$$

where $f(\overrightarrow{x})$ is a continuous real-valued objective function to be minimized, and s is the solution space. Finally, lb_i and ub_i are the box constraints corresponding to the ith dimension of the search space, that is, $\forall i \in \{1, \ldots, D\}, -\infty < l_i < u_i < \infty$.

4.4.1 *Behavior Control in DE with Variable-Structure Learning Automaton*

Two approaches for selecting the value of a parameter are applied to the DE algorithm using variable-structure LA. In both approaches, the LA is responsible for selecting the value of a parameter from a permissible range, allowing a parameter to change drastically from one end to the other end of its range in two consecutive iterations. To this end, a permissible range for parameters F and CR is defined. In the first approach, called PADE, all individuals of the population share the same values for DE parameters (i.e., F and CR) which are adaptively set at each iteration using two learning automata LA_F and LA_{CR}, respectively. In contrast, in the second approach, the so-called IADE, every individual can adjust its parameters independently. Therefore, in IADE, each individual i has two learning automata LA_F^i and LA_{CR}^i which enables it to decide how to change the value of its parameters. The learning structure in PADE and IADE is illustrated in Fig. 4.1.

4.4.1.1 PADE

The working mechanism of PADE is as follows. At each iteration t, the value of each parameter is selected using LA_F and LA_{CR}. Then the success of the parameter selection is evaluated to update the two learning automata. In PADE, the *success rate* (ϕ) of the population is used to measure the progress of the population. A high success rate indicates that F and CR's selected values were favorable to the DE

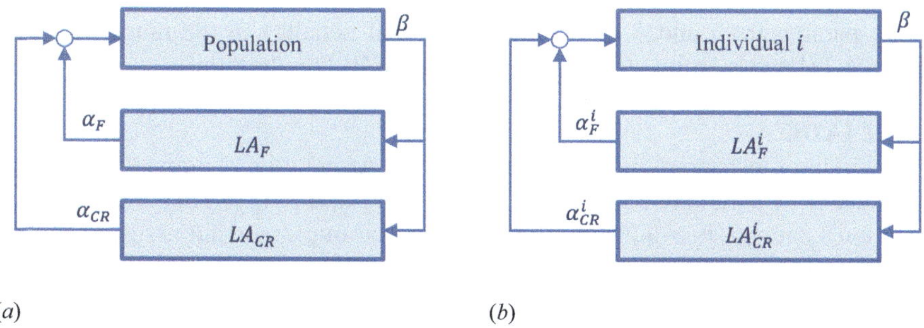

Fig. 4.1 Schematic representation of learning structure for selecting the parameters of DE in (a) PADE and (b) IADE

performance. Similarly, a low success rate indicates that F and CR's selected values were not contributing to the population's performance during the last iteration.

The success rate of the population is computed based on the success and failure of the individuals. The success of individual i at iteration t in a minimization problem is defined as follows:

$$\eta_i(t) = \begin{cases} 1\ if\ \left(f\left(\vec{x}_{i,t}\right) < f\left(\vec{x}_{i,t+1}\right)\right) \\ 0\ if\ \left(f\left(\vec{x}_{i,t}\right) = f\left(\vec{x}_{i,t+1}\right)\right) \end{cases} \tag{4.16}$$

The success rate of the population is then calculated as follows:

$$\phi = \frac{\sum_{i=1}^{NP} \eta_i(t)}{NP} \tag{4.17}$$

The success rate is then used for computing the reinforcement signal β to update the probability vectors of the LA_F and LA_{CR}. In this work, a linear function is used to map the values of ϕ to the possible range of β as follows:

$$\beta = 1 - \phi \tag{4.18}$$

Then, the automata LA_F and LA_{CR} update their action probabilities according to the following equations (Mahdaviani et al. 2015):

$$p_i(n+1) = p_i(n) - \beta(n) \cdot b \cdot p_i(n) + [1 - \beta(n)] \cdot a \cdot (1 - p_i(n)) \quad if\ i = j \tag{4.19}$$

$$p_j(n+1) = p_j(n) - \beta(n) \cdot \left[\frac{b}{(r-1)} - b \cdot p_j(n)\right] + [1 - \beta(n)] \cdot a \cdot p_j(n) \quad if\ i \neq j \tag{4.20}$$

where parameters a and b determine reward and penaltsy in the range of [0, 1]. Figure 4.2 shows step-by-step pseudocode for the PADE algorithm.

4.4.1.2 IADE

The working mechanism of IADE is as follows. At iteration t, individual i selects the value of its parameters with using its own LA_F^i and LA_{CR}^i. Then the parameter selection's success is evaluated to update two learning automata assigned for the individual. In IADE, parameter selection for individual i is considered as "successful" when the fitness of individual i improves.

$$Reinforcement\ signal\ \beta = \begin{cases} 0i\ if\ f\left(\vec{X}_{i,G}\right) < \left(\vec{X}_{i,G+1}\right) \\ 1\ \ otherwise \end{cases} \tag{4.21}$$

where f is the fitness function, and $\vec{X}_{i,G}$ is the i-th individual of the algorithm at generation G. If the parameter selection is successful, the two respective learning automata will receive a favorable response and be rewarded according to Eq. (4.22); otherwise, an unfavorable response will be generated, and learning automata will be penalized to Eq. (4.23).

Algorithm 4-1. PADE algorithm

define
Initialize the algorithm and problem parameters: population size *NP* fitness evaluations counter *fe* = 0, maximum fitness evaluations FE_{max}, current generation *G* = 0.
Initialize LA_F with parameters: action set *a* = {*F* = 0.1, *F* = 0.3, *F* = 0.5, *F* = 0.7, *F* = 0.9}, action probability vector *p* = {(1/5), (1/5), (1/5), (1/5), (1/5)}, alpha = 0.1, beta = 0.01.
Initialize LA_{CR} with parameters: action set *a* = {*CR* = 0.1, *CR* = 0.3, *CR* = 0.5, *CR* = 0.7, *CR* = 0.9}, action probability vector *p* = {(1/5), (1/5), (1/5), (1/5), (1/5)}, alpha = 0.1, beta = 0.01.
Let $\alpha_j \mid j \in \{1,2,3,4,5\}$ be the selected action of LA_F.
Let $\alpha_k \mid k \in \{1,2,3,4,5\}$ be the selected action of LA_{CR}.
Generate an initial population $P_0 = \{\vec{X}_{1,0}, \ldots, \vec{X}_{NP,0}\}$ in the *D*-dimensional search space according to Eq. (4-10).
Let *f* be the fitness value.
Evaluate the objective function values $f(\vec{X}_1), \ldots, f(\vec{X}_{NP})$.
fe := *NP*;
while (*fe* < FE_{max})
  $P_{G+1} = \emptyset$ // to determine the population of the next generation
  Select the actions α_j and α_k for LA_F and LA_{CR} from their action set based on their probability vector *p*;
  for each genome $i \in [1, \ldots, NP]$ in current population (P_G) **do**
    Generate the donor vector $\vec{v}_i$ using selected *F* and DE/rand/bin/1; // mutation step
    Repair $\vec{v}_i$ according to Eq. (4-12). // repair operator
    Generate trial vector $\vec{u}_i$ using selected crossover rate *CR* and Eq. (4-13); // crossover step
    Evaluate the objective function value $f(\vec{u}_i)$.
    Add an individual to P_{G+1} using Eq. (4-14). // selection and replacement step
  end-for
  Calculate the reinforcement signal β according to Eq. (4-18);
  Update the probability vector *p* for LA_F and LA_{CR} according to Eq. (4-19), Eq. (4-20);
  G := *G* + 1;
  fe := *fe* + *NP*;
end-while

Fig. 4.2 Pseudocode of PADE

$$p_j(n+1) = \begin{cases} p_j(n) + a.\big(1 - p_j(n)\big) & if\ i = j \\ p_j(n).(1-a) & if\ i \neq j \end{cases} \tag{4.22}$$

$$p_j(n+1) = \begin{cases} p_j(n).(1-b) & if\ i = j \\ \frac{b}{r-1} + (1-b).p_j(n) & if\ i \neq j \end{cases} \tag{4.23}$$

where parameters a and b determine reward and penalty in the range of [0, 1]. Figure 4.3 shows step-by-step pseudocode for the IADE algorithm.

Algorithm 4-2. IADE algorithm

```
define
Initialize the algorithm and problem parameters: population size NP fitness evaluations counter fe = 0, maximum
   fitness evaluations FE_max, current generation G = 0.
Initialize the set of LA_F = {LA_F (1), ..., LA_F (NP)} with parameters: action set a = {F = 0.1, F = 0.3, F = 0.5, F = 0.7,
   F = 0.9}, action probability vector p = {(1/5), (1/5), (1/5), (1/5), (1/5)}, alpha = 0.1, beta = 0.01.
Initialize the set of LA_CR = {LA_CR (1), ..., LA_CR (NP)} with parameters: action set a = {CR = 0.1, CR = 0.3, CR =
   0.5, CR = 0.7, CR = 0.9}, action probability vector p = {(1/5), (1/5), (1/5), (1/5), (1/5)}, alpha = 0.1, beta = 0.01.
Let α_j(M) | j ϵ{1,2,3}, Mϵ[1,...,NP] be the selected action of LA_F (M).
Let α_k(M) | k ϵ{1,2}, Mϵ[1,...,NP] be the selected action of LA_CR (M).
Generate an initial population P_0 = {X⃗_1,0, ..., X⃗_NP,0} in the D-dimensional search space according to Eq. (4-10).
Let f be the fitness value.
while (fe < FE_max)
|  P_G+1 = ∅; // to determine the population of the next generation
|  for each genome i ϵ [1,...,NP] in current population (P_G) do
|  |  LA_F (i), and LA_CR (i) select an action from their action set based on their probability
|  |     vector p, according to a random uniform distribution.
|  |  Generate the donor vector v⃗_i using selected F and DE/rand/bin/1; // mutation step
|  |  Repair v⃗_i according to Eq. (4-12); // repair operator
|  |  Generate trial vector u⃗_i using selected crossover rate CR and Eq. (4-13); // crossover step
|  |  Evaluate the objective function value f(u⃗_i);
|  |  Add an individual to P_G+1 using Eq. (4-14); // selection and replacement step
|  |  Calculate the reinforcement signal β according to Eq. (4-21);
|  |  if (β==0) then
|  |  |  Reward LA_F (i), and LA_CR (i) using Eq. (4-22);
|  |  else
|  |  |  Penalize LA_F (i), and LA_CR (i) using Eq. (4-23);
|  |  end-if
|  end-for
|  G := G + 1;
|  fe := fe + NP;
end-while
```

Fig. 4.3 Pseudocode of IADE

4.4.2 Behavior Control in DE with Fixed-Structure Learning Automaton

Apart from the above algorithms for behavior control in DE using variable-structure LA, we propose a scheme for controlling the DE using the $L_{2,2}$ automaton proposed by Tsetlin (Narendra and Thathachar 2012). The $L_{2,2}$ automaton has two states, Φ_1 and Φ_2 and two actions α_1 and α_2. The automaton accept input from a set of {0, 1} and switches its states upon encountering an input 1 (unfavorable response) and remains in the same state on receiving an input 0 (favorable response). Figure 4.4 illustrates the state transition and action selection in the $L_{2,2}$ automaton.

Here, we propose two approaches for choosing the search strategy, i.e., global search and local search, in the DE algorithm using the $L_{2,2}$ automaton. In the first approach, called PSADE, all individuals of the population follow the same strategy (i.e., global search or local search), which are adaptively set at each iteration using one learning automaton LA_S. In contrast, in the second approach, the so-called ISADE, every individual can independently adjust its search strategy. Therefore, in ISADE, each individual i has one learning automata LA_S^i which enables it to decide how to change its search strategy. The learning structure in PSADE and ISADE is illustrated in Fig. 4.5.

4.4.2.1 PSADE

The proposed PSADE works as follows. Initially, the $L_{2,2}$ automaton is in state $i = 1$. Therefore, the automaton selects the action α_1, i.e., global search, which is related to evolving DE with parameters $F = 0.5$ and $CR = 0.9$. Similarly, when the $L_{2,2}$ automaton is in state $i = 2$, the automaton selects the action α_2, i.e., local search, which is related to evolving DE with parameters $F = 0.1$ and $CR = 0.2$. Afterward, the $L_{2,2}$ receives an input β according to Eq. (4.24).

$$Reinforcement\ signal\ \beta = \begin{cases} 0\ if\ f\left(\vec{X}_{P,G}\right) < \left(\vec{X}_{P,G+1}\right) \\ 1\ otherwise \end{cases} \tag{4.24}$$

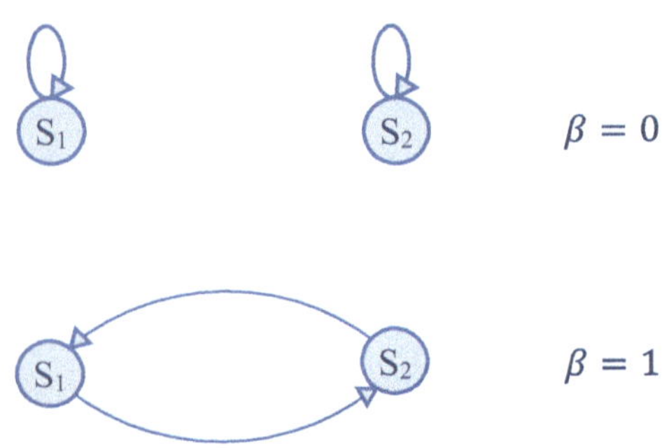

Fig. 4.4 The state transition graph for $L_{2,2}$ automaton

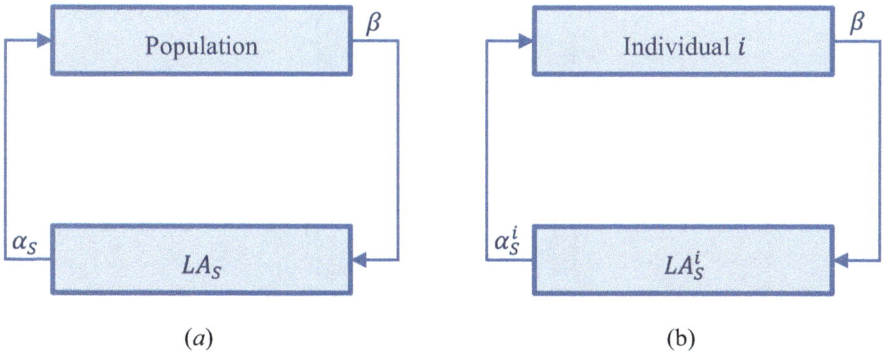

Fig. 4.5 Schematic representation of learning structure for selecting the strategy of DE in (*a*) PSADE and (*b*) ISADE

where f is the fitness function, and $\vec{X}_{P,G}$ is the global best individual of the whole population of the algorithm at generation G. If $\beta = 0$, $L_{2,2}$ stays in the same state, otherwise it switches to the other state.

The simple strategy used by $L_{2,2}$ implies that the DE algorithm continues to execute the whole population with the search strategy it was evolving earlier as long as the response of the environment is favorable but changes to another strategy as soon as the response is unfavorable. Figure 4.6 shows the general procedure used in PSADE.

Finally, Fig. 4.7 shows the details of the proposed PSADE method.

4.4.2.2 ISADE

As the final method, we combine the general idea of IADE and PSADE in a method called ISADE. In ISADE, each individual $\boldsymbol{i}$ of the population is equipped with a Tsetlin automaton $\boldsymbol{LA_s^i}$ which is used to control its search strategy during the run. Figure 4.8 shows the general procedure of ISADE.

4.5 Experimental Setup

4.5.1 Benchmark Functions

To study the performance of the proposed adaptive DEs, a set of experiments was carried out using five classical numerical benchmark functions. For all test problems, the dimension of the search space is 30. Table 4.2 illustrates the benchmark functions with their names, formula, search space ranges, dimensions, optimum global values $f_{min,}$ and optimization goal. Meanwhile, their two-dimensional views are illustrated in Fig. 4.9.

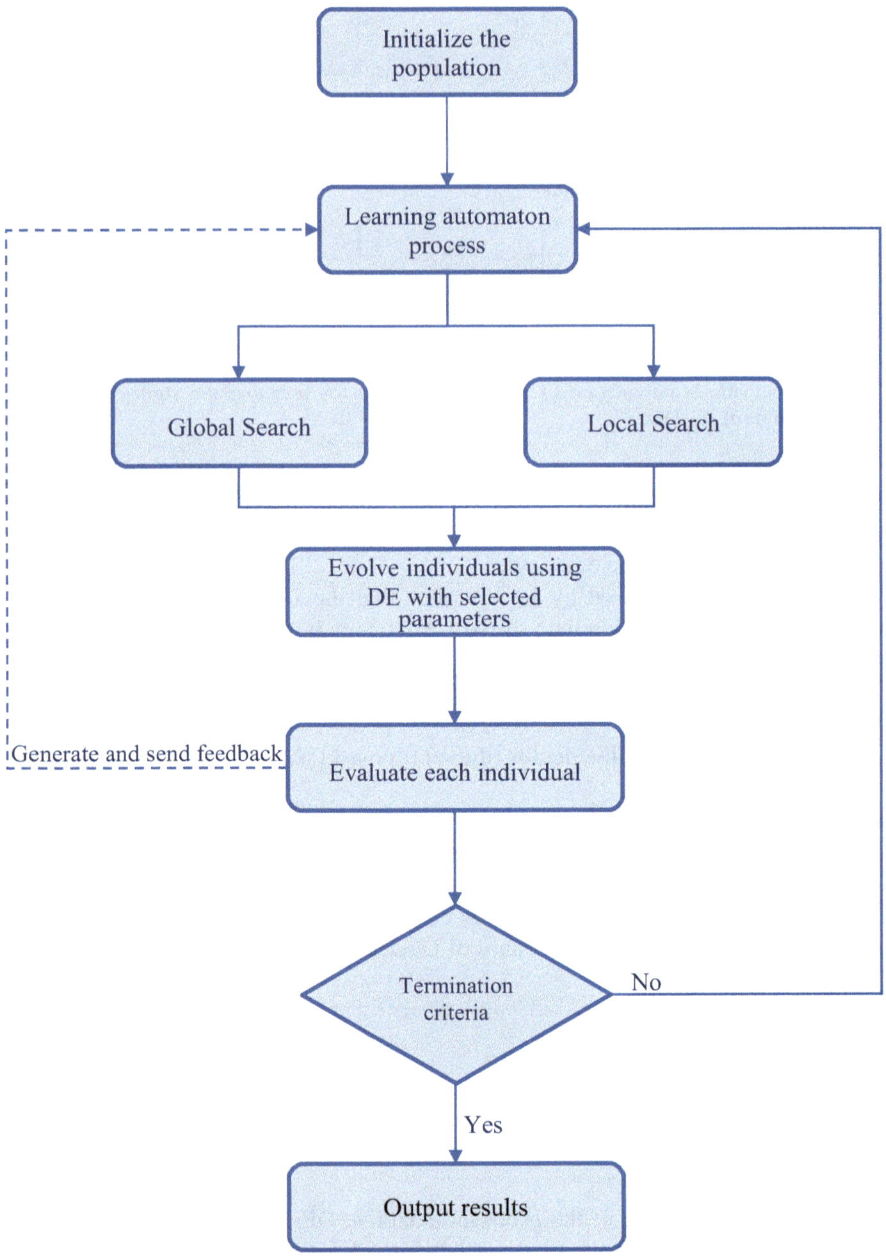

Fig. 4.6 General procedure for PSADE

Algorithm 4-3. PSADE algorithm

```
define
Initialize the algorithm and problem parameters: population size NP fitness evaluations counter fe = 0, maximum
fitness evaluations FEmax, current generation G = 0.
Initialize the LAs with Φ = (Φ1, Φ2) and α = (α1, α2) where α1={F = 0.5,CR=0.9} and α2={F = 0.1,CR=0.2}
Set the current state of LAs to k = 1;
Let αj | j ∈{1,2} be the selected action of LAs.
Generate an initial population P0 = {X1,0, ..., XNP,0} in the D-dimensional search space according to Eq. (4-10).
Let f be the fitness value.
Evaluate the objective function values f(X1), ..., f(XNP).
fe := NP;
while (fe < FEmax)
    PG+1 = ∅ // to determine the population of the next generation
    Select the action αk for LAs from the action set based on the current state k;
    for each genome i ∈ [1, ..., NP] in current population (PG) do
        Generate the donor vector vi using selected F and DE/rand/bin/1; // mutation step
        Repair vi according to Eq. (4-12). // repair operator
        Generate trial vector ui using selected crossover rate CR and Eq. (4-13); // crossover step
        Evaluate the objective function value f(ui).
        Add an individual to PG+1 using Eq. (4-14). // selection and replacement step
    end-for
    Calculate the reinforcement signal β according to Eq. (4-24);
    if (β==0) then //favorable response
        //do nothing and stay in the current state
    else if (β==1) then //unfavorable response
        k = (k%2) + 1; //go to next state
    end-if
    G := G + 1;
    fe := fe + NP;
end-while
```

Fig. 4.7 Pseudocode of PSADE

The benchmark functions include two unimodal functions, Rosenbrock and Sphere, and three multimodal functions, Rastrigin, Griewank, and Ackley. The Rastrigin function has many local optima around the global optima and no correlation among its variables. The Ackley function is multimodal at low resolution. Finally, the Griewank function is the only function, which introduces correlation between its variables.

4.5.2 Algorithm's Configuration

For all DE variants, DE/rand/1/bin with a population size of $NP = 100$ is used. For DE, the scale factor F is set to 0.5, and the crossover rate CR is set to 0.9. For PADE

Algorithm 4-4. ISADE algorithm

define

Initialize the algorithm and problem parameters: population size *NP* fitness evaluations counter *fe* = 0, maximum fitness evaluations FE_{max}, current generation *G* = 0.

Initialize the set of LA_s={LA_s(1), ..., LA_s(*NP*) } with $\Phi = (\Phi_1, \Phi_2)$ and $\alpha = (\alpha_1, \alpha_2)$ where α_1={*F* = 0.5,*CR*=0.9} and α_2={*F* = 0.1,*CR*=0.2}.

Let $k(M) \mid M\epsilon[1, \ldots, NP]$ be the current state of LA_s (*M*).

Set the current state of all LA_s to 1, i.e., $k(1) = 1, k(2) = 1, \ldots, k(NP) = 1$.

Let $\alpha_j(M) \mid j\, \epsilon\{1,2\}, M\epsilon[1, \ldots, NP]$ be the selected action of LA_s (*M*).

Generate an initial population $P_0 = \{\vec{X}_{1,0}, \ldots, \vec{X}_{NP,0}\}$ in the *D*-dimensional search space according to Eq. (4-10).

Let *f* be the fitness value.

while (*fe* < FE_{max})

  $P_{G+1} = \emptyset$; // to determine the population of the next generation

  for each genome $i\, \epsilon\, [1, \ldots, NP]$ in current population (P_G) **do**

    LA_s (i) selects an action from its action set based on its current state $k(i)$; Generate the donor vector $\vec{v}_i$ using selected *F* and DE/rand/bin/1; // mutation step

    Repair $\vec{v}_i$ according to Eq. (4-12); // repair operator

    Generate trial vector $\vec{u}_i$ using selected crossover rate *CR* and Eq. (4-13); // crossover step

    Evaluate the objective function value $f(\vec{u}_i)$;

    Add an individual to P_{G+1} using Eq. (4-14); // selection and replacement step

    Calculate the reinforcement signal β according to Eq. (4-21);

    if (β==0) then

      //do nothing and stay in the current state

    else

      $k(i) = (k(i)\%2) + 1$; //go to next state

    end-if

  end-for

  G := G + 1;

  fe := *fe* + *NP*;

end-while

Fig. 4.8 Pseudocode of ISADE

and IADE, the permissible value for parameters *F* and *CR* is chosen {0.1, 0.3, 0.5, 0.7, 0.9}. Moreover, in the proposed PADE, all learning automata use $L_{R\varepsilon P}$ with learning parameters $a = 0.01$ and $b = 0.001$, whereas in IADE, all learning automata use $L_{R\varepsilon P}$ with learning parameters $a = 0.1$ and $b = 0.01$. For PSADE and ISADE, the actions of the applied Tsetlin automata is $\alpha_1 = \{F = 0.5, CR = 0.9\}$ and $\alpha_2 = \{F = 0.1, CR = 0.2\}$.

4.5.3 Simulation Settings and Results

The following settings are adopted for all experiments of this chapter. For a fair comparison among DE variants, the maximum number of function evaluations was set to 100,000. All experiments on each function were run 30 times. All the algorithms

Table 4.2 Benchmark functions used in the experiments of this chapter

Name	Formula	Search domain	D	f_{min}	Goal
Sphere	$\sum_{i=1}^{D} x_i^2$	$[-100,100]^D$	30	0	1E–06
Rosenbrock	$-\cos\left(2\pi\sqrt{\sum_{i=1}^{D} x_i^2}\right) + \frac{1}{10}\sqrt{\sum_{i=1}^{D} x_i^2} + 1$	$[-2.00,2.00]^D$	30	0	1E–06
Rastrigin	$10D + \sum_{i=1}^{D} (x_i^2 - 10\cos(2\pi x_i))$	$[-5.12,5.12]^D$	30	0	1E–06
Ackley	$-20\exp\left(-0.2\sqrt{\frac{1}{D}\sum_{i=1}^{D} x_i^2}\right) - \exp\left(\frac{1}{D}\sum_{i=1}^{D} \cos 2\pi x_i\right) + 20 + e$	$[-32,32]^D$	30	0	1E–06
Griewank	$\sum_{i=1}^{D} -x_i \sin\left(\sqrt{\lvert x_i \rvert}\right)$	$[-600,600]^D$	30	0	1E–06

were executed on an Intel dual-core i7-6500U with a 2.5 GHz CPU and 8 GB of memory.

The average, best, worst, median and standard deviation of the function error values $f(\vec{x}_{best}) - f(\vec{x}_{opt})$ among 30 independent runs recorded for each benchmark function, where $f(\vec{x}_{best})$ is the best solution found by the algorithm in a typical run and $f(\vec{x}_{opt})$ is the optimum value of the test function.

4.5.4 Experimental Results

4.5.4.1 Experiment 1: Effect of the LA-Based Parameter Control on the Performance of DE

This experiment investigates the performance of LA-based DEs in reaching global optimum, success rate, and speed of convergence. Table 4.3 summarizes the optimization results regarding the mean and standard deviation of the function error values of the best solutions found by different approaches over 30 independent trials. In Table 4.3, the best performing algorithm results, which are significantly superior to the others with a Wilcoxon rank-sum test at 0.05 significant level, are highlighted in gray.

From Table 4.3, the first thing that stands out is that the results obtained by PADE, IADE, PSADE, and ISADE are better than those achieved by DE. The reason is mainly that the adaptive mechanism used in the mentioned methods can balance the diversification and intensification capability of the DE. Another observation from

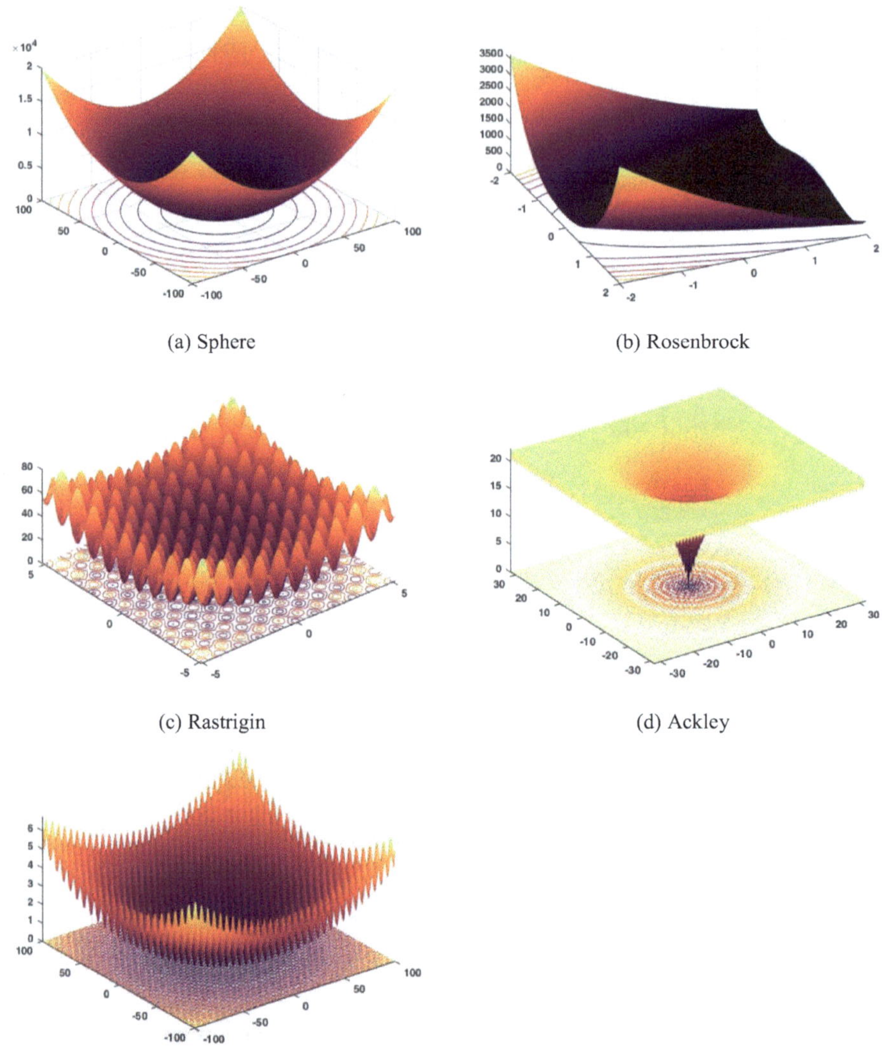

(a) Sphere (b) Rosenbrock

(c) Rastrigin (d) Ackley

(e) Griewank

Fig. 4.9 Test functions ($D = 2$)

Table 4.3 is that PSADE is superior to the other DE-based algorithms. In other words, PSADE is the best-performing DE variant on 3 out of 5 benchmark functions.

For each algorithm, the average number of function evaluations required to reach the specified optimization goal and the given problems' success rate is shown in Table 4.4. Considering the Rosenbrock benchmark function, none of the compared algorithms have reached the specified goal. Table 4.4 shows that IADE, PSADE, and ISADE are the only algorithms that can effectively solve most problems. A

Table 4.3 Numerical results obtained by different DE variants

Method	Stats.	Sphere	Rosenbrock	Rastrigin	Ackley	Griewank
DE	Mean	4.04E-08	1.94E+01	1.88E+02	6.55E-05	1.34E-07
	Best	7.16E-09	1.68E+01	1.56E+02	3.57E-05	1.42E-08
	Worst	1.72E-07	2.06E+01	2.03E+02	1.35E-04	7.62E-07
	Median	3.10E-08	1.95E+01	1.89E+02	6.40E-05	1.09E-07
	Std.	3.24E-08	8.12E-01	1.10E+01	2.10E-05	1.34E-07
PADE	Mean	3.40E-19	2.56E+01	1.23E+01	5.27E-11	0.00E+00
	Best	1.16E-22	2.31E+01	8.02E-02	1.01E-12	0.00E+00
	Worst	4.19E-18	2.76E+01	2.19E+01	3.14E-10	0.00E+00
	Median	2.04E-20	2.59E+01	1.17E+01	3.25E-11	0.00E+00
	Std.	9.89E-19	1.06E+00	5.45E+00	7.28E-11	0.00E+00
IADE	Mean	7.44E-19	2.52E+01	1.14E-14	1.58E-10	0.00E+00
	Best	2.15E-19	2.38E+01	0.00E+00	8.83E-11	0.00E+00
	Worst	3.13E-18	2.59E+01	5.68E-14	3.22E-10	0.00E+00
	Median	5.32E-19	2.54E+01	0.00E+00	1.50E-10	0.00E+00
	Std.	5.82E-19	5.38E-01	2.31E-14	5.15E-11	0.00E+00
PSADE	Mean	5.48E-19	2.31E+01	3.32E-02	1.23E-10	0.00E+00
	Best	7.28E-20	1.95E+01	2.84E-13	5.45E-11	0.00E+00
	Worst	1.69E-18	2.57E+01	9.95E-01	2.80E-10	0.00E+00
	Median	5.33E-19	2.36E+01	1.22E-11	1.14E-10	0.00E+00
	Std.	3.45E-19	1.63E+00	1.82E-01	4.63E-11	0.00E+00
ISADE	Mean	8.92E-18	2.35E+01	6.63E-02	5.35E-10	1.85E-17
	Best	1.76E-18	2.09E+01	1.42E-12	2.34E-10	0.00E+00
	Worst	2.15E-17	2.62E+01	9.95E-01	9.58E-10	5.55E-16
	Median	8.48E-18	2.34E+01	3.64E-11	4.89E-10	0.00E+00
	Std.	4.71E-18	9.96E-01	2.52E-01	1.86E-10	1.01E-16

closer look at Table 4.4 reveals that while DE fails to reach the optimization goal on Rastrigin and Ackley, even for one trial, IADE, PSADE, and ISADE can reach a 100% success rate these problems.

Based on Table 4.4, we also provide the success rate and the average number of function evaluation ranks of each method in Table 4.5. As shown in Table 4.5, PADE is the best performing algorithm in terms of approaching speed the global optima. On the other hand, IADE, PSADE, and ISADE are the most successful algorithms among the other contestant methods.

To further validate the superiority of the LA-enhanced DEs over the basic method, we performed a Friedman test with a 0.05 level of significance over all instances of all algorithms. Figure 4.10 illustrates the Friedman rank obtained by each method. It is important to note that the lower the rank, the better is the algorithm performance.

Table 4.4 The average number of function evaluations to reach the optimization goal for successful runs (mean success rate) obtained by different DE variants

Function	DE	PADE	IADE	PSADE	ISADE
Sphere	91,303 (1)	44,346 (1)	50,733 (1)	48,510 (1)	50,753 (1)
Rosenbrock	−(0)	−(0)	−(0)	−(0)	−(0)
Rastrigin	−(0)	−(0)	69,280 (1)	79,733 (1)	82,243 (1)
Ackley	−(0)	63,800 (0.96)	70,806 (1)	68,103 (1)	72,050 (1)
Griewank	95,003 (0.9)	48,096 (1)	55,906 (1)	50,690 (1)	53,536 (1)

Table 4.5 Average function evaluation ranks (success rate ranks) for different DE variants

Function	DE	PADE	IADE	PSADE	ISADE
Sphere	5 (1)	1 (1)	3 (1)	2 (1)	4 (1)
Rosenbrock	1 (1)	1 (1)	1 (1)	1 (1)	1 (1)
Rastrigin	4 (2)	4 (2)	1 (1)	2 (1)	3 (1)
Ackley	5 (3)	1 (2)	3 (1)	2 (1)	4 (1)
Griewank	5 (2)	1 (1)	4 (1)	2 (1)	3 (1)
Overall rank	4 (1.80)	1.60 (1.40)	3.40 (1)	1.80 (1)	3 (1)

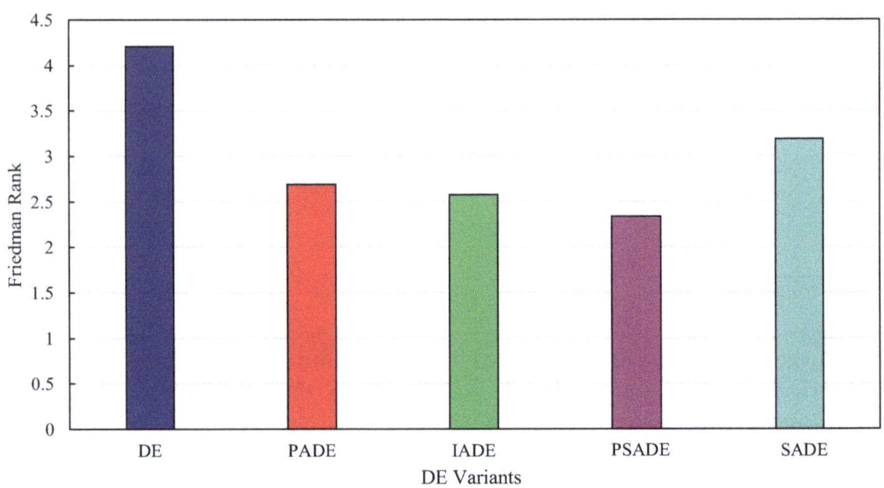

Fig. 4.10 Ranking of different methods using Friedman test in all experiments

Several conclusions can be drawn from Fig. 4.10 as follows:

- It is observed that controlling the behavior of DE has a significant influence on its performance.

- The performance of LA-enhance methods is significantly better than the basic method.
- PSADE is the best-performing DE variant.

Apart from the above numerical results, Fig. 4.11 depicts the convergence curves (evolution of the mean error values versus the number of iterations) for DE versus PSADE on different benchmark functions. A closer look at the convergence curves indicates that the PSADE has a faster convergence rate to the optimum.

To see the adaptive nature of LA-based methods, Fig. 4.12 shows how F and CR's average values are changed over time in PADE applied to the Sphere test problem. The results show that F and CR's value is larger in the first iterations and decreases as the algorithm reaches its final stages, where exploitation is desirable.

4.5.4.2 Experiment 2: Classical Engineering Design Problem

In addition to the well-known test functions used in the previous experiment, we also study the LA-based DE's effectiveness on a real-world engineering problem, namely the pressure vessel design. The pressure vessel design is a classical engineering problem in which the goal is to minimize the cost of material, forming, and welding of a cylindrical pressure vessel. The problem contains four design variables which are shell's thickness (x_1), spherical head thickness (x_2), the inner radius of the cylindrical shell (x_3) and length of the cylindrical section (x_4) and can be formulated as follows:

$$\text{Minimize } f(\vec{x}) = 0.6224x_1x_3x_4 + 1.7781x_2x_3^2 + 3.1661x_1^2x_4 + 19.84x_1^2x_3 \tag{4.25}$$

$$\text{Subject to: } g_1(\vec{x}) = -x_1 + 0.0193x_3 \leq 0 \tag{4.26}$$

$$g_2(\vec{x}) = -x_2 + 0.00954x_3 \leq 0 \tag{4.27}$$

$$g_3(\vec{x}) = -\pi x_3^2 x_4 - \frac{4}{3}\pi x_3^3 + 1296000 \leq 0 \tag{4.28}$$

$$g_4(\vec{x}) = x_4 - 240 \leq 0 \tag{4.26}$$

The range of design variables is also as follows:

$$0 \leq x_1 \leq 100, \quad 0 \leq x_2 \leq 100, \quad 10 \leq x_3 \leq 200, \quad 10 \leq x_4 \leq 200 \tag{4.27}$$

The structure of the pressure vessel design problem is shown in Fig. 4.13.

Table 4.6 summarizes PSADE and CDE's optimization results (Huang et al. 2007), another well-known DE-based method for constrained design problems. For each method, the best-found solution ($\vec{x}$) and the corresponding optimal cost (f_{min}) are reported. The reported results of PSADE are the best result of 10 independent runs. Moreover, each run was continued up to 50,000 fitness evaluations.

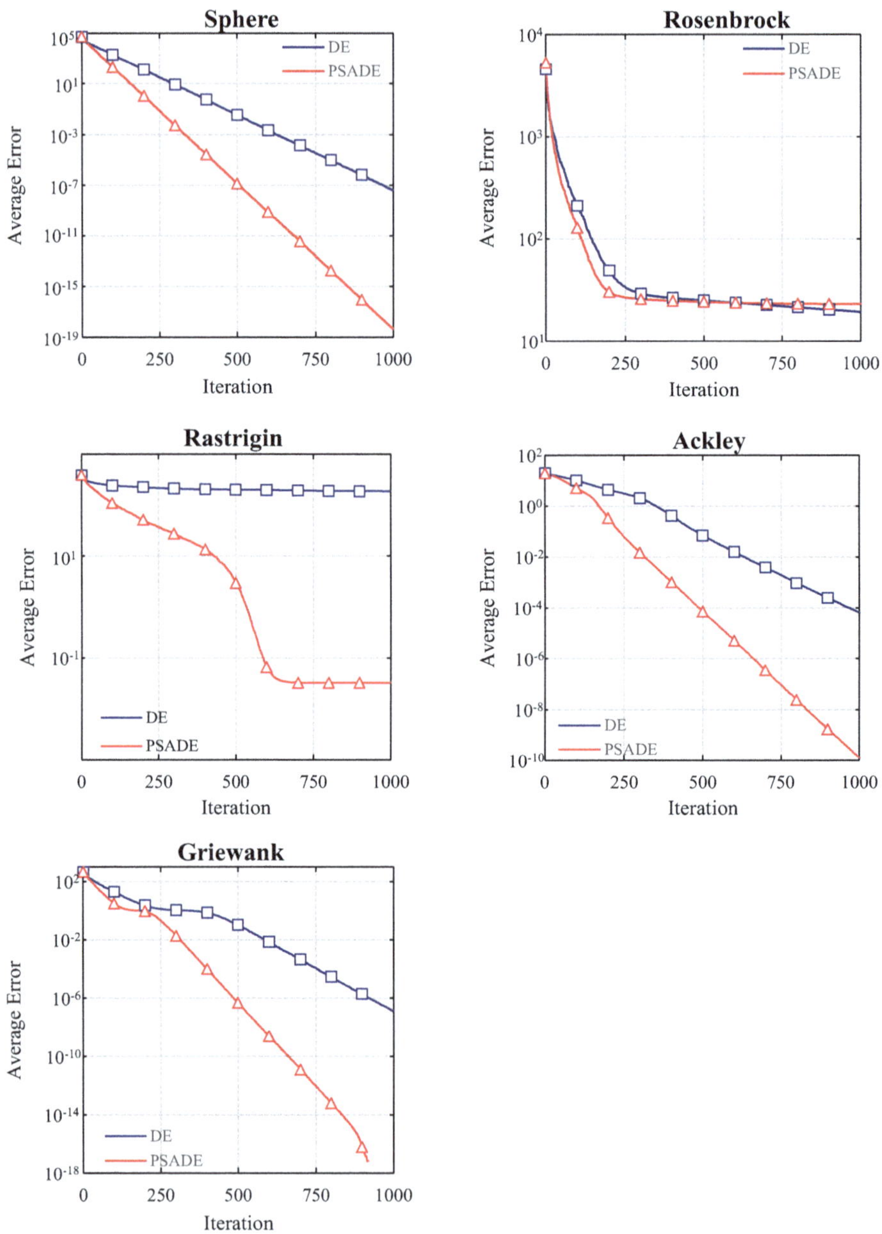

Fig. 4.11 Evolution of the mean error values versus the number of iterations for different DE variants on various test problems

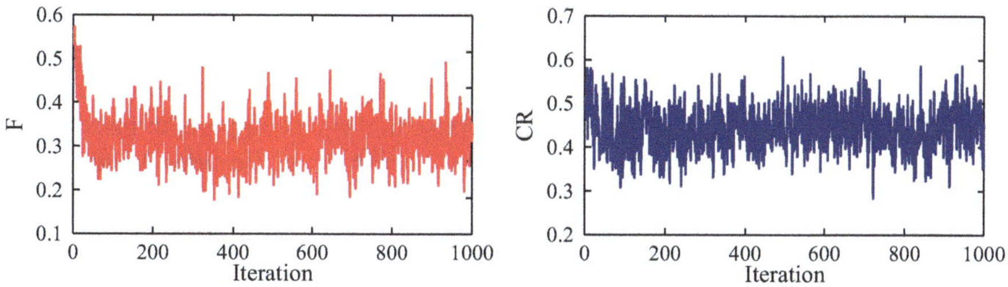

Fig. 4.12 Evolution of the mean value of F and CR over time for Sphere function

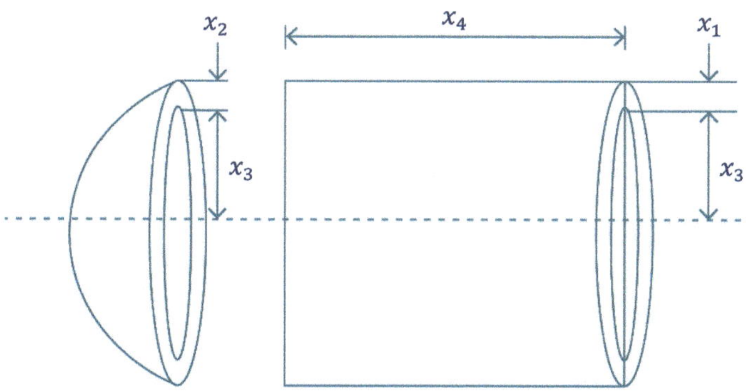

Fig. 4.13 Schematic of the pressure vessel design problem (Coello Coello 2000)

Table 4.6 Optimization results for pressure vessel design problem

Methods	x_1	x_2	x_3	x_4	f_{min}
CDE	0.812500	0.437500	42.098411	176.63769	6059.7340
PSADE	0.808487	0.443152	42.098445	176.63659	6059.7143

From the results in Table 4.6, it is observed that PSADE can find better results.

4.5.4.3 Experiment 3: Computational Complexity of the Proposed Methods

In this experiment, the involved computational complexity of LA-enhance DEs is analyzed.

Regarding the first approach, PADE involves four major steps:

1) Selecting two parameters F and CR, using the probability vector of LA_F and LA_{CR};
2) Evolve the population using DE,
3) Computing the β parameter, and
4) Update the probability vector of LA_F and LA_{CR}.

Since running the population depends on the selected inner optimizer, the population evolution (step 2) will be excluded from our analysis.

For executing step 1, we require a cycle over the probability vector of LA_F and LA_{CR}, which have a size of 2×5. In step 3 and step 4, it is clear from the explanation given in Sect. 4.4.1.1 that they are single operations related to the selected parameters. Thus, these steps do not rely on any loop. Suppose that each comparison involved in step 1 (for finding the maximum probability) has a cost C_1; while steps 3 and 4 have the costs C_2 and C_3, respectively. Then, the PADE approach has linear complexity, given as:

$$C_{PADE} = 2 \cdot C_1 + C_2 + 2 \cdot C_3 \tag{4.31}$$

In summary, the instantiations of the PADE introduce only a marginal complexity to the base algorithm.

A similar analysis for IADE shows that it also has linear complexity, given as:

$$C_{IADE} = N \cdot 2 \cdot C_1 + N \cdot C_2 + N \cdot 2 \cdot C_3 \tag{4.32}$$

Although this is indeed a higher complexity than the DE and PADE, the performance of IADE is superior.

Regarding the PSADE and ISADE methods, a similar analysis shows that they have even a lower complexity than the PADE and IADE.

4.6 Conclusion

This chapter studied the usefulness of learning automaton (LA) for controlling the behavior of evolutionary computation methods during the run. Various conclusions can be drawn from the present study. Firstly, The LA-enhanced differential evolution (DE) algorithms could outperform the DE on a set of well-known benchmark functions in terms of optimality. Secondly, the convergence speed of the LA-enhanced DE was faster than that of DE. Thirdly, the LA-enhanced algorithms do not impose very substantial computational complexity on the base algorithm. Several research directions can be pursued as future works. First, the value of the parameters for reward and penalty rates in PADE and IADE were set based on preliminary experiments. Therefore, a comprehensive sensitivity analysis on these parameters' effect could be a direction for future research. Second, the community lacks a comprehensive study on the effect of different methods for calculating reinforcement signals. Therefore, an empirical study on the effect of different methods for generating reinforcement signals is another direction for future research. Third, the proposed approach can be used to improve the performance of other evolutionary computation methods. Forth, it is also very valuable to apply the proposed approach to other real-world optimization problems.

References

Abedi Firouzjaee, H., Kazemi Kordestani, J., Meybodi, M.R.: Cuckoo search with composite flight operator for numerical optimization problems and its application in tunnelling. Eng. Optim. **49**, 597–616 (2017). https://doi.org/10.1080/0305215X.2016.1206535

Abshouri, A.A., Meybodi, M.R., Bakhtiary, A.: New firefly algorithm based on multi swarm & learning automata in dynamic environments. In: IEEE Proceedings, pp. 989–993 (2011)

Alirezanejad, M., Enayatifar, R., Motameni, H., Nematzadeh, H.: GSA-LA: gravitational search algorithm based on learning automata. 1–17 (2020). https://doi.org/10.1080/0952813X.2020.1725650

Arora, S., Anand, P.: Learning automata-based butterfly optimization algorithm for engineering design problems. Int. j. Comput. Mater. Sci. Eng. **07**, 1850021 (2018). https://doi.org/10.1142/S2047684118500215

Brest, J., Greiner, S., Boskovic, B., Mernik, M., Zumer, V.: Self-adapting control parameters in differential evolution: a comparative study on numerical benchmark problems. IEEE Trans. Evol. Comput. **10**, 646–657 (2006). https://doi.org/10.1109/TEVC.2006.872133

Clerc, M.: The swarm and the queen: towards a deterministic and adaptive particle swarm optimization. In: Proceedings of the 1999 Congress on Evolutionary Computation-CEC99 (Cat. No. 99TH8406), vol. 3, pp. 1951–1957 (1999)

Coello Coello, C.A.: Use of a self-adaptive penalty approach for engineering optimization problems. Comput. Ind. **41**, 113–127 (2000). https://doi.org/10.1016/S0166-3615(99)00046-9

Das, S., Suganthan, P.N.: Differential evolution: a survey of the state-of-the-art. IEEE Trans. Evol. Comput. **15**, 4–31 (2011). https://doi.org/10.1109/TEVC.2010.2059031

Das, S., Konar, A., Chakraborty, U.K.: Two improved differential evolution schemes for faster global search. In: Proceedings of the 2005 conference on genetic and evolutionary computation, Washington, DC, USA, pp. 991–998. ACM (2005)

Eberhart, R.C., Shi, Y.: Tracking and optimizing dynamic systems with particle swarms. In: Evolutionary Computation, 2001. Proceedings of the 2001 Congress on. IEEE, pp 94–100 (2001)

Enayatifar, R., Yousefi, M., Abdullah, A.H., Darus, A.N.: LAHS: a novel harmony search algorithm based on learning automata. Commun. Nonlinear Sci. Numer. Simul. **18**, 3481–3497 (2013). https://doi.org/10.1016/j.cnsns.2013.04.028

Fan, S.-K.S., Chiu, Y.-Y.: A decreasing inertia weight particle swarm optimizer. Eng. Optim. **39**, 203–228 (2007). https://doi.org/10.1080/03052150601047362

Gong, W., Fialho, Á., Cai, Z.: Adaptive strategy selection in differential evolution. In: Proceedings of the 12th Annual Conference on Genetic and Evolutionary Computation, Portland, Oregon, USA, pp. 409–416. ACM (2010)

Hashemi, A.B., Meybodi, M.R.: A note on the learning automata based algorithms for adaptive parameter selection in PSO. Appl. Soft Comput. **11**, 689–705 (2011)

Huang, F., Wang, L., He, Q.: An effective co-evolutisonary differential evolution for constrained optimization. Appl. Math. Comput. **186**, 340–356 (2007). https://doi.org/10.1016/j.amc.2006.07.105

Juang, C.-F., Chang, Y.-C.: Evolutionary-group-based particle-swarm-optimized fuzzy controller with application to mobile-robot navigation in unknown environments. IEEE Trans. Fuzzy Syst. **19**, 379–392 (2011)

Kazemi Kordestani, J., Ahmadi, A., Meybodi, M.R.: An improved differential evolution algorithm using learning automata and population topologies. Appl. Intell. **41**, 1150–1169 (2014). https://doi.org/10.1007/s10489-014-0585-2

Kazemi Kordestani, J., Abedi Firouzjaee, H., Meybodi, M.R.: An adaptive bi-flight cuckoo search with variable nests for continuous dynamic optimization problems. Appl. Intell. **48**, 97–117 (2018). https://doi.org/10.1007/s10489-017-0963-7

Kordestani, J.K., Rezvanian, A., Meybodi, M.R.: An efficient oscillating inertia weight of particle swarm optimisation for tracking optima in dynamic environments. J. Exp. Theor. Artif. Intell. **28**, 137–149 (2016). https://doi.org/10.1080/0952813X.2015.1020521

Li, J., Tang, Y., Hua, C., Guan, X.: An improved krill herd algorithm: krill herd with linear decreasing step. Appl. Math. Comput. **234**, 356–367 (2014). https://doi.org/10.1016/j.amc.2014.01.146
Liu, J., Lampinen, J.: A fuzzy adaptive differential evolution algorithm. Soft Comput. **9**, 448–462 (2005). https://doi.org/10.1007/s00500-004-0363-x
Mahdaviani, M., Kazemi Kordestani, J., Rezvanian, A., Meybodi, M.R.: LADE: learning automata based differential evolution. Int. j. Artif. Intell. Tools **24**, 1550023 (2015). https://doi.org/10.1142/S0218213015500232
Mallipeddi, R., Suganthan, P.N., Pan, Q.K., Tasgetiren, M.F.: Differential evolution algorithm with ensemble of parameters and mutation strategies. Appl. Soft Comput. **11**, 1679–1696 (2011). https://doi.org/10.1016/j.asoc.2010.04.024
Narendra, K.S., Thathachar, M.A.L.: Learning automata: an introduction. Courier Corporation (2012)
Qin, A.K., Huang, V.L., Suganthan, P.N.: Differential evolution algorithm with strategy adaptation for global numerical optimization. IEEE Trans. Evol. Comput. **13**, 398–417 (2009). https://doi.org/10.1109/TEVC.2008.927706
Rezvanian, A., Meybodi, M.R.: An adaptive mutation operator for artificial immune network using learning automata in dynamic environments. In: 2010 Second World Congress on Nature and Biologically Inspired Computing (NaBIC), pp. 479–483. IEEE (2010a)
Rezvanian, A., Meybodi, M.R.: Tracking extrema in dynamic environments using a learning automata-based immune algorithm. In: Communications in Computer and Information Science, pp. 216–225. Springer, Heidelberg (2010)
Rezvanian, A., Meybodi, M.R.: Tracking extrema in dynamic environments using a learning automata-based immune algorithm. In: Grid and Distributed Computing, Control and Automation, pp. 216–225. Springer (2010c)
Rezvanian, A., Saghiri, A.M., Vahidipour, S.M., Esnaashari, M., Meybodi, M.R.: Recent Advances in Learning Automata. Springer (2018a)
Rezvanian, A., Saghiri, A.M., Vahidipour, S.M., Esnaashari, M., Meybodi, M.R.: Learning automata theory. In: Recent Advances in Learning Automata, pp. 3–19. Springer (2018b)
Rezvanian, A., Moradabadi, B., Ghavipour, M., Daliri Khomami, M.M., Meybodi, M.R.: Introduction to learning automata models. In: Learning Automata Approach for Social Networks, pp. 1–49. Springer (2019)
Shi, Y., Eberhart, R.: A modified particle swarm optimizer. In: The 1998 IEEE International Conference on Evolutionary Computation Proceedings, 1998. IEEE World Congress on Computational Intelligence, pp. 69–73. IEEE (1998)
Shi, Y., Eberhart, R.C.: Empirical study of particle swarm optimization. In: Proceedings of the 1999 Congress on Evolutionary Computation, 1999, CEC 99, IEEE (1999)
Storn, R., Price, K.: Differential evolution-a simple and efficient adaptive scheme for global optimization over continuous spaces. International Computer Science Institute, Berkeley (1995)
Storn, R., Price, K.: Differential evolution - a simple and efficient heuristic for global optimization over continuous spaces. J. Global Optim. **11**, 341–359 (1997). https://doi.org/10.1023/A:1008202821328
Vafashoar, R., Morshedlou, H., Rezvanian, A., Meybodi, M.R.: Applications of cellular learning automata and reinforcement learning in global optimization. In: Cellular Learning Automata: Theory and Applications, pp. 157–224. Springer (2021)
Wang, Y., Cai, Z., Zhang, Q.: Differential evolution with composite trial vector generation strategies and control parameters. IEEE Trans. Evol. Comput. **15**, 55–66 (2011). https://doi.org/10.1109/TEVC.2010.2087271
Yousri, D., Allam, D., Eteiba, M.B.: Chaotic whale optimizer variants for parameters estimation of the chaotic behavior in Permanent Magnet Synchronous Motor. Appl. Soft Comput. **74**, 479–503 (2019). https://doi.org/10.1016/j.asoc.2018.10.032
Zhang, J., Sanderson, A.C.: JADE: adaptive differential evolution with optional external archive. IEEE Trans. Evol. Comput. **13**, 945–958 (2009). https://doi.org/10.1109/TEVC.2009.2014613

Zheng, Y., Ma, L., Zhang, L., Qian, J.: Empirical study of particle swarm optimizer with an increasing inertia weight. In: The 2003 Congress on Evolutionary Computation, 2003. CEC 2003, vol. 1, pp. 221–226 (2003a)

Zheng, Y.-L., Ma, L.-H., Zhang, L.-Y., Qian, J.-X.: On the convergence analysis and parameter selection in particle swarm optimization. In: Proceedings of the 2003 International Conference on Machine Learning and Cybernetics (IEEE Cat. No. 03EX693), vol. 3, pp. 1802–1807 (2003b)

Chapter 5
A Memetic Model Based on Fixed Structure Learning Automata for Solving NP-Hard Problems

Javidan Kazemi Kordestani, Mehdi Razapoor Mirsaleh, Alireza Rezvanian, and Mohammad Reza Meybodi

Abstract Combing a genetic algorithm (GA) with a local search method produces a type of evolutionary algorithm (EA) known as a memetic algorithm (MA). This chapter presents a new memetic algorithm called OMA-MA (object migration automaton-based memetic algorithm) that behaves according to learning the automata-based memetic algorithm (LA-MA) model. Like LA-MA, OMA-MA has composed of two parts: a genetic section and a memetic section. Evolution is performed in the genetic section, and a local search is performed in the memetic section. The genetic section consists of a population of chromosomes, mutation operator, crossover operator, and a chromosome's fitness function. The memetic section consists of a meme that corresponds to a local search method. The meme saves the effect (history) of its corresponding local search method. The meme is represented by an object migration automaton (OMA) whose states keep information about the local search process's history. Based on the relationship between the genetic section and the memetic section, three versions of OMA–MA called LOMA–MA, BOMA-MA, and HOMA–MA are presented.

J. Kazemi Kordestani
Department of Computer Engineering, Science and Research Branch, Islamic Azad University, Tehran, Iran

M. R. Mirsaleh
Department of Computer Engineering and Information Technology, Payame Noor University (PNU), P.O. BOX 19395-3697, Tehran, Iran
e-mail: mrezapoorm@aut.ac.ir

A. Rezvanian (✉)
Department of Computer Engineering, University of Science and Culture, Tehran, Iran
e-mail: rezvanian@usc.ac.ir

M. R. Meybodi
Computer Engineering Department, Amirkabir University of Technology (Tehran Polytechnic), Tehran, Iran
e-mail: mmeybodi@aut.ac.ir

J. Kazemi Kordestani et al. (eds.), *Advances in Learning Automata and Intelligent Optimization*, Intelligent Systems Reference Library 208, https://doi.org/10.1007/978-3-030-76291-9_5

5.1 Introduction

Exploration and exploitation are two main search goals. Exploration is important for ensuring global reliability: the whole search space needs to be searched to provide a reliable estimate of the global optimum (De Jong 1975) Exploitation is important because it focuses the search effort around the best solutions by searching their neighborhoods to find more accurate solutions (Weber et al. 2009). Many search algorithms use a global search method and a local search method to achieve their goal. These algorithms are known as hybrid methods. Combining a traditional genetic algorithm (GA) with local search methods that incorporate local improvement procedures can improve the performance of GAs. These hybrid methods are commonly known as memetic algorithms (MAs), or Baldwinian (Ku and Mak 1998), or Lamarckian (Morris et al. 1998) evolutionary algorithms. The particular local search method employed is the important aspect of these algorithms. In the Lamarckian approach, the local search method is used as a refinement genetic operator that modifies an individual's genetic structure and places it back in the genetic population (Chen et al. 2011). Lamarckian evolution can increase the speed of search processes in genetic algorithms. However, it can damage schema processing by changing individuals' genetic structure, leading to premature convergence (Downing 2001; Krasnogor and Smith 2005).

The Baldwinian learning approach improves an individual's fitness by applying a local search; however, individual genotypes remain unchanged. Thus, it increases the individual's chances of remaining in subsequent generations. Similar to natural evolution, Baldwinian learning does not modify an individual's genetic structure, but it does increase its chances of survival. Unlike the Lamarckian learning model, the Baldwinian approach does not allow parents to transfer what they have learned to their children (Rezapoor Mirsaleh and Meybodi 2005). The local search method is used as a part of the individual's evaluation process in the Baldwinian approach. The local search method uses local knowledge to create a new fitness that the global genetic operators can use to improve an individual's capability. In this method, one or more individuals of a population that are similar in genotype gain similar fitness. These individuals are probably near each other in search space and equal in fitness after applying the local search. Therefore, the new search space will be a smooth surface and cover many of the new search space's local minima. This fitness modification is known as the smoothing effect. The Baldwinian learning approach can be more effective, albeit slower, than Lamarckian approaches since it does not alter gas's global search process (Krasnogor and Smith 2005).

GALA which is obtained from a combination of genetic algorithm and learning automata (LA) (Rezvanian et al. 2018c), is a type of memetic algorithm which was first reported by Rezapoor and Meybodi in (Rezapoor Mirsaleh and Meybodi 2005). GALA represents chromosomes as object migration automata (OMAs) (Rezvanian et al. 2018b), whose states represent the local search process's history. In GALA, GA performs the global search (Exploration), and a set of object migrating learning automaton performs a local search (Exploitation). Each state in an OMA has two attributes: the gene's value and the degree of association with its value. Information about the local search process's history shows the degree of association between genes and their values. The local search changes the degree of association between genes and their values. In GALA, a chromosome's fitness is computed using only the value of the genes. According to a Lamarckian learning model, GALA performs because it modifies the genotype and only uses a chromosome's fitness to fitness function computation. In another words, it passes on the learned traits acquired by its local search method to offspring by a modification of the genotype.

In this chapter, a modified version of GALA called OMA-MA that behaves according to the LA-MA model defined in the previous chapter will be presented. Like LA-MA, OMA-MA has composed of two parts: a genetic section and a memetic section. Evolution is performed in the genetic section, and a local search is performed in the memetic section. The genetic section consists of a population of chromosomes, mutation operator, crossover operator, and a chromosome's fitness function. The memetic section consists of a meme that corresponds to a local search method. The meme saves the effect (history) of its corresponding local search method. Like GALA, the meme is represented by an object migration automaton (OMA) whose states keep information about the local search process's history. Based on the genetic section's relationship and the memetic section, three OMA-MA versions are introduced. LOMA–MA behaves according to the Lamarckian learning model, BOMA-MA behaves according to the Baldwinian learning model, and HOMA–MA behaves according to both Baldwinian and Lamarckian learning models. In the remaining part of this chapter, at first, fixed structure learning automata and object migrating automata are described, and then GALA is presented in more detail, and finally, the modified version of GALA, called OMA-MA and its variants called LOMA–MA, BOMA-MA, and HOMA-MA are proposed.

5.2 Fixed Structure Learning Automata and Object Migrating Automata

5.2.1 Fixed Structure Learning Automata

A fixed structure learning automaton (Rezvanian et al. 2019b) is represented by a quintuple $<\underline{\alpha}, \underline{\Phi}, \underline{\beta}, F, G>$ where:

- $\underline{\alpha} = \{\alpha_1, \ldots, \alpha_r\}$ is the set of actions that it must choose from.
- $\underline{\Phi} = \{\varphi_1, \ldots, \varphi_s\}$ is the set of internal states.
- $\underline{\beta} = \{0, 1\}$ is the set of inputs where 1 represents a penalty and 0 represents a reward.
- $F : \underline{\Phi} * \underline{\beta} \rightarrow \underline{\Phi}$ is a function that maps the current state and current input into the next state.

- $G : \underline{\Phi} \rightarrow \underline{\alpha}$ is a function that maps the current state into the current output. In other words, G determines the action taken by the automaton.

The operation of fixed learning automata could be described as follows:

At the first step, the selected action $\alpha(n) = G[\Phi(n)]$ serves as the input to the environment, which in turn emits a stochastic response $\beta(n)$ at the time n. $\beta(n)$ is an element of $\underline{\beta} = \{0, 1\}$ and is the feedback response of the environment to the automaton. In the second step, the environment penalizes (i.e., $\beta(n) = 1$) the automaton with the penalty probability c_i, which is action-dependent. Based on the response $\beta(n)$, the state of the automaton is updated by $\Phi(n + 1) = F[\Phi(n), \beta(n)]$. This process continues until the desired result is obtained.

We describe some of the fixed structure learning automata such as Tsetline, Krinsky, and Krylov automata in the following paragraphs.

The Two-State Automata ($L_{2,2}$): This automaton has two states, φ_1 and φ_2 and two actions α_1 and α_2. The automaton accepts input from a set of $\{0, 1\}$ and switches its states upon encountering an input 1 (unfavorable response) and remains in the same state on receiving an input 0 (favorable response). An automaton that uses this strategy is referred as $L_{2,2}$ where the first subscript refers to the number of states and the second subscript to the number of actions.

The Tsetline Automata (The Two-Action Automata with Memory $L_{2N,2}$): Tsetline suggested a modification of $L_{2,2}$ denoted by $L_{2N,2}$. This automaton has *2N* states and two actions and attempts to incorporate the system's past behavior in its decision rule for choosing the sequence of actions. While the automaton $L_{2,2}$ switches from one action to another on receiving a failure response from the environment, $L_{2N,2}$ keeps an account of the number of successes and failures received for each action. Only when the number of failures exceeds the number of successes or some maximum value *N*; the automaton switches from one action to another. The procedure described above is one convenient method of keeping track of the performance of the actions α_1 and α_2. *N* is called depth of memory associated with each action, and automaton is said to have a total memory of *2N*. For every favorable response, the automaton's state moves deeper into the memory of the corresponding action, and for an unfavorable response, moves out of it. This automaton can be extended to multiple action automata. The state transition graph of $L_{2N,2}$ automaton is shown in Fig. 5.1.

The Krinsky Automata: This automaton behaves exactly like $L_{2N,2}$ automaton when the response of the environment is unfavorable, but for a favorable response, any state φ_i (for $i = 1, \ldots N$) passes to the state φ_1 and any state φ_i ($i = N + 1$, *2N*) passes to the state φ_{N+1}. This implies that a string of *N* consecutive unfavorable responses is needed to change from one action to another. The state transition graph of the Krinsky automaton is shown in Fig. 5.2.

The Krylov Automata: This automaton has state transitions that are identical to the $L_{2N,2}$ automaton when the output of the environment is favorable. However, when the response of the environment is unfavorable, a state $\varphi_i (i \neq 1, N, N + 1, 2N)$ passes

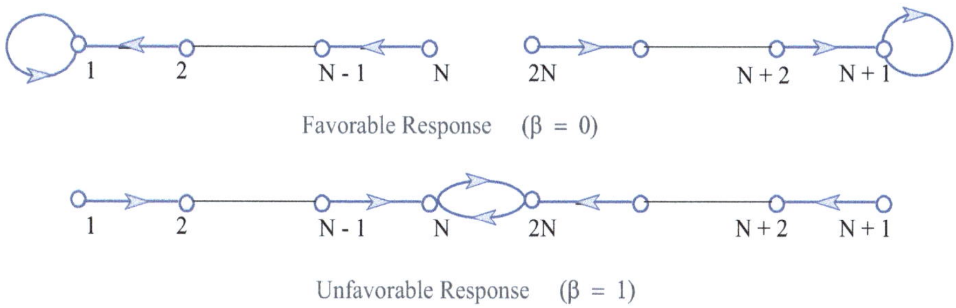

Fig. 5.1 The state transition graph for $L_{2N,2}$ (Tsetline Automaton)

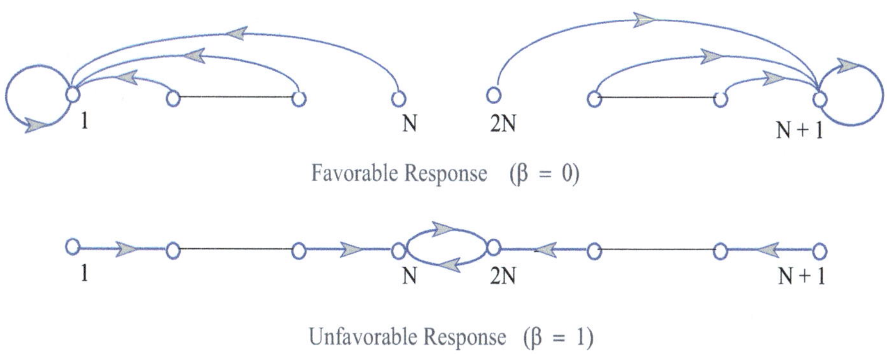

Fig. 5.2 The state transition graph for Krinsky Automaton

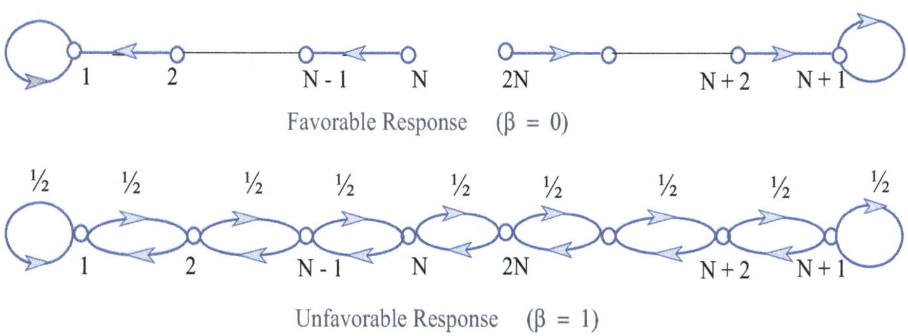

Fig. 5.3 The state transition graph for Krylov Automaton

to a state φ_{i+1} with probability 1/2 and to state φ_{i-1} with probability 1/2, as shown in Fig. 5.3. When $i = 1$ or $N + 1$, φ_i stays in the same state with probability 1/2 and moves to φ_{i+1} with the same probability. When $i = N$, φ_N moves to φ_{N-1} and φ_{2N} each with probability 1/2 and similarly, when $i = 2N$, φ_{2N} moves to φ_{2N-1} and

φ_N each with probability 1/2. The state transition graph of the Krylov automaton is shown in Fig. 5.3.

Object migration automaton (OMA) is an example of fixed structure learning automata described in the next section. Learning automata have a vast variety of applications in combinatorial optimization problems (Soleimani-Pouri et al. 2014), Vahidipour et al. 2019b; Khomami et al. 2020b), global optimization (Vafashoar et al. 2021c), channel assignment (Vafashoar et al. 2021e), influence maximization (Khomami et al. 2021), recommender systems (Rezvanian et al. 2019d), peer-to-peer networks (Rezvanian et al. 2018f), network sampling (Rezvanian et al. 2019i), computer networks (Torkestani and Meybodi 2010), queuing theory (Vahidipour et al. 2015), image processing (Hasanzadeh Mofrad et al. 2015), information retrieval (Torkestani 2012b), Internet of Things based on blockchain (Saghiri et al. 2018), neural networks engineering (Beigy and Meybodi 2009), cloud computing (Vafashoar et al. 2021f; Vafashoar et al. 2021g), social networks (Mirsaleh and Meybodi 2016a; Rezvanian and Meybodi 2017a), community detection (Daliri Khomami et al. 2020a, 2020b; Khomami et al. 2020a), and pattern recognition (Mahmoudi et al. 2021).

5.2.2 Object Migration Automata

Object migration automata were first proposed by Oommen and Ma (Oommen and Ma 1988). OMAs are a type of fixed structure learning automata and are defined by a quintuple $<\underline{\alpha}, \underline{\Phi}, \underline{\beta}, F, G>$.$\underline{\alpha} = \{\alpha_1, \ldots, \alpha_r\}$. is the set of allowed actions for the automaton. For each action α_k, there is a set of states $\{\varphi_{(k-1)N+1}, \ldots, \varphi_{kN}\}$, where N is the depth of memory. The states $\varphi_{(k-1)N+1}$ and φ_{kN} are the most internal state and the boundary state of action α_k, respectively. The set of all states is represented by $\underline{\Phi} = \{\varphi_1, \ldots, \varphi_s\}$, where $s = N * r$. $\underline{\beta} = \{0, 1\}$ is the set of inputs, where 1 represents an unfavorable response, and 0 represents a favorable response. $F : \underline{\Phi} * \underline{\beta} \to \underline{\Phi}$ is a function that maps the current state and current input into the next state, and $G : \underline{\Phi} \to \underline{\alpha}$ is a function that maps the current state into the current output. In other words, G determines the action taken by the automaton. W objects are assigned to actions in an OMA and moved around the automaton states, as opposed to general learning automata, in which the automaton can move from one action to another by the environmental response. The state of objects is changed based on the feedback response from the environment. If the object w_i is assigned to action α_k (i.e., w_i is in state ξ_i, where $\xi_i \in \{\varphi_{(k-1)N+1}, \ldots, \varphi_{kN}\}$), and the feedback response from the environment is 0, α_k is rewarded, and w_i is moved toward the most internal state ($\varphi_{(k-1)N+1}$) of that action. If the feedback from the environment is 1, then α_k is penalized, and w_i is moved toward the boundary state (φ_{kN}) of action α_k. The variable γ_k denotes the reverse of the state number of the object assigned to action α_k (i.e., degree of association between action α_k and its assigned object). By rewarding an action, the degree of association between that action and its assigned object will be increased. Conversely, penalizing an action causes the degree of association between that action and its assigned object to be decreased. An object associated with state $\varphi_{(k-1)N+1}$ has the most degree of association with action α_k, and an object associated with state φ_{kN} has the least degree of association with action α_k.

5.3 GALA

GALA, a hybrid model, based on a GA and an LA, was introduced by Rezapoor and Meybodi (Rezapoor Mirsaleh and Meybodi 2005). OMAs in this model represent chromosomes. In the OMA-based representation, there are n actions in each automaton corresponding to n genes in each chromosome. Furthermore, for each action, there are a fixed number of states N. The value of each gene, as a migratory object in the automata, is selected from the set $W = \{w_1, \ldots, w_m\}$ and assigned to states of the corresponding action. After applying a local search, if the assignment of an object to the states of action is promising, then the automaton is rewarded, and the assigned object moves toward the most internal state of that action; otherwise, the automaton is penalized, and the assigned object moves toward the boundary state of that action. The rewarding and penalizing of action changes the degree of association between an object and its action. Figure 5.4 shows a representation of chromosome "dfabec" using the Tsetline automaton-based OMA with six actions and a depth of memory of 5.

In Fig. 5.4, there are six actions (genes), denoted by α_1, α_2, α_3, α_4, α_5, and α_6. Genes 1, 2, and 6 possess values 'd,' 'f,' and 'c,' located at internal states 2, 3, and 4 of their actions. The value of genes 3 and 5 are 'a' and 'e,' respectively, and both of them are located at the boundary states of their actions. Consequently, there is a minimum degree of association between these actions and their corresponding object. The remaining action, gene 4, has a value of 'b' and is located at the most internal state of its action. That is, it has the maximum degree of association with action 4. Representation of chromosomes based on other fixed structure learning automata is also possible. In a Krinsky-based OMA representation, as shown in Fig. 5.5, the object will be associated with the most internal state (i.e., it gets the highest degree of association with the corresponding action) rewarded and moves according to the Tsetline automaton-based OMA when it is penalized. In the representation based on the Krylov OMA shown in Fig. 5.6, the object moves either toward the most internal state, or toward the boundary state, with a probability of 0.5 toward penalty, and moves according to the Tsetline automaton-based OMA upon reward.

5.3.1 Global Search in GALA

The global search in GALA is based on a traditional genetic algorithm. An OMA represents a population of chromosomes. Chromosome i is denoted by $CR_i = [(CR_i.Action(1), CR_i.Object(1), CR_i State(1)), \ldots, (CR_i.Action(n),$

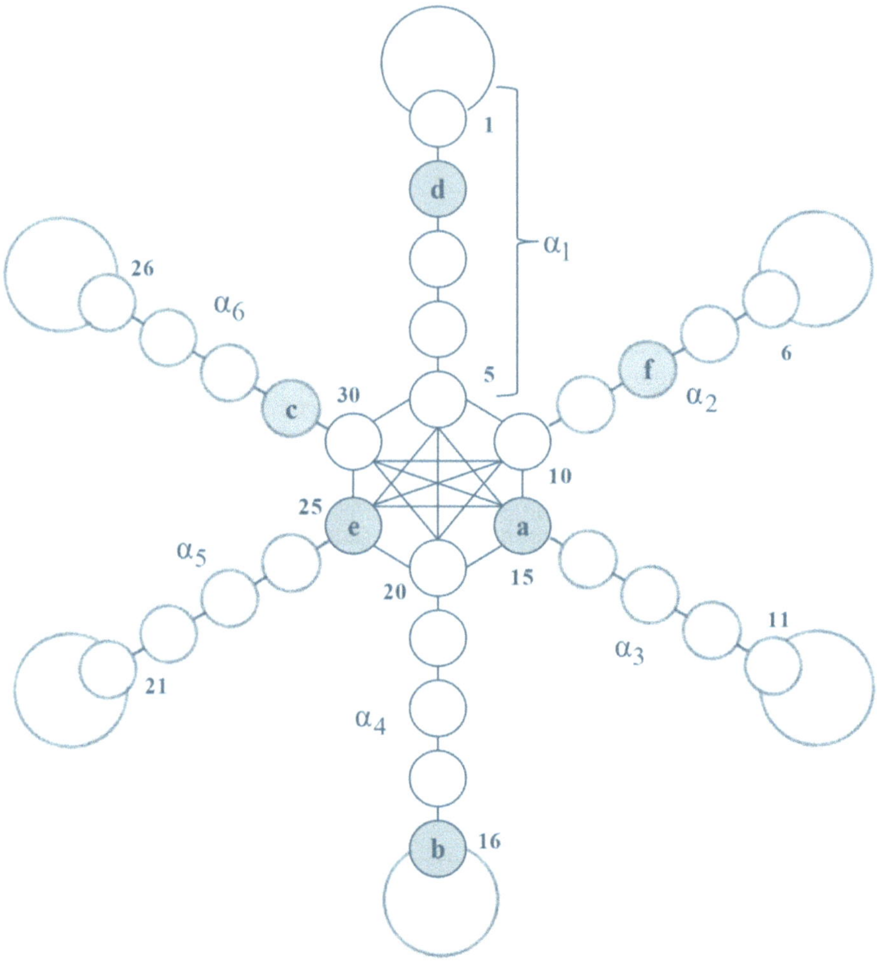

Fig. 5.4 The state transition graph of a Tsetline-based OMA

$CR_i.Object(n), CR_i.State(n))]$, where $CR_i.Action(k)$ is the k_{th} action of CR_i, $CR_i.Object(k)$ is the object assigned to the k_{th} action (the value of the k_{th} gene), and $CR_i.State(k)$ is the state of the object assigned to the k_{th} action (the degree of association between gene k and its value), specifying $1 \leq k \leq n$, $1 \leq CR_i.Object(k) \leq m$, and $(k-1)N+1 \leq CR_i.State(k) \leq kN$. The initial population is created randomly, and objects are located at the boundary state of their actions. At the beginning of each generation, the best chromosome from the previous generation is moved to the current generation population. Next, the crossover operator is applied to the parent chromosomes at the rate r_c (parents are selected according to the chromosome's fitness using a tournament mechanism), and then the mutation operator is applied at the rate r_m.

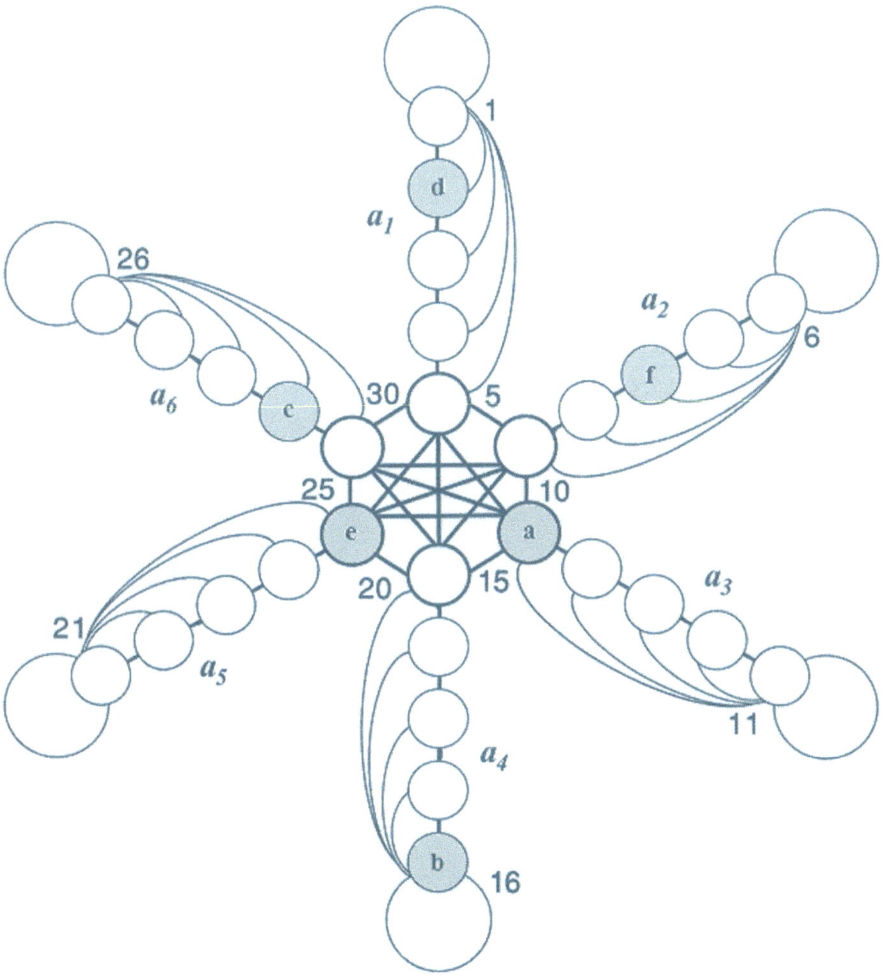

Fig. 5.5 The state transition graph of a Krinsky-based OMA

5.3.2 *Crossover Operator*

The crossover operator in GALA is applied as follows: Two chromosomes, CR_1 and CR_2, are selected by the selection mechanism as parent chromosomes. Two actions, r_1 and r_2, are also randomly selected from CR_1 and CR_2, respectively. Then for each action in the range of $[r_1, r_2]$ of CR_1, the assigned object is exchanged with an assigned object of the same action in chromosome CR_2. In the crossover operator, the previous states of the selected actions in CR_1 and CR_2 are transferred to the child chromosomes. The pseudo-code for the crossover operator is shown in Fig. 5.7.

Figure 5.8 illustrates an example of a crossover operator. First, two actions are randomly selected in the parent chromosomes (e.g., actions 2 and 4 here), and then

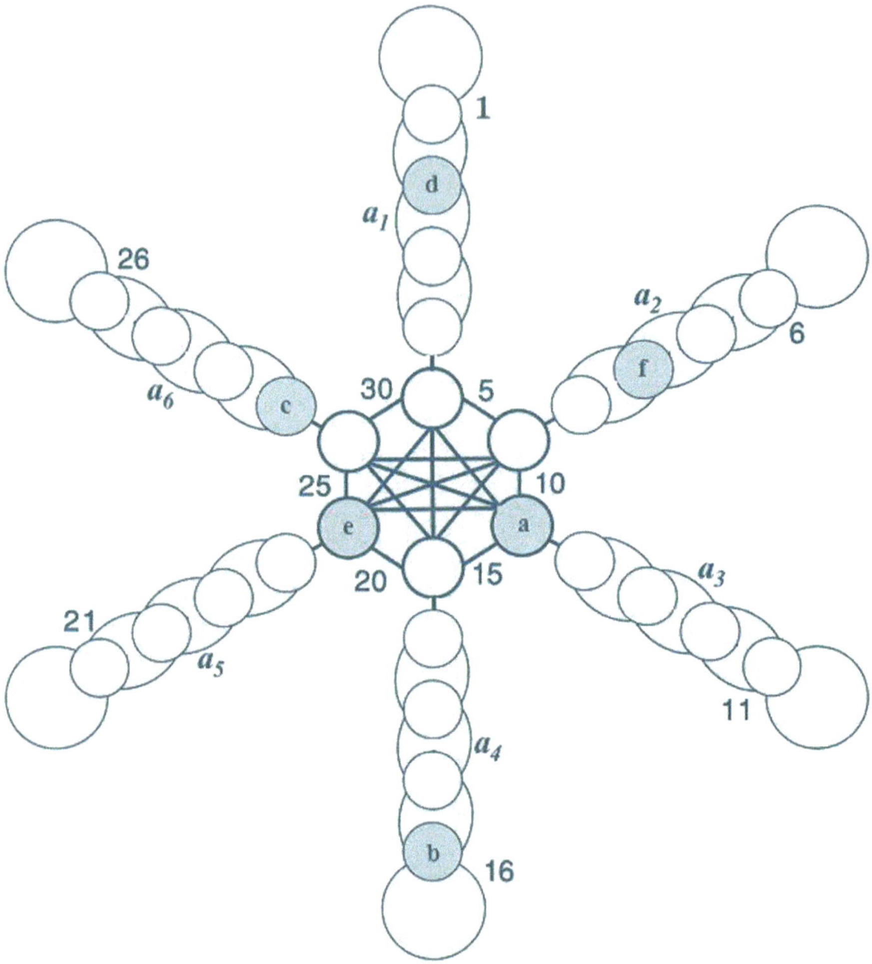

Fig. 5.6 The state transition graph of a Krylov-based OMA

Algorithm 5-1. crossover operator

Procedure Crossover (CR_1, CR_2)
    Generate two random numbers r_1 and r_2 in $[1, N]$ where $r_1 < r_2$;
    For i = r_1 to r_2 **do**
        Swap(*CR₁.Object(CR₁.Action(i)), CR₂.Object(CR₂.Action(i))*);
        Swap(*CR₁.State(CR₁.Action(i)), CR₂.State(CR₂.Action(i))*);
    End For
End Crossover

Fig. 5.7 Pseudocode for a crossover operator

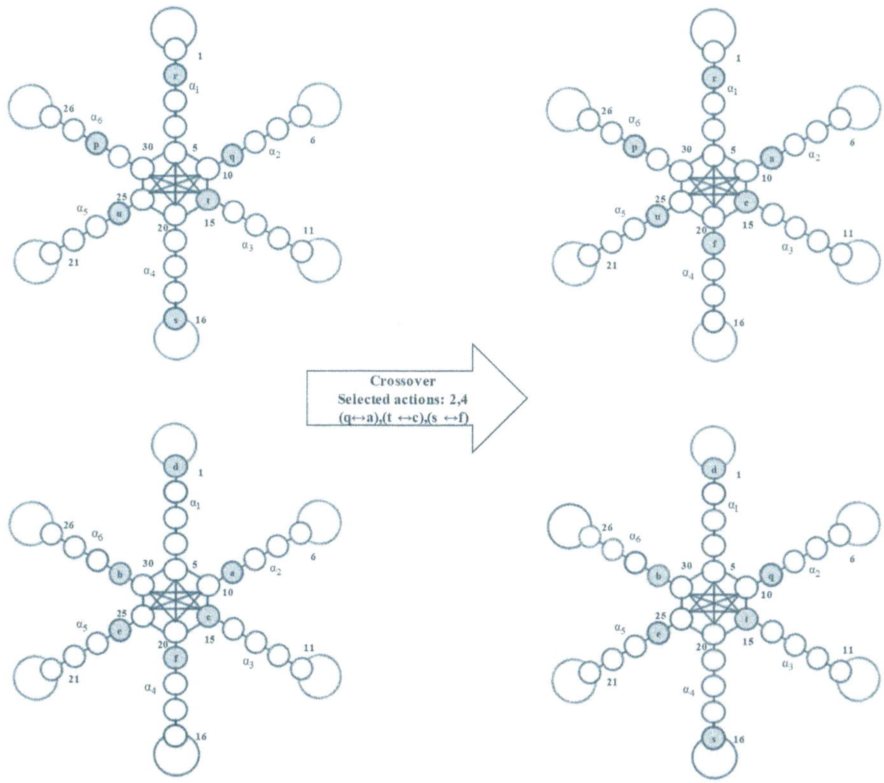

Fig. 5.8 An example of crossover operator

objects are assigned to all actions in the range of [2, 4] in CR_1, and are exchanged with the objects of the corresponding actions in CR_2.

5.3.3 *Mutation Operator*

The mutation operator is the same as in a traditional genetic algorithm. Two actions are selected randomly in the parent chromosome, and then their assigned objects are exchanged. The previous states of selected parent chromosome actions are transferred to the child chromosomes in this operator. Pseudocode for the mutation operator is shown in Fig. 5.9.

Figure 5.10 illustrates an example of a mutation operator. The first two actions in the parent chromosome are randomly selected (e.g., actions 1 and 2 here). The mutation operator exchanges both the state and the object assigned to action 1 with the state and the object assigned to action 2.

```
Algorithm 5-2. Mutation
Procedure Mutation (CR)
    Generate two random numbers r1 and r2 in [1,n] where r1 < r2;
    Swap(CR.Object(CR.Action(r1)), CR.Object(CR.Action(r2)));
    Swap(CR.State(CR.Action(r1)), CR.State(CR.Action(r2)));
End Mutation
```

Fig. 5.9 Pseudocode for a mutation operator

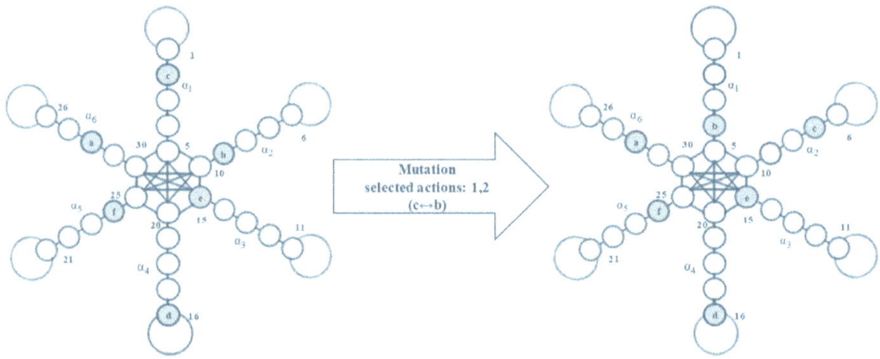

Fig. 5.10 A sample of the mutation operator

```
Algorithm 5-3. Reward
Procedure Reward( CR, u )
    If (CR.State(u)-1) mod N <> 0 then
        Dec(CR.State(u)); //Decrement the state number of action (the state of action move toward the most internal
state)
    End If
End Reward
```

Fig. 5.11 Pseudocode for a reward function

5.3.4 Local Learning in GALA

Local learning in GALA is done using OMA representations of chromosomes. If the objects assigned to action are the same before and after applying a given local search, then that action will be rewarded; otherwise, it will be penalized. It is worth noting that a local search only changes the state of the actions (according to the OMA connections), not the objects assigned to the actions (i.e., the action associated with an object will not change). By rewarding an action, the state of that action will move toward the most internal state according to the OMA connection. This causes the degree of association between an object, and its corresponding action, to be increased. The state of action remains unchanged if the object is located at its internal state, such as the state of object d in action 4, shown in Fig. 5.12.

Figure 5.11 provides a pseudo-code for rewarding an action. Figure 5.12 illustrates an example of rewarding an action.

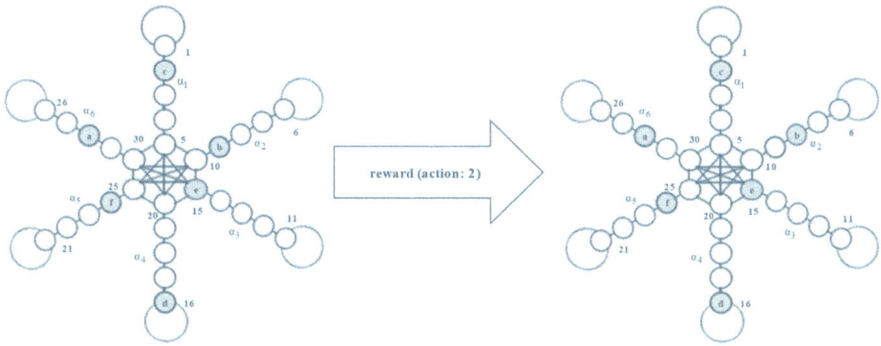

Fig. 5.12 An example of a reward function

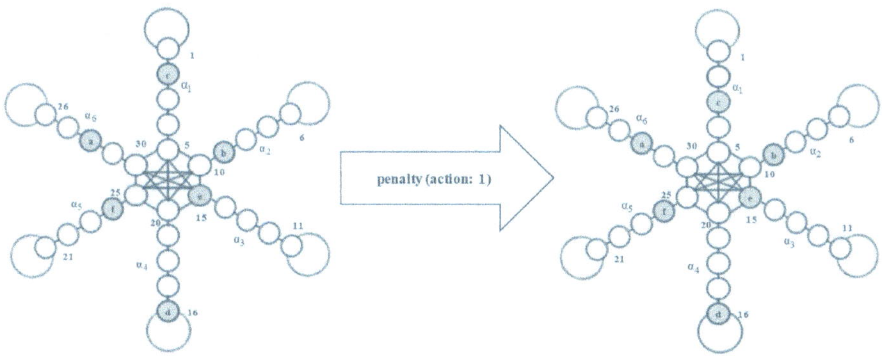

Fig. 5.13 An example of a penalty function

Penalizing an action causes the degree of association between an object and its corresponding action to be decreased. If an object is not in the boundary state of its action, it is then penalizing causes the object assigned to the action to move toward the boundary state. This means that the degree of association between the action and the corresponding object will be decreased (Fig. 5.13). If an object is in the boundary state of its action, then penalizing the action causes the object assigned to that action to change and create a new chromosome. How a new chromosome is created depends on the application. A new chromosome is always created so that its fitness becomes greater than the old chromosome's fitness. Figure 5.14 shows the effect of the penalty function on action 3 of a sample chromosome (assuming that chromosome "cbadfe" has better fitness than chromosome "cbedfa"). The pseudo-code for the penalty function is shown in Fig. 5.15.

The pseudo-code for GALA is shown in Fig. 5.16.

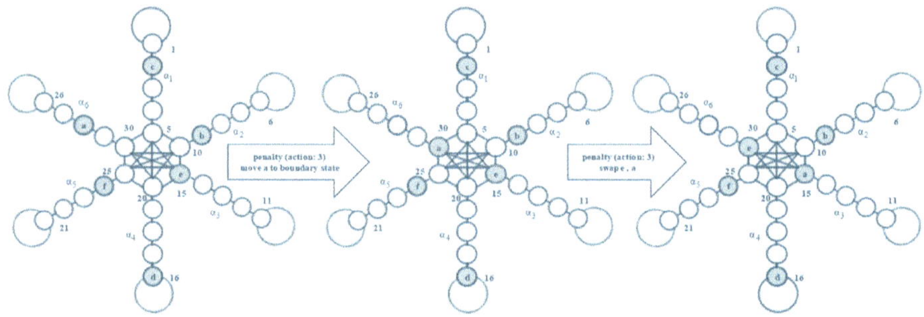

Fig. 5.14 Another example of a penalty function

```
Algorithm 5-4. Penalize
Procedure Penalize(CR, u)
  If (CR.State(u)) mod N <> 0 then
    Inc(CR.State(u)); //Increment the state number of action (the state of action move toward the boundary state)
  Else
    Find action U of chromosome CR so that swapping the object assigned to u and object assigned to U causes
CR to be minimized.
    CR.State(U) = CR.Action(U)*N; //the state of action changed to the boundary state
    CR.State(u) = CR.Action(u)*N; //the state of action changed to the boundary state
    Swap(CR.Object(CR.Action(u)), CR.Object(CR.Action(U)));
  End If
End Penalize
```

Fig. 5.15 Pseudocode for the penalty function

```
Algorithm 5-5. GALA
Function GALA
t←0;
Init population P(t) with size ps; //Create an initial population P(0) of chromosomes P(0).CR1 ... P(0).CRps;
EvaluateFitness (); // Evaluate fitness of all chromosomes of initial population
While (while termination criteria are not satisfied) do
  P(t+1).CR1 ← Best chromosome of P(t); //the best chromosome move to the population of next generation
  For i = 2 to ps do
    Select parents CR1, CR2 based on tournament mechanism from P(t);
    NewCR←Crossover(CR1, CR2);
    NewCR←Mutation(NewCR);
    TempCR←LocalSearch(NewCR);
    For j=1 to n do
      If (TempCR.Object(TempCR.Action (j)) = NewCR.Object (NewCR.Action (j))) then
        Reward(NewCR, NewCR.Action (j));
      Else
        Penalize(NewCR, NewCR.Action (j));
      End If
      P(t+1).CRi ← NewCR;
    End For
  End For
t←t + 1;
End While
End GALA
```

Fig. 5.16 Pseudocode for GALA

5.3.5 Applications of GALA

GALA has been used in a variety of applications, including the graph isomorphism problem (Rezapoor Mirsaleh and Meybodi 2006), join ordering problems in database queries (Asghari et al. 2001, 2007; Mamaghani et al. 2007; Safari Mamaghani et al. 2008), the traveling salesman problem (Zaree and Meybodi 2007a; Zaree et al. 2008), the problem of the Hamiltonian cycle (Asghari and Meybodi 2008), sorting problems in graphs (Zaree and Meybodi 2007b), the graph bandwidth minimization problem (Mamaghani and Meybodi 2008a; Mamaghani and Meybodi 2011; Isazadeh et al. 2012), software clustering problems (Mamaghani and Meybodi 2008b; Mamaghani and Meybodi 2009), the single machine total weighted tardiness scheduling problem (Asghari and Meybodi 2009), data allocation problems in distributed database systems (Mamaghani et al. 2010a, 2010b), and the task graph scheduling problem (Nezhad et al. 2011; Bansal and Kaur 2012).

In (Rezapoor Mirsaleh and Meybodi 2006), a hybrid algorithm based on GALA has been reported to solve graph isomorphism. It was shown that the proposed algorithm performs better than the GA method (Wang et al. 1997) and also the LA-based method (Rezapour 2004) for weighted graphs. This algorithm has the highest convergence rate when the depth of memory is 9, 3, and 1 when chromosomes are represented using Tsetline, Krinsky, and Krylov-based OMA representations.

In (Asghari et al. 2001; Asghari et al. 2007; Mamaghani et al. 2007; Safari Mamaghani et al. 2008), GALA has been used for solving Join ordering problem in database queries (Asghari et al. 2001; Asghari et al. 2007; Mamaghani et al. 2007; Safari Mamaghani et al. 2008). Query optimization aims to select the most efficient execution plan for achieving relevant data and responding to data queries (Safari Mamaghani et al. 2008). Because of the join operator's associative and commutative features, the number of execution plans for a query increases exponentially when the number of joins between relations increases. In GALA based solution (Mamaghani et al. 2007), the join operators order has been represented by OMA, where each join operator (an object) is located in a state of one of the actions. OMA representation of chromosomes based on Krinsky, Krylov, and Tsetline learning automata have been tested and shown that Krinsky-based OMA representation performs better than other representations. It has also been shown that GALA results in a higher rate of convergence and also prevents premature convergence.

In (Zaree and Meybodi 2007a; Zaree et al. 2008), GALA has been the basis for designing an algorithm for the traveling salesman problem (TSP). In this algorithm, tours are represented by OMA in which each city that plays the role of the object is located in a state of one of the actions of OMA. It was shown that this hybrid algorithm performs better than genetic and greedy methods and converges faster to the global solution when Krylov-based OMA representation has been used. An algorithm based on GALA for TSP problem for large graphs has also been reported (Zaree et al. 2008). This algorithm consists of two phases. In the first phase, using a clustering technique, the graph is partitioned into several smaller subgraphs, and in the second phase, the subgraphs are searched using GALA.

A GALA-based algorithm for solving the Hamiltonian cycle problem has been reported (Asghari and Meybodi 2008). A variety of mutation and crossover operators, such as Order Based, Sub List, Based, Insertion and Scramble operators for mutation operator and Ordered, Reverse Ordered, Partially Mapped Cycle operators for crossover operator which are adopted for GALA have been tested in this algorithm. Simulation results have shown that a sublist-based operator for mutation produces a faster convergence rate than genetic, greedy, and learning automata-based methods (Asghari and Meybodi 2008).

In (Zaree and Meybodi 2007b), GALA has been used to design a hybrid algorithm for sorting problems in graphs. Two new operators, one for mutation and one for crossover, were introduced. The algorithm has been tested on several data sets, and the results are compared with the results obtained from genetic and greedy methods. The comparison showed the superiority of the proposed algorithm.

In (Mamaghani and Meybodi 2008a; Mamaghani and Meybodi 2011; Isazadeh et al. 2012), the graph bandwidth minimization problem has been solved using four algorithms: genetic algorithm, an algorithm based on OMA, a hybrid algorithm based on variable structure LA and GALA. It has been shown that GALA performs the best.

In (Mamaghani and Meybodi 2008b; Mamaghani and Meybodi 2009), GALA application in the software clustering problem has been reported. The proposed GALA-based algorithm has been compared with two other algorithms: an OMA-based algorithm and a GA-based algorithm. In order to examine these algorithms, two groups of module dependency graphs are used. The first group includes graphs that are created randomly. The second group includes software graphs obtained by CIA code analyzer tools (Chen et al. 1998). The comparison has shown the superiority of GALA based algorithm.

In (Asghari and Meybodi 2009), a single machine total weighted tardiness scheduling problem has been solved using a GALA-based solution. The algorithm was tested on a benchmark with 40, 50, and 100 OR library jobs (Beasley 1990). The results obtained were compared with the results obtained for four other methods: iterative, greedy, descent, and genetic algorithms (Grosso et al. 2004; Huegler and Vasko 1997, Madureira et al. 2001). The comparison results showed that GALA performs best in terms of the solution's quality and convergence rate.

In (Mamaghani et al. 2010a, 2010b), several algorithms based on GALA are designed for solving data allocation problems in distributed database systems. Experimental results showed that the proposed algorithms have superiority over several well-known methods such as the neighborhood random search algorithm (RS), Corcoran Genetic Algorithm, the Ishfaq Genetic Algorithm OMA-based algorithm, in terms of quality of solution and the rate of convergence. The proposed algorithms were tested with Tsetline, Krinsky, and Krylov-based OMA representations. The results indicated that solutions generated when Tsetline based OMA representation is used lower transmission cost.

In (Nezhad et al. 2011; Bansal and Kaur 2012), two new algorithms based on GALA for task graph scheduling problems have been reported. The algorithm proposed in (Nezhad et al. 2011), unlike other GALA based algorithm which maps the chromosomes to the OMAs at the beginning of the algorithm, it starts with an initial random population and then after several generations, mapping of chromosome to OMAs are done. The algorithms are compared with modified critical path (MCP), dominant sequence clustering (DSC) (Yang and Gerasoulis 1994), mobility directed (MD), dynamic critical path (DCP), PMC_GA, first come-first serve (FCFS), and minimum execution time (MET) methods (Hwang et al. 2008; Kwok and Ahmad 1996; Wu and Gajski 1990; Yang and Gerasoulis 1994). The results of comparisons indicated that both algorithms perform better than the existing algorithms.

5.4 The New Memetic Model Based on Fixed Structure Learning Automata

This section introduces a new memetic model based on fixed structure learning automata called OMA-MA that behaves according to LA-MA. The genetic section performs the global search and consists of a population of chromosomes with size n, mutation operator, crossover operator, and a chromosome's fitness function. The memetic section consists of a meme that corresponds to a local search method. The meme saves the effect (history) of its corresponding local search method by an OMA. In the OMA-based representation, n actions are corresponding to n genes in each chromosome. Furthermore, a fixed number of states N keep information about each action's local search process's history. For simplicity, we can use the GALA-based representation by combining genetic information and memetic information. In this representation, each state has two attributes: the genes' values (allele) and the degree of association between genes and their allele. As migratory objects in the automata, the genes' values are assigned to states of the corresponding action. The hybrid fitness function is computed using the local search history kept in the OMA states and the chromosome's fitness. In OMA-MA, the local search history is used as memetic fitness, and chromosome fitness is used as genetic fitness. OMA-MA uses all the information in the genetic section and memetic section to compute the hybrid fitness function. The relationship between the genetic section and the memetic section in OMA-MA is shown in Fig. 5.17.

5.4.1 Hybrid Fitness Function

The hybrid fitness function in OMA-MA is not only dependent on genotype information but also on phenotype information. We use the hybrid fitness function $f'(CR) = \sum_{i=1}^{n} f_i(1+\gamma_i)$ for maximization problems, and $f'(CR) = \sum_{i=1}^{n} f_i(1-\gamma_i)$ for minimization problems in the selection of chromosomes. In these functions, f_i is the fitness of the i_{th} gene, which is referred to as genetic fitness, and γ_i is the degree of association between action α_i and its assigned object is referred to as memetic fitness.

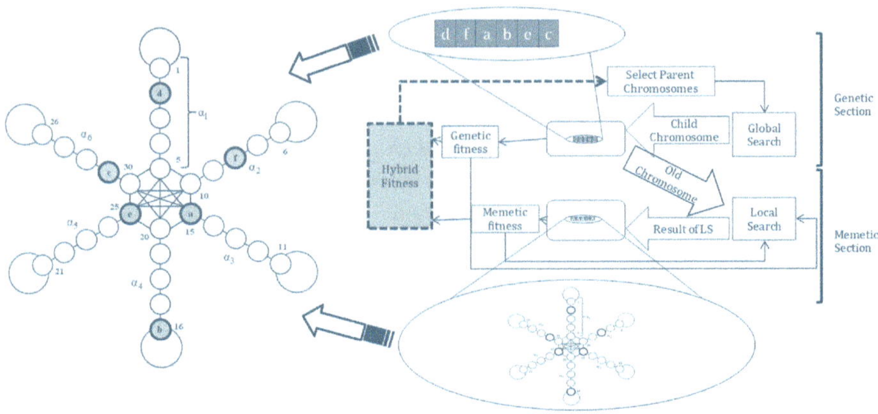

Fig. 5.17 The relationship between the genetic section and memetic section in OMA-MA

The depth of memory is N, so we have $\frac{1}{N} \leq \gamma_i \leq 1$. The parent chromosomes and the next generation's chromosomes are selected based on the defined hybrid fitness function using a tournament mechanism.

5.4.2 Mutation Operators

Whether the state of the selected actions changes or not, we define two types of mutation operators in OMA-MA. In the first type, the states of the selected actions remain unchanged (i.e., the degree of association between actions and their assigned objects are saved). In the second class, the states of the selected actions are changed (i.e., the degree of association between actions and their assigned objects are lost).

OMA-MA has three mutation operators defined in this section: SS-Mutation, XS-Mutation, and LS-Mutation. Specifically:

- The mutation operator in which the previous state of selected actions can be saved is referred to as the SS-Mutation.
- The mutation operator in which the previous state of selected actions can be exchanged is referred to as the XS-Mutation.
- The mutation operator in which the previous state of selected actions can be lost is referred to as the LS-Mutation.

The SS-Mutation and XS-Mutation are examples of the first type of mutation operator described above, and the LS-Mutation is an example of the second type. These mutation operators are described in more detail below.

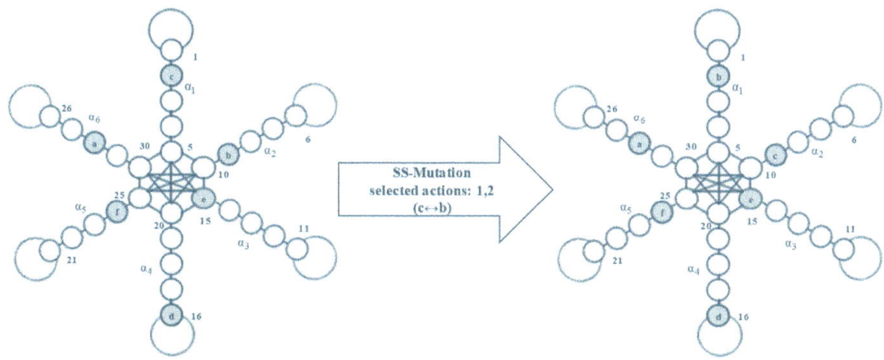

Fig. 5.18 An example of an SS-Mutation

Algorithm 5-6. SS-Mutation

Procedure SS-Mutation (*CR*)
    Generate two random numbers r_1 and r_2 in *[1, n]* where $r_1 < r_2$;
    Swap(*CR.Object(CR.Action(r₁)), CR.Object(CR.Action(r₂))*);
End SS-Mutation

Fig. 5.19 Pseudocode for the SS-Mutation

5.4.2.1 SS-Mutation

Assuming actions 1 and 2 are the selected actions; the SS-Mutation exchanges the objects assigned to the states of actions 1 and 2. Figure 5.18 shows an example of an SS-Mutation.

Pseudocode for our SS-Mutation is shown in Fig. 5.19.

5.4.2.2 XS-Mutation

The XS-Mutation is the same as the mutation that GALA uses. The states of the actions, along with their assigned objects, are exchanged. Figure 5.20 shows an example of an XS-Mutation. Assuming actions 1 and 2 are the selected actions, the XS-Mutation exchanges the object and the state of action 1 with those of action 2.

Pseudocode for our XS-Mutation is shown in Fig. 5.21.

5.4.2.3 LS-Mutation

The LS-Mutation is the same as that used in GALA, except that each action's state is changed to become its corresponding boundary state. Figure 5.22 shows an example of an LS-Mutation operator. Assuming that actions 1 and 2 are the selected actions, the LS-Mutation operator causes 1) The object assigned to the state of action 1 is

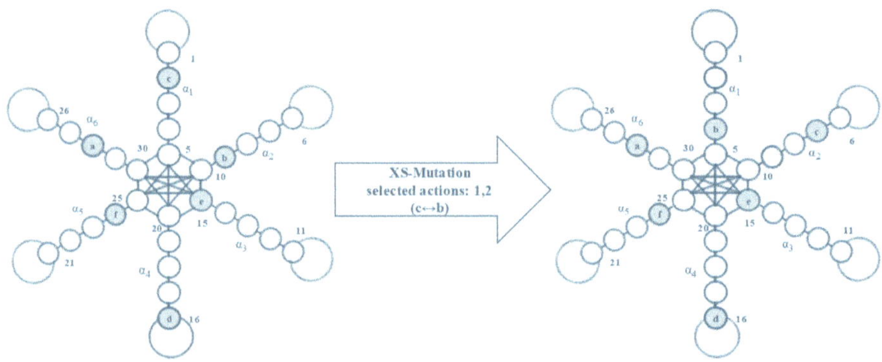

Fig. 5.20 An example of an XS-Mutation

```
Algorithm 5-7. XS-Mutation
Procedure XS-Mutation (CR)
    Generate two random numbers r1 and r2 in [1,n] where r1 < r2;
    Swap(CR.Object(CR.Action(r1)), CR.Object(CR.Action(r2)));
    Swap(CR.State(CR.Action(r1)), CR.State(CR.Action(r2)));
End XS-Mutation
```

Fig. 5.21 Pseudocode for the XS-Mutation

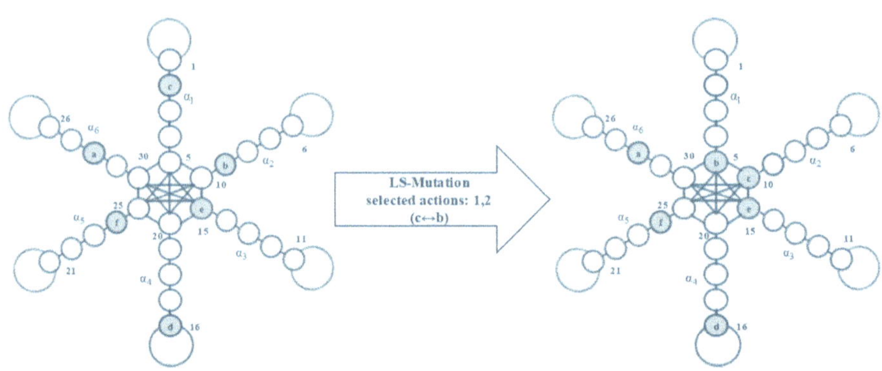

Fig. 5.22 An example of an LS-Mutation

exchanged with the assigned object of the action; and 2) The state of each action changes to become its corresponding boundary state.

Pseudocode for our LS-Mutation operator is given in Fig. 5.23.

Algorithm 5-8. LS-Mutation

```
Procedure LS-Mutation (CR)
    Generate two random numbers r1 and r2 in [1,n] where r1 < r2;
    Swap(CR.Object(CR.Action(r1)), CR.Object(CR.Action(r2)));
    CR.State(CR.Action(r1)) = CR.Action(r1)*N; //the state of action changed to the boundary state
    CR.State(CR.Action(r2)) = CR.Action(r2)*N; //the state of action changed to the boundary state
End LS-Mutation
```

Fig. 5.23 Pseudocode for the LS-Mutation

5.4.3 Crossover Operators

Similar to the mutation operators, we define two different types of crossover operators for OMA-MA. In the first type, the states of selected actions remain unchanged, and in the second type, the states of the actions change to the boundary states.

Three crossover operators, SS-Crossover, XS-Crossover, and LS-Crossover, are defined in OMA-MA:

- The crossover operator in which the previous state of selected actions can be saved is referred to as the SS-Crossover.
- The crossover operator in which the previous state of selected actions can be exchanged is referred to as the XS-Crossover.
- The crossover operator in which the previous state of selected actions can be lost is referred to as the LS-Crossover.

The SS-Crossover and XS-Crossover are examples of crossover operators of the first type, and the LS-Crossover is an example of the second type. These crossover operators are described in further detail below.

5.4.3.1 SS-Crossover

Figure 5.24 shows an example of an SS-Crossover. Assuming actions 2 and 4 are selected randomly from the parent chromosomes, the SS-Crossover exchanges the assigned object of each action in the range of [2, 4] of CR_1, with the assigned object of the same action in the range of [2, 4] of CR_2. Note that in the SS-Crossover the states of the actions remain unchanged.

Pseudocode for the SS-Crossover is shown in Fig. 5.25.

5.4.3.2 XS-Crossover

The XS-Crossover is the same as the crossover used in GALA. Figure 5.26 shows an example of an XS-Crossover. Assuming actions 2 and 4 are selected randomly from the parent chromosomes, the XS-Crossover exchanges both the assigned object and

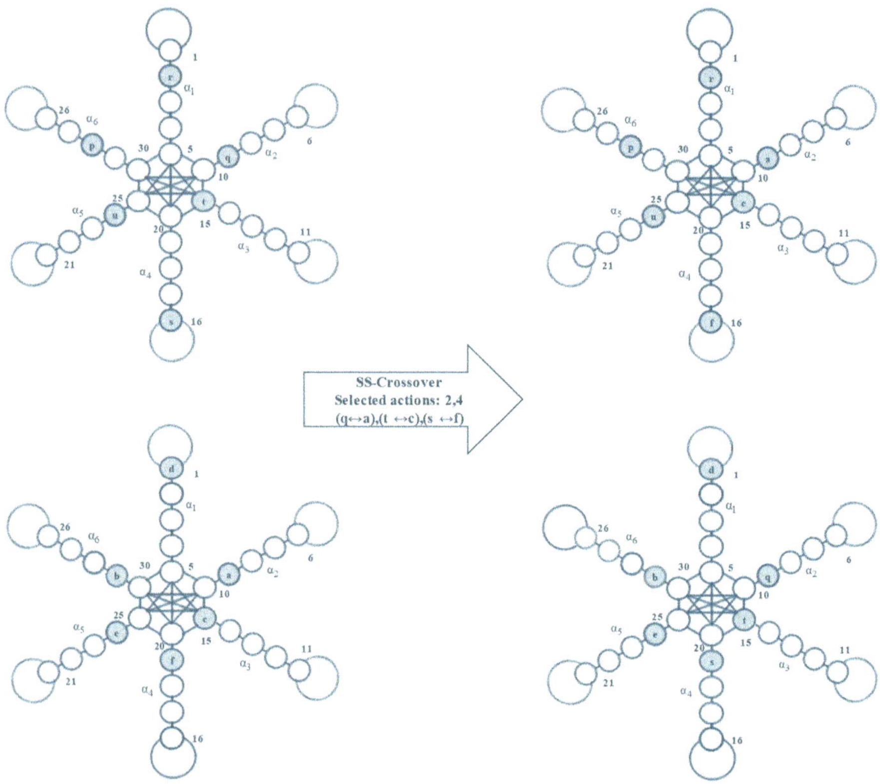

Fig. 5.24 An example of an SS-Crossover

Algorithm 5-9. SS-Crossover

Procedure SS-Crossover (*CR₁*, *CR₂*)
  Generate two random numbers *r₁* and *r₂* in *[1,n]* where *r₁ < r₂*;
  For i = *r₁* to *r₂* **do**
    Swap(*CR₁.Object(CR₁.Action(i)), CR₂.Object(CR₂.Action(i))*);
  End For
End SS-Crossover

Fig. 5.25 Pseudocode for the SS-Crossover

the state of each action in the range of [2, 4] of CR_1, with the assigned object and the state of the same action in the range of [2, 4] of CR_2.

Figure 5.27 shows the pseudo-code for the XS-Crossover.

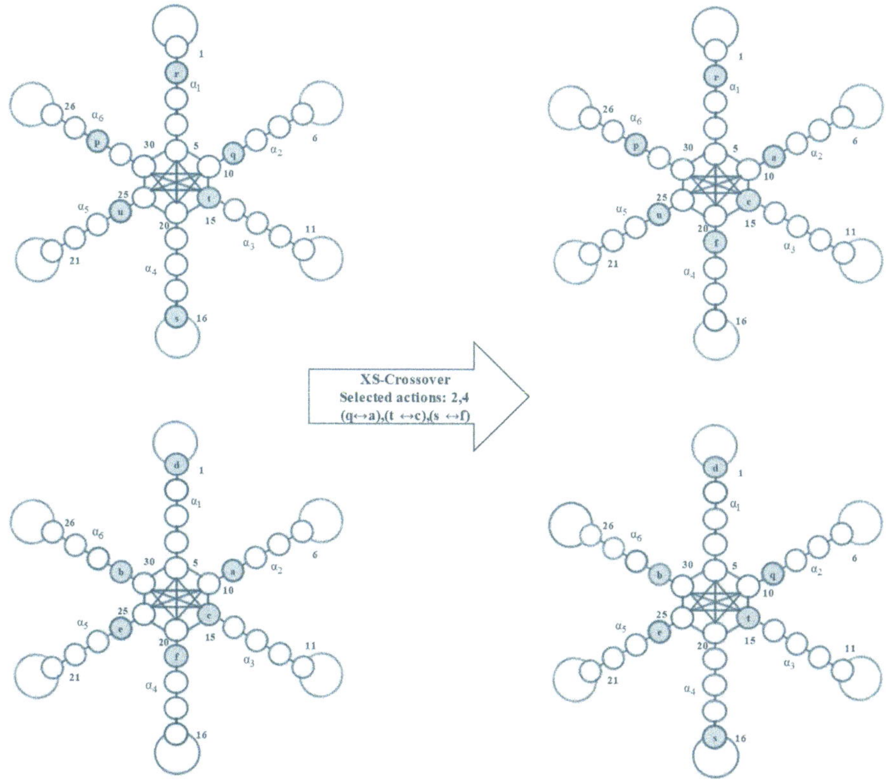

Fig. 5.26 An example of an XS-Crossover

```
Algorithm 5-10. XS-Crossover
Procedure XS-Crossover ( CR1, CR2 )
    Generate two random numbers r1 and r2 in [1,n] where r1 < r2;
    For i = r1 to r2 do
        Swap(CR1.Object(CR1.Action(i)), CR2.Object(CR2.Action(i)));
        Swap(CR1.State(CR1.Action(i)), CR2.State(CR2.Action(i)));
    End For
End XS-Crossover
```

Fig. 5.27 Pseudocode for the XS-Crossover

5.4.3.3 LS-Crossover

The LS-Crossover is the same as the crossover used in GALA, except that each action's state is changed to become its corresponding boundary state. Figure 5.28 shows an example of an LS-Crossover. Assume that actions 2 and 4 are randomly selected in the parent chromosomes, then the LS-Crossover causes: 1) Each object assigned to the actions in the range of [2, 4] of CR_1 is exchanged with the assigned

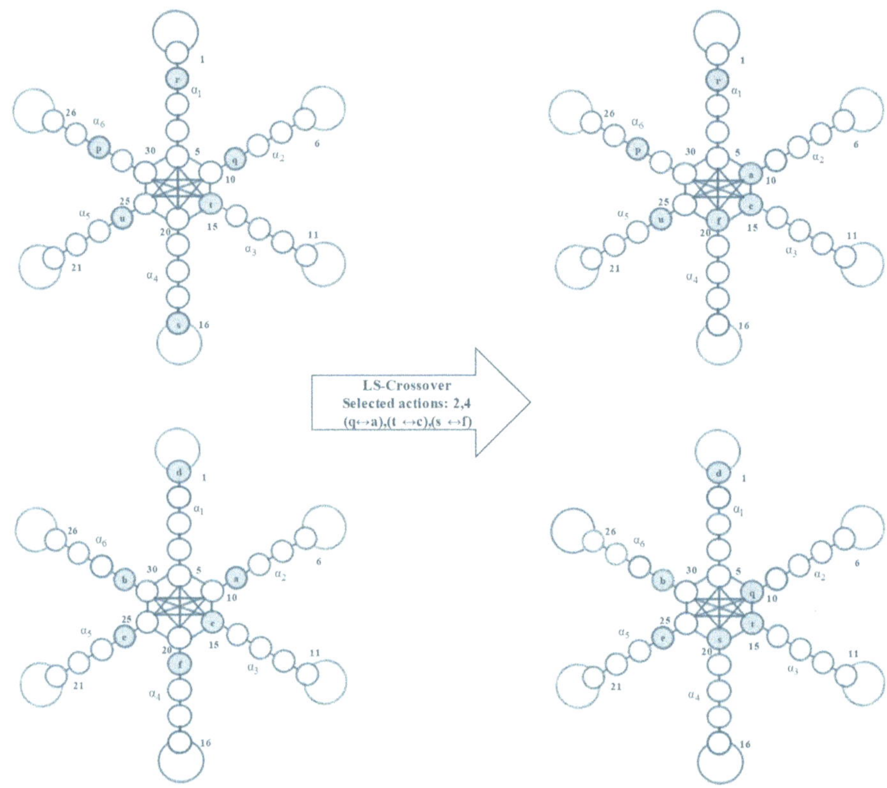

Fig. 5.28 An example of an LS-Crossover

```
Algorithm 5-11. LS-Crossover
Procedure LS-Crossover ( CR1, CR2 )
    Generate two random numbers r1 and r2 in [1,n] where r1 < r2;
    For i = r1 to r2 do
        Swap(CR1.Object(CR1.Action(i)), CR2.Object(CR2.Action(i)));
        CR.State(CR.Action(i)) = CR.Action(i)*N; //the state of action changed to the boundary state
    End For
End LS-Crossover
```

Fig. 5.29 Pseudocode for the LS-Crossover

object of the same action in the range of [2, 4] of CR_2; and 2) The state of each action changes to become its boundary state.

Figure 5.29 shows the pseudo-code of the LS-Crossover.

Depending on whether the memetic information changes or not, we introduce three learning models for each generation.

- Lamarckian learning model (LOMA-MA)
- Baldwinian learning model (BOMA-MA)
- Hybrid (Lamarckian-Baldwinian) learning model (HOMA-MA)

To emphasize the differences between these learning models, we list each of these models' main characteristics.

Main characteristics of LOMA-MA:

1. Mutation and crossover operators perform based on the XS-Mutation and XS-Crossover operators, respectively.
2. At each generation, the memetic information is exchanged between parents and children (i.e., the degree of association between actions and their assigned objects is exchanged).

Main characteristics of BOMA-MA:

1. Mutation and crossover operators perform based on the SS-Mutation and SS-Crossover operators, respectively.
2. The memetic information remains unchanged (i.e., the degree of association between actions and their assigned objects are saved).

Main characteristics of HOMA-MA:

1. Mutation and crossover operators perform based on the LS-Mutation and LS-Crossover operators, respectively.
2. At each generation, the memetic information is changed (i.e., the degree of association between actions and their assigned objects are lost).

5.5 The OneMax Problem

In this section, we have considered solving the OneMax problems in order to justify the approach. For solving the OneMax problem $X = \{x_1, x_2, \ldots, x_n\}$, with $x_i \in \{0, 1\}$ by OMA-MA, we define chromosomes to have n genes, and the value of genes is selected from {0,1} in the genetic section. Also, an OMA with n actions corresponding to n genes in each chromosome is created in the memetic section to save the local search method's effect. Furthermore, a fixed number of states N keep information about each action's local search process's history. Combing the genotype information kept in the genetic section with phenotype information kept in the memetic section produces a GALA-based representation shown in Fig. 5.30. This figure illustrates a representation based on a Tsetline OMA with $N = 5$ for the OneMax problem with 6 bits. The representation in Fig. 5.30 corresponds to the following situation:

- Value 1 is assigned to genes 1, 3, and 6, and value 0 is assigned to genes 2, 4, and 5,
- There are six actions denoted by α_1, α_2, α_3, α_4, α_5, and α_6 corresponding to 6 genes,
- The value of gene 1 is assigned to the internal state 2 of action α_1,
- The value of gene 2 is assigned to the internal state 3 of action α_2,

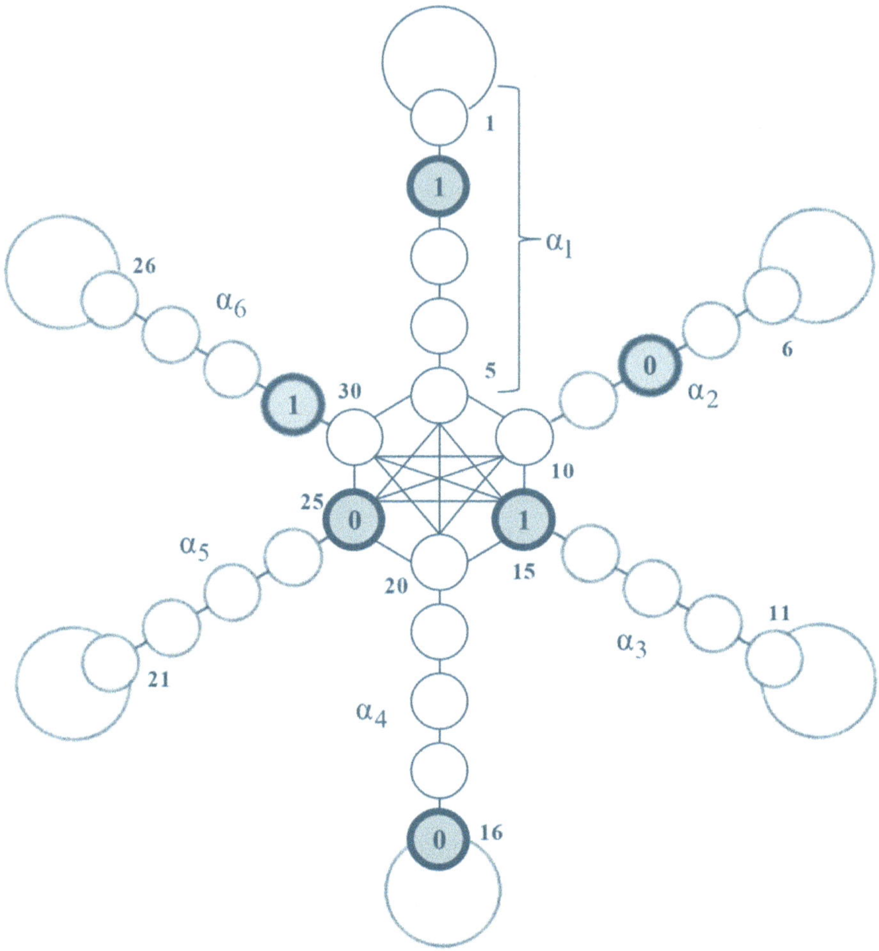

Fig. 5.30 A GALA based representation for the OneMax problem with *6* bits and $N = 5$

- The value of gene 3 is assigned to the boundary state of action α_3,
- The value of gene 4 is assigned to the most internal state of action α_4,
- The value of gene 5 is assigned to the boundary state of action α_5,
- The value of gene 6 is assigned to the internal state 4 of action α_6,
- There is a minimum degree of association between actions α_3 and α_5 and their corresponding values,
- There is a maximum degree of association between actions α_4 and its corresponding value,

5.5.1 Local Search for OneMax

The proposed algorithm for the OneMax problem includes a local search that is applied to all genes in each chromosome. Specifically, for a certain gene k with value s in chromosome X, a temporary chromosome X' is created based on the chromosome X in which the value assigned to gene k is replaced with another value $s' \neq s$ of set $\{0,1\}$.

If the hybrid fitness of a temporary chromosome X' is less than hybrid fitness of current chromosome X, the assignment of the value s to action α_k corresponding to gene k is rewarded. Otherwise, it is penalized. By rewarding an action, the state of that action will move toward the most internal state according to the OMA connection. This causes the degree of association between action, and its corresponding value, to be increased. If the value is located at the most internal state of its action, the state of action remains unchanged.

Penalizing an action causes the degree of association between a value and its corresponding action to be decreased. If a value is not in the boundary state of its action, it is then penalizing causes the action's value to move toward the boundary state. This means that the degree of association between the action and the corresponding value will be decreased. If a value is in the boundary state of its action, then penalizing the action causes the value to be replaced with another value of set $\{0,1\}$, and results in creating a new chromosome.

5.5.2 Experimental Results

To study the efficiency of OMA-MA in solving the OneMax problem, we have conducted two experiments. For both experiments, an initial population of chromosomes of size ten was randomly created, mutation rate and crossover rate are set to the 0.05 and selection mechanism $(\mu + \lambda)$ based on the hybrid fitness using a tournament mechanism of size two, and the depth of memory was 2. The algorithm terminates when all the values are located in the most internal state of their actions in at least a population's chromosome. For both experiments, a Tsetline-based OMA was used for chromosome representation. Each reported result was averaged over 30 runs. OMA-MA was tested with three different mutation operators: SS-Mutation, XS-Mutation, and LS-Mutation. We performed a parametric test (t-test) at the 95% significance level to provide statistical confidence. The t-tests were performed after ensuring that the data followed a normal distribution (by using the Kolmogorov–Smirnov test).

5.5.2.1 Experiment 1

Experiment 1 aimed to find the optimal memory depth for the OMA-MA-based algorithm in different versions of the OneMax problem. For this purpose, we studied the effect of parameter N (depth of memory) on the number of hybrid fitness evaluations. The depth of memory was varied from 2 to 10 by increments of 2. Table 5.1 lists the number of hybrid fitness evaluations and the standard deviation for the different depths of memory employed. From the results, we conclude the following:

- For all versions of the OneMax problem, the minimum value for the number of hybrid fitness evaluations is obtained when $N = 2$.
- For all versions of the OneMax problem, the number of hybrid fitness evaluations increases as memory depth increases.

Table 5.2 shows that, according to the T-test results, the OMA-MA-based algorithm with a memory depth of $N = 2$ performs significantly better than the OMA-MA-based algorithm for $N \neq 2$ for different versions OneMax problem (p-value $<$ 0.05).

5.5.2.2 Experiment 2

In this experiment, we compared the results obtained from OMA-MA based algorithm with the results of two other algorithms, an OMA-based algorithm reported in (Oommen and Ma 1988) and the GALA version of the algorithm, for the OneMax problem, in terms of the number of evaluations required by the algorithm. For this experiment, the depth of memory was set to 2. Table 5.3 presents the different algorithms for three different versions of OneMax concerning the average number of hybrid fitness evaluations and their standard deviation. From the results reported in Table 5.3, we report the following:

- The OMA-MA-based algorithm outperforms both the OMA and GALA algorithms.
- For most cases, OMA-MA based algorithm with the SS-Mutation performs the best.
- The OMA algorithm displays the worst performance compared with the GALA and OMA-MA-based algorithms.
- The difference between the OMA performance and the other algorithms' performance is statistically significant (p-value < 0.05) in most cases.

Table 5.1 Performance of OMA-MA based algorithm concerning the depth of memory

	Algorithm	Length of chromosome	Number of hybrid fitness evaluations	Depth of memory (*N*)				
				2	4	6	8	10
OMA-MA	LS-Mutation	64	Std.	5.23E+01	1.45E+01	1.95E+01	3.52E+01	3.61E+01
			Avg.	889	1791	2769	3769	4689
		32	Std.	3.50E+00	2.16E+01	2.26E+01	3.67E+01	5.24E+01
			Avg.	552	1253	1765	2356	2896
		16	Std.	3.70E+00	1.43E+01	1.94E+01	2.54E+01	3.64E+01
			Avg.	123	263	389	496	652
	XS-Mutation	64	Std.	6.54E+01	2.63E+01	3.54E+01	3.68E+01	4.56E+01
			Avg.	898	1896	2569	3367	4658
		32	Std.	2.53E+01	2.46E+01	3.65E+01	3.94E+01	4.53E+01
			Avg.	655	1263	1986	2245	3125
		16	Std.	5.60E+00	1.73E+01	2.43E+01	2.36E+01	3.51E+01
			Avg.	125	268	358	512	689
	SS-Mutation	64	Std.	4.23E+01	2.18E+01	3.22E+01	1.25E+01	2.44E+02
			Avg.	864	1569	2569	3201	4236
		32	Std.	1.43E+01	2.32E+01	1.86E+01	1.76E+01	1.94E+01
			Avg.	548	1125	1652	2235	2765
		16	Std.	3.40E+00	1.46E+01	2.53E+01	1.86E+01	2.64E+01
			Avg.	121	256	346	442	635

Table 5.2 The results of statistical test for OMA-MA based algorithm with the depth of memory $N = 2$ vs. OMA-AM based algorithm with other depths of memory

		LS-Mutation			XS-Mutation			SS-Mutation		
Length of chromosome		64	32	16	64	32	16	64	32	16
Algorithm	*N*	P value	P value	P value	P value	P value	P value	P value	P value	P value
OMA-MA	4	3.43E−37	1.92E−45	3.66E−30	3.51E−35	1.21E−37	7.68E−28	9.47E−36	3.11E−40	1.59E−29
	6	4.51E−46	8.67E−52	1.48E−34	5.56E−41	1.33E−44	5.51E−30	1.86E−45	2.79E−50	2.94E−29
	8	6.57E−50	8.96E−51	1.65E−35	8.89E−46	3.56E−46	1.11E−36	8.94E−52	4.77E−56	1.85E−37
	10	2.68E−53	1.29E−49	1.91E−35	2.61E−50	1.99E−50	1.30E−36	1.03E−34	1.01E−58	4.47E−39

Table 5.3 Comparison of the number of fitness evaluations and results of statistical test for different algorithms

Length of chromosome	OMA-MA									GALA			OMA	
	LS-Mutation			XS-Mutation			SS-Mutation							
	P-value	Std.	Avg.	P-value	Std.	Avg.	P-value	Std.	Avg.	P-value	Std.	Avg.	Std.	Avg.
16	1.00E−13	3.70E+00	123	1.76E−12	5.60E+00	125	1.77E−14	3.40E+00	121	1.10E−03	5.60E+00	145	1.24E+01	154
32	1.70E−30	3.50E+00	552	2.37E−17	2.53E+01	655	1.26E−28	1.43E+01	548	3.00E−05	2.53E+01	735	2.16E+01	765
64	7.67E−07	5.23E+01	889	1.70E−05	6.54E+01	898	2.35E−09	4.23E+01	864	2.04E−01	6.54E+01	963	6.56E+01	985

5.6 Conclusion

In this chapter, a new memetic model called OMA-MA is introduced for designing an algorithm for optimization purposes. OMA-MA, a fixed structure learning automata-based model, is obtained from a combination of a memetic algorithm and learning automata, in which the LA plays the role of providing the local search. According to the relationship between the genetic section and memetic section, three versions of OMA-MA, called LOMA-MA, BOMA-MA, and HOMA-MA were also introduced.

References

Asghari, K., Meybodi, M.R.: Searching for Hamiltonian cycles in graphs using evolutionary methods. Iran Joint Congress on Fuzzy Intell. Syst. (2008)

Asghari, K., Meybodi, M.R.: Solving single machine total weighted tardiness scheduling problem using learning automata and combination of it with genetic algorithm. In: Proceedings of the 3rd Iran Data Mining Conference. Tehran University of Science and Technology. Iran (2009)

Asghari, K, Safari, M.A., Mahmoudi, F.: A relational databases query optimization using hybrid evolutionary algorithm (2001)

Asghari, K., Mamaghani, A.S., Meybodi, M.R.: An evolutionary approach for query optimization problem in database. In: Proceeding of International Joint Conference on Computers, Information and System Sciences, and Engineering (CISSE2007), University of Bridgeport, England (2007)

Bansal, A., Kaur, R.: Task graph scheduling on multiprocessor system using genetic algorithm. Int. J. Eng. Res. Technol. (IJERT) (2012), ISSN 2278-0181

Beasley, J.: OR-library: distributing test problems by electronic mail. J. Oper. Res. **41**, 1069–1072 (1990)

Beigy, H., Meybodi, M.R.: A learning automata-based algorithm for determination of the number of hidden units for three-layer neural networks. Int. J. Syst. Sci. **40**, 101–118 (2009)

Chen, Y.-F., Gansner, E.R., Koutsofios, E.: A C++ data model supporting reachability analysis and dead code detection. IEEE Trans. Software Eng. **24**, 682–694 (1998)

Chen, X., Ong, Y.-S., Lim, M.-H., Tan, K.C.: A multi-facet survey on memetic computation. IEEE Trans. Evol. Comput. **15**, 591–607 (2011)

Daliri Khomami, M.M., Rezvanian, A., Saghiri, A.M., Meybodi, M.R.: Utilizing cellular learning automata for finding communities in weighted networks. In: 2020 6th International Conference on Web Research (ICWR), pp. 325–329 (2020a)

Daliri Khomami, M.M., Rezvanian, A., Saghiri, A.M., Meybodi, M.R.: SIG-CLA: a significant community detection based on cellular learning automata. In: 2020 8th Iranian Joint Congress on Fuzzy and intelligent Systems (CFIS), pp. 039–044 (2020b)

De Jong, K.A.: Analysis of the behavior of a class of genetic adaptive systems (1975)

Downing, K.L.: Reinforced genetic programming. Genet. Program Evolvable Mach. **2**, 259–288 (2001)

Grosso, A., Della Croce, F., Tadei, R.: An enhanced dynasearch neighborhood for the single-machine total weighted tardiness scheduling problem. Oper. Res. Lett. **32**, 68–72 (2004)

Hasanzadeh Mofrad, M., Sadeghi, S., Rezvanian, A., Meybodi, M.R.: Cellular edge detection: combining cellular automata and cellular learning automata. AEU – Int. J. Electron. Commun. **69**, 1282–1290 (2015). https://doi.org/10.1016/j.aeue.2015.05.010

Huegler, P.A., Vasko, F.J.: A performance comparison of heuristics for the total weighted tardiness problem. Comput. Ind. Eng. **32**, 753–767 (1997)

Hwang, R., Gen, M., Katayama, H.: A comparison of multiprocessor task scheduling algorithms with communication costs. Comput. Oper. Res. **35**, 976–993 (2008)

Isazadeh, A., Izadkhah, H., Mokarram, A.H.: A learning based evolutionary approach for minimization of matrix bandwidth problem. Appl. Math. **6**, 51–57 (2012)

Khomami, M.M.D., Rezvanian, A., Saghiri, A.M., Meybodi, M.R.: Distributed learning automata-based algorithm for finding K-Clique in complex social networks. In: 2020 11th International Conference on Information and Knowledge Technology (IKT), pp. 139–143. IEEE (2020a)

Khomami, M.M.D., Rezvanian, A., Saghiri, A.M., Meybodi, M.R.: Overlapping community detection in social networks using cellular learning automata. In: 2020 28th Iranian Conference on Electrical Engineering (ICEE), pp. 1–6. IEEE (2020b)

Khomami, M.M.D., Rezvanian, A., Meybodi, M.R., Bagheri, A.: CFIN: a community-based algorithm for finding influential nodes in complex social networks. J. Supercomp. 2207–2236 (2021). https://doi.org/10.1007/s11227-020-03355-2

Krasnogor, N., Smith, J.: A tutorial for competent memetic algorithms: model, taxonomy, and design issues. IEEE Trans. Evol. Comput. **9**, 474–488 (2005)

Ku, K.W., Mak, M-.W.: Empirical analysis of the factors that affect the Baldwin effect. In: International Conference on Parallel Problem Solving from Nature, pp. 481–490. Springer (1998)

Kwok, Y.-K., Ahmad, I.: Dynamic critical-path scheduling: an effective technique for allocating task graphs to multiprocessors. IEEE Trans. Parallel Distrib. Syst. **7**, 506–521 (1996)

Madureira, A., Ramos, C., do Carmo Silva, S.: A GA based scheduling system for dynamic single machine problem. In: Proceedings of the 2001 IEEE International Symposium on Assembly and Task Planning (ISATP2001). Assembly and Disassembly in the Twenty-first Century.(Cat. No. 01TH8560), pp. 262–267. IEEE (2001)

Mahmoudi, F., Razmkhah, M., Oommen, B.J.: Nonparametric "anti-Bayesian" quantile-based pattern classification. Pattern Anal. Appl. **24**, 75–87 (2021)

Mamaghani, A.S., Meybodi, M.R.: Hybrid algorithms (learning automata+ genetic algorithm) for solving graph bandwidth minimization problem. In: Proceedings of the second Joint Congress on Fuzzy and Intelligent Systems (2008a)

Mamaghani, A.S., Meybodi, M.R.: Hybrid evolutionary algorithms for solving software clustering problem. In: Proceedings of the second Joint Congress on Fuzzy and Intelligent Systems (2008b)

Mamaghani, A.S., Meybodi, M.R.: Clustering of software systems using new hybrid algorithms. In: 2009 Ninth IEEE International Conference on Computer and Information Technology, pp. 20–25. IEEE (2009)

Mamaghani, A.S., Meybodi, M.R.: A learning automaton based approach to solve the graph bandwidth minimization problem. In: 2011 5th International Conference on Application of Information and Communication Technologies (AICT), pp. 1–5. IEEE (2011)

Mamaghani, A.S., Asghari, K., Mahmoudi, F., Meybodi, M.R.: A novel hybrid algorithm for join ordering problem in database queries. In: Proceedings of the 6th WSEAS International Conference on Computational Intelligence, Man-Machine Systems and Cybernetics, Tenerife. Citeseer (2007)

Mamaghani, A.S., Mahi, M., Meybodi, M.R.: A learning automaton based approach for data fragments allocation in distributed database systems. In: 2010 10th IEEE International Conference on Computer and Information Technology, pp. 8–12. IEEE (2010a)

Mamaghani, A.S., Mahi, M., Meybodi, M.R., Moghaddam, M.H.: A novel evolutionary algorithm for solving static data allocation problem in distributed database systems. In: 2010 Second International Conference on Network Applications, Protocols and Services, pp. 14–19. IEEE (2010b)

Mirsaleh, M.R., Meybodi, M.R.: A Michigan memetic algorithm for solving the community detection problem in complex network. Neurocomputing **214**, 535–545 (2016)

Morris, G.M., Goodsell, D.S., Halliday, R.S., Huey, R., Hart, W.E., Belew, R.K., Olson, A.J.: Automated docking using a Lamarckian genetic algorithm and an empirical binding free energy function. J. Comput. Chem. **19**, 1639–1662 (1998)

Nezhad, V.M., Gader, H.M., Efimov, E.: A novel hybrid algorithm for task graph scheduling. IJCSI 32 (2011)

Oommen, B.J., Ma, D.C.Y.: Deterministic learning automata solutions to the equipartitioning problem. IEEE Trans. Comput. **37**, 2–13 (1988)

Rezapoor Mirsaleh, M., Meybodi, M.R.: A hybrid algorithm for solving graph isomorphism problem. In: Proceedings of the Second International Conference on Information and Knowledge Technology (IKT2005), Tehran, Iran (2005)
Rezapoor Mirsaleh, M., Meybodi, M.R.: Improving GA+ LA algorithm for solving graph isomorphic problem. In: 11th Annual CSI Computer Conference of Iran, Tehran, Iran, pp. 474–483 (2006)
Rezapour, M.: Solving Graph Isomorphism problem Using Learning Automata. PhD Thesis, M. Sc. Thesis, Amirkabir University of technology, Tehran, Iran (2004)
Rezvanian, A., Meybodi, M.R.: Sampling algorithms for stochastic graphs: a learning automata approach. Knowl.-Based Syst. **127**, 126–144 (2017). https://doi.org/10.1016/j.knosys.2017.04.012
Rezvanian, A., Saghiri, A.M., Vahidipour, S.M., Esnaashari, M., Meybodi, M.R.: Recent advances in learning automata. Springer (2018a)
Rezvanian, A., Saghiri, A.M., Vahidipour, S.M., Esnaashari, M., Meybodi, M.R.: Learning automata theory. In: Recent Advances in Learning Automata, pp. 3–19. Springer (2018b)
Rezvanian, A., Saghiri, A.M., Vahidipour, S.M., Esnaashari, M., Meybodi, M.R.: Learning automata for cognitive peer-to-peer networks. In: Recent Advances in Learning Automata, pp. 221–278 (2018c)
Rezvanian, A., Moradabadi, B., Ghavipour, M., Daliri Khomami, M.M., Meybodi, M.R.: Introduction to learning automata models. In: Learning Automata Approach for Social Networks, pp. 1–49. Springer (2019a)
Rezvanian, A., Moradabadi, B., Ghavipour, M., Daliri Khomami, M.M., Meybodi, M.R.: Social recommender systems. In: Learning Automata Approach for Social Networks, pp. 281–313. Springer (2019b)
Rezvanian, A., Moradabadi, B., Ghavipour, M., Khomami, M.M.D., Meybodi, M.R.: Social network sampling. In: Learning Automata Approach for Social Networks, pp. 91–149. Springer (2019c)
Safari Mamaghani, A., Asghari, K., Meybodi, M.R., Mahmoodi, F.: A new method based on genetic algorithm for minimizing join operations cost in database. In: 13th Annual CSI Computer Conference of Iran, Kish Island, Iran (2008)
Saghiri, A.M., Vahdati, M., Gholizadeh, K., Meybodi, M.R., Dehghan, M., Rashidi, H.: A framework for cognitive Internet of Things based on blockchain. In: 2018 4th International Conference on Web Research (ICWR), pp. 138–143. IEEE (2018)
Soleimani-Pouri, M., Rezvanian, A., Meybodi, M.R.: Distributed learning automata based algorithm for solving maximum clique problem in stochastic graphs. Int. J. Comput. Inf. Syst. Ind. Manage. Appl. **6**, 484–493 (2014)
Torkestani, J.A.: An adaptive focused web crawling algorithm based on learning automata. Appl. Intell. **37**, 586–601 (2012)
Torkestani, J.A., Meybodi, M.R.: An efficient cluster-based CDMA/TDMA scheme for wireless mobile ad-hoc networks: a learning automata approach. J. Netw. Comput. Appl. **33**, 477–490 (2010)
Vafashoar, R., Morshedlou, H., Rezvanian, A., Meybodi, M.R.: Applications of cellular learning automata and reinforcement learning in global optimization. In: Cellular Learning Automata: Theory and Applications, pp. 157–224. Springer (2021a)
Vafashoar, R., Morshedlou, H., Rezvanian, A., Meybodi, M.R.: Applications of multi-reinforcement cellular learning automata in channel assignment. In: Cellular Learning Automata: Theory and Applications, pp. 225–254. Springer (2021b)
Vafashoar, R., Morshedlou, H., Rezvanian, A., Meybodi, M.R.: Cellular learning automata for competitive loss sharing. In: Cellular Learning Automata: Theory and Applications, pp. 285–333. Springer (2021c)
Vafashoar, R., Morshedlou, H., Rezvanian, A., Meybodi, M.R.: Cellular learning automata for collaborative loss sharing. In: Cellular Learning Automata: Theory and Applications, pp. 255–284. Springer (2021d)

Vahidipour, S.M., Meybodi, M.R., Esnaashari, M.: Learning automata-based adaptive Petri net and its application to priority assignment in queuing systems with unknown parameters. IEEE Trans. Syst. Man, Cybern. Syst. **45**, 1373–1384 (2015)

Vahidipour, S.M., Esnaashari, M., Rezvanian, A., Meybodi, M.R.: GAPN-LA: a framework for solving graph problems using Petri nets and learning automata. Eng. Appl. Artif. Intell. **77**, 255–267 (2019). https://doi.org/10.1016/j.engappai.2018.10.013

Wang, Y.-K., Fan, K.-C., Horng, J.-T.: Genetic-based search for error-correcting graph isomorphism. IEEE Trans. Syst. Man Cybern. Part B (cybernetics) **27**, 588–597 (1997)

Weber, M., Neri, F., Tirronen, V.: Distributed differential evolution with explorative–exploitative population families. Genet. Program Evolvable Mach. **10**, 343 (2009)

Wu, M.-Y., Gajski, D.D.: HyperTool: a programming aid for message-passing systems. IEEE Trans. Parallel Distrib. Syst. **1**, 330–343 (1990)

Yang, T., Gerasoulis, A.: DSC: scheduling parallel tasks on an unbounded number of processors. IEEE Trans. Parallel Distrib. Syst. **5**, 951–967 (1994)

Zaree, B., Asghari, K., Meybodi, M.R.: A hybrid method based on clustering for solving large traveling salesman problem. In: 13th Annual CSI Computer Conference of Iran, Kish Island, Iran (2008)

Zaree, B., Meybodi, M.R.: An evolutionary method for solving symmetric TSP. In: Third International Conference on Information and Knowledge Technology (IKT2007), Mashhad, Iran (2007a)

Zaree, B., Meybodi, M.R.: A hybrid method for sorting problem. In: Third International Conference on Information and Knowledge Technology (IKT2007), Mashhad, Iran (2007b)

Chapter 6
The Applications of Object Migration Automaton (OMA)-Memetic Algorithm for Solving NP-Hard Problems

Javidan Kazemi Kordestani, Mehdi Razapoor Mirsaleh, Alireza Rezvanian, and Mohammad Reza Meybodi

Abstract In this chapter, to evaluate the efficiency of the object migration automaton (OMA)-memetic algorithm (OMA-MA), it has been used for designing memetic algorithms to solve the equipartitioning problem (EPP), the graph isomorphism problem (GIP), and the assignment of cells to switches problem in cellular mobile network (ACTSP). The results of computer experimentations have shown that all the OMA-MAs outperform the others in terms of quality of solution and rate of convergence.

6.1 Introduction

In this chapter, to evaluate the efficiency of OMA-MA, it has been used for designing memetic algorithms to solve the equipartitioning problem (EPP), the graph isomorphism problem (GIP), and the assignment of cells to switches problem in cellular mobile network (ACTSP). The results of computer experimentations have shown that all the proposed algorithms outperform the others in terms of quality of solution and rate of convergence.

J. Kazemi Kordestani
Department of Computer Engineering, Science and Research Branch, Islamic Azad University, Tehran, Iran

M. Razapoor Mirsaleh
Department of Computer Engineering and Information Technology, Payame Noor University (PNU), P.O. BOX 19395-3697, Tehran, Iran
e-mail: mrezapoorm@aut.ac.ir

A. Rezvanian (✉)
Department of Computer Engineering, University of Science and Culture, Tehran, Iran
e-mail: rezvanian@usc.ac.ir

M. R. Meybodi
Computer Engineering Department, Amirkabir University of Technology (Tehran Polytechnic), Tehran, Iran
e-mail: mmeybodi@aut.ac.ir

J. Kazemi Kordestani et al. (eds.), *Advances in Learning Automata and Intelligent Optimization*, Intelligent Systems Reference Library 208, https://doi.org/10.1007/978-3-030-76291-9_6

6.2 The Equipartitioning Problem

Let $\underline{A} = \{A_1, \ldots, A_W\}$ be a set of W objects. We want to partition $\underline{A}$ into R classes $\{P_1, \ldots, P_R\}$ such that the objects used more frequently are located together in the same class. We assume that the joint access probabilities of the objects are unknown. This problem is called the object partitioning problem. A special case of the object partitioning problem referred to as the equal partitioning problem (EPP), is where the objects are equipartitioned. In an EPP, each class has exactly $M = W/R$ objects. For solving the EPP with OMA-MA, we define a chromosome to have W genes (actions), and the value of the genes are selected from a set of classes $\{P_1, \ldots, P_R\}$ (as migratory objects in the OMA) such that each class is assigned to W/R of genes (actions). Objects are initially assigned to the boundary state of actions. Figure 6.1 shows a chromosome based on a Tsetline OMA representation for six objects and two classes (called class α and class β) with $N = 5$. In this figure, objects 1, 3, and 4 are assigned to class α, and objects 2, 5, and 6 are assigned to class β.

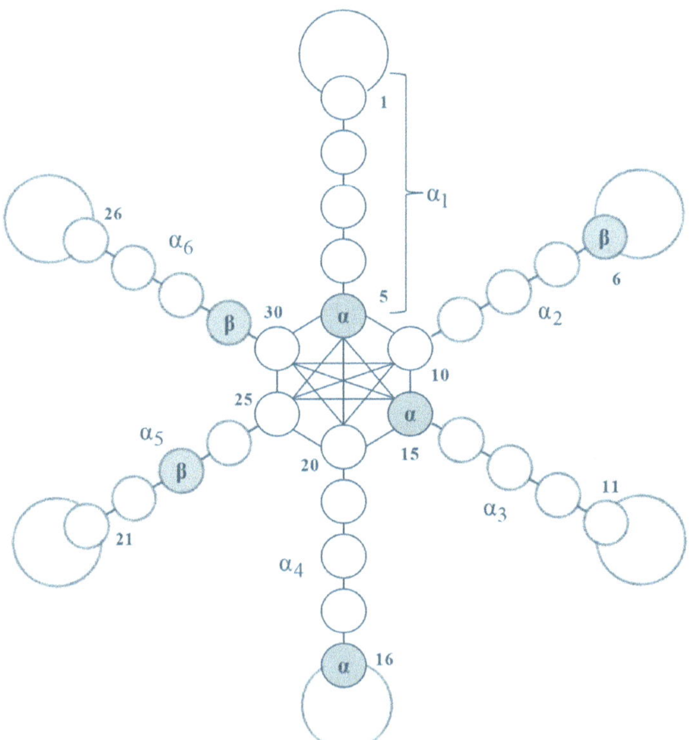

Fig. 6.1 A chromosome representation of the EPP with $W = 6$, $R = 2$, and $N = 5$

Fig. 6.2 Pseudocode for our local search of the EPP

Algorithm 6-1. Local search for EPP

```
Procedure LocalSearch( CR )
    Access query (A_i, A_j) from the pool of queries;
    If ( CR.Object(CR.Action(i)) = CR.Object(CR.Action(j)) )
        Reward(CR, CR.Action(i));
        Reward(CR, CR.Action(j));
    Else
        Penalty(CR, CR.Action(i));
        Penalty(CR, CR.Action(j));
    End if
End LocalSearch;
```

6.2.1 *Local Search for EPP*

Suppose a query (which is a pair of objects (A_i, A_j)) has been accessed. If the assigned objects of actions α_i and α_j are the same, then both actions α_i and α_j are rewarded, and their states change according to OMA connections. If the assigned objects of actions α_i and α_j are different, then they are penalized, and their states change according to OMA connections. The pseudo-code for our local search of the EPP is shown in Fig. 6.2.

6.2.2 *Experimental Results*

We studied the efficiency of our OMA-MA algorithm in solving the EPP by comparing its results to those obtained for an OMA method reported in Oommen and Ma (1988) and to the GALA algorithm. Queries were chosen randomly from a pool of queries for all experiments. The pool of queries was generated in such a way that the sum of probabilities that object A_i in partition π_i is jointly accessed with other objects in partition π_i is p, and with objects in partition $\pi_j (j \neq i)$ is $1 - p$, that is:

$$\sum_{A_j \in \pi_i} Pr\left[A_i, A_j \ accessed\ together\right] = p \tag{6.1}$$

Therefore, if $p = 1$, then queries will only involve objects in the same partition. As the value of p decreases, the queries will become decreasingly informative about the EPP solution (Oommen and Ma 1988). For all experiments, an initial population of chromosomes of size one was randomly created, the size of the chromosome was set equal to the number of objects, the mutation rate was 0.05, the selection mechanism was (1,1), p was 0.9, and the depth of memory was 2. The algorithm terminates when all the objects in the only chromosome are located in the most internal state of their actions. Each reported result was averaged over 30 runs. We performed a parametric test (t-test) and two non-parametric tests (Wilcoxon rank-sum test and permutation test) at the 95% significance level to provide statistical confidence. The t-tests were performed after ensuring that the data followed a normal distribution (by using the Kolmogorov–Smirnov test).

6.2.2.1 Experiment 1

In this experiment, we compared the results obtained from OMA-MA based algorithm with the results of two other algorithms, an OMA-based algorithm reported in Oommen and Ma (1988) and the GALA version of the algorithm, for the EPP, in terms of the number of iterations (number of accessed queries) required by the algorithm. The OMA-MA-based algorithm was tested with three different mutation operators: SS-Mutation, XS-Mutation, and LS-Mutation. Table 6.1 presents the different algorithms for 14 different cases concerning the average number of iterations and their standard deviation. From the results reported in Table 6.1, we report the following:

- The OMA-MA-based algorithm outperforms both the OMA and GALA algorithms.
- For cases ($W = 9$ and $R = 3$), ($W = 12$ and $R = 2$) and ($W = 8$ and $R = 2$), OMA-MA based using the LS-Mutation performs the best, and for the other cases, OMA-MA based with the SS-Mutation displays the best performance.
- The OMA algorithm displays the worst performance compared with the GALA and OMA-MA-based algorithms.
- As the number of classes (R) decreases, the number of iterations required by all algorithms increases. This is because a low value for R means a higher number of objects will be placed in each class, leading to a situation where more actions have the same class number (migratory objects). This causes the probability that a mutation operator swaps two objects between two actions to decrease.
- OMA-MA with the XS-Mutation displays the same performance as GALA. This is because 1) In the OMA-MA based algorithm and GALA algorithms, the selection mechanism is considered to be (1,1), that is, OMA-MA based algorithm, like GALA, has no selection mechanism in this mode; and 2) the XS-Mutation operator used by OMA-MA based algorithm is the same as the mutation operator used by GALA.

Table 6.2 shows the p-values of the two-tailed t-test, the p-values of the two-tailed Wilcoxon rank-sum test, and the p-values of the two-tailed permutation test. From the results reported in Table 6.2, we report the following:

- For all three kinds of statistical tests (Wilcoxon, permutation, and T-test), the difference between the performance of the OMA and the performance of the other algorithms is statistically significant (p-value < 0.05) in most cases.

Table 6.1 Comparison of the number of iterations required by different algorithms

W	R	OMA		GALA		OMA-MA					
						SS-Mutation		XS-Mutation		LS-Mutation	
		Avg.	Std.	Avg.	Std.	Avg.	Std.	Avg.	Std.	Avg.	Std.
4	2	18	8.50E+00	10	1.50E+00	10	2.10E+00	10	1.50E+00	12	2.00E+00
6	3	33	1.63E+01	18	2.20E+00	18	1.80E+00	18	2.20E+00	21	3.50E+00
	2	115	5.63E+01	98	6.30E+00	85	6.50E+00	98	6.30E+00	102	9.80E+00
9	3	209	9.86E+01	194	2.16E+01	186	1.26E+01	194	2.16E+01	180	1.72E+01
12	6	70	2.56E+01	36	6.50E+00	34	6.70E+00	36	6.50E+00	41	1.29E+01
	4	243	5.86E+01	181	1.23E+01	179	9.50E+00	181	1.23E+01	190	2.18E+01
	3	569	1.26E+02	500	2.53E+01	486	2.56E+01	500	2.53E+01	512	3.13E+01
	2	1993	3.65E+02	1884	1.52E+01	1869	8.56E+01	1884	1.52E+01	1823	9.84E+01
15	5	317	9.53E+01	198	1.95E+01	188	1.53E+01	198	1.95E+01	212	6.84E+01
	3	2000	1.56E+02	1902	1.25E+02	1892	8.86E+01	1902	1.25E+02	1911	1.09E+02
18	9	111	3.43E+01	62	6.20E+00	58	9.50E+00	62	6.20E+00	62	1.67E+01
	6	334	1.25E+02	211	1.52E+01	201	1.25E+01	211	1.52E+01	227	5.23E+01
	3	5169	5.69E+02	4443	4.52E+01	3896	9.64E+01	4443	4.52E+01	4563	1.73E+02
	2	10096	2.36E+02	9898	6.85E+01	9865	1.22E+02	9898	6.85E+01	9755	1.32E+02

Table 6.2 The results of statistical tests for the OMA algorithm and other algorithms

W	R	GALA		OMA-MA									
				SS-Mutation				XS-Mutation			LS-Mutation		
		P-value		P-value				P-value			P-value		
		T-Test	Permutation	Wilcoxon	T-Test	Permutation	Wilcoxon	T-Test	Permutation	Wilcoxon	T-Test	Permutation	Wilcoxon
4	2	2.05E−05	0.00E+00	1.49E−04	2.50E−05	0.00E+00	1.78E−04	2.05E−05	0.00E+00	1.17E−04	7.57E−04	0.00E+00	1.77E−03
6	3	2.57E−05	0.00E+00	6.37E−04	2.47E−05	0.00E+00	6.37E−04	2.57E−05	0.00E+00	6.04E−04	4.68E−04	0.00E+00	3.76E−03
	2	1.11E−01	1.10E−01	4.25E−01	7.06E−03	5.00E−03	1.03E−02	1.11E−01	1.18E−01	4.51E−01	2.23E−01	2.21E−01	5.89E−01
9	3	4.22E−01	4.20E−01	5.74E−01	2.15E−01	1.98E−01	3.33E−01	4.22E−01	4.35E−01	5.84E−01	1.23E−01	1.06E−01	2.04E−01
12	6	9.36E−08	0.00E+00	8.35E−08	3.27E−08	0.00E+00	3.35E−08	9.36E−08	0.00E+00	1.31E−07	5.65E−06	0.00E+00	3.83E−06
	4	3.94E−06	0.00E+00	1.17E−05	2.07E−06	0.00E+00	5.46E−06	3.94E−06	0.00E+00	7.74E−06	6.84E−05	0.00E+00	4.49E−05
	3	6.23E−03	1.00E−02	6.67E−03	1.35E−03	0.00E+00	7.30E−04	6.23E−03	6.00E−03	7.79E−03	2.24E−02	1.30E−02	3.64E−02
	2	1.13E−01	9.80E−02	8.11E−02	8.06E−02	7.30E−02	1.01E−01	1.13E−01	9.40E−02	8.37E−02	2.00E−02	2.00E−02	3.58E−02
15	5	2.38E−07	0.00E+00	1.20E−07	4.60E−08	0.00E+00	8.70E−08	2.38E−07	0.00E+00	1.31E−07	3.32E−05	0.00E+00	3.06E−05
	3	1.19E−02	1.00E−03	2.71E−02	2.62E−03	0.00E+00	1.72E−03	1.19E−02	1.90E−02	2.76E−02	1.61E−02	1.50E−02	1.70E−02
18	9	1.72E−08	0.00E+00	3.66E−07	5.40E−09	0.00E+00	1.25E−07	1.72E−08	0.00E+00	4.44E−07	9.75E−08	0.00E+00	3.13E−07
	6	9.93E−06	0.00E+00	1.05E−05	2.88E−06	0.00E+00	2.08E−06	9.93E−06	0.00E+00	8.29E−06	1.68E−04	0.00E+00	9.79E−05
	3	1.15E−07	0.00E+00	8.12E−09	7.55E−13	0.00E+00	1.77E−09	1.15E−07	0.00E+00	8.12E−09	5.01E−06	0.00E+00	2.47E−07
	2	1.29E−04	0.00E+00	1.37E−04	4.86E−05	0.00E+00	3.71E−05	1.29E−04	0.00E+00	1.68E−04	1.38E−07	0.00E+00	2.11E−07

6.2.2.2 Experiment 2

This experiment's goal was to evaluate the accuracy of the solution produced by OMA-MA based algorithm. Before we introduce the concept of accuracy for OMA-MA, we will provide some preliminaries. For this purpose, we use an example of equipartitioning with four objects A_1, A_2, A_3, and A_4, and two classes, α and β. A Tsetline based on an OMA with a depth of memory 2 will be used for the chromosome representation shown in Fig. 6.3.

We also assume that the initial population is of size one and that the only chromosome in the initial population is created randomly and has its migratory objects in the boundary state of the actions. For this example, there are the three possible object equipartitioning schemes specified below:

- $\alpha\alpha\beta\beta$: Objects A_1, A_2 are in class α and objects A_3, A_4 are in class β.
- $\alpha\beta\alpha\beta$: Objects A_1, A_3 are in class α and objects A_2, A_4 are in class β.
- $\alpha\beta\beta\alpha$: Objects A_1, A_4 are in class α and objects A_2, A_3 are in class β.

The chromosome in Fig. 6.3 can also be represented by $\left(\overline{\alpha}\underline{\alpha\beta}\overline{\beta}\right)$ which corresponds to the following situation:

- Object A_1 is in partition α, and the migratory object is located in the boundary state of action 1.
- Object A_2 is in partition α, and the migratory object is located in the internal state of action 2.
- Object A_3 is in partition β, and the migratory object is located in the internal state of action 3.

Fig. 6.3 A representation of the EPP with $W = 4$, $R = 2$, and $N = 2$

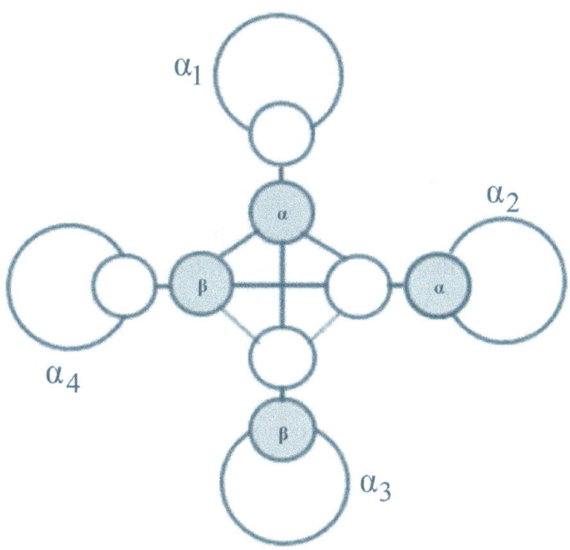

- Object A_4 is in partition β, and the migratory object is located in the boundary state of action 4.

Each possible equipartitioning scheme may correspond to any of the 16 possible chromosomes out of all 48 possible chromosomes. For example, chromosomes $(\overline{\alpha\alpha}\,\overline{\beta\beta})$, $(\overline{\alpha\alpha}\,\overline{\beta}\underline{\beta})$, $(\overline{\alpha\alpha}\,\underline{\beta}\overline{\beta})$, $(\overline{\alpha\alpha}\,\underline{\beta\beta})$, $(\overline{\alpha}\underline{\alpha}\,\overline{\beta\beta})$, $(\overline{\alpha}\underline{\alpha}\,\overline{\beta}\underline{\beta})$, $(\overline{\alpha}\underline{\alpha}\,\underline{\beta}\overline{\beta})$, $(\overline{\alpha}\underline{\alpha}\,\underline{\beta\beta})$, $(\underline{\alpha}\overline{\alpha}\,\overline{\beta\beta})$, $(\underline{\alpha}\overline{\alpha}\,\overline{\beta}\underline{\beta})$, $(\underline{\alpha}\overline{\alpha}\,\underline{\beta}\overline{\beta})$, $(\underline{\alpha}\overline{\alpha}\,\underline{\beta\beta})$, $(\underline{\alpha\alpha}\,\overline{\beta\beta})$, $(\underline{\alpha\alpha}\,\overline{\beta}\underline{\beta})$, $(\underline{\alpha\alpha}\,\underline{\beta}\overline{\beta})$, and $(\underline{\alpha\alpha}\,\underline{\beta\beta})$ show the equipartitioning $\alpha\alpha\beta\beta$. We group chromosomes corresponding to a given equipartitioning into two sets: converged chromosomes set (CCS) and non-converged chromosomes (NCCS). The CCS includes all the chromosomes for which all objects of class α or class β are in the internal states. All other chromosomes are in the NCCS. For example, chromosomes $(\overline{\alpha\alpha}\,\overline{\beta\beta})$, $(\overline{\alpha\alpha}\,\overline{\beta}\underline{\beta})$, $(\overline{\alpha\alpha}\,\underline{\beta}\overline{\beta})$, $(\overline{\alpha}\underline{\alpha}\,\overline{\beta\beta})$, $(\overline{\alpha}\underline{\alpha}\,\overline{\beta}\underline{\beta})$, $(\overline{\alpha}\underline{\alpha}\,\underline{\beta}\overline{\beta})$, $(\underline{\alpha}\overline{\alpha}\,\overline{\beta\beta})$, $(\underline{\alpha}\overline{\alpha}\,\overline{\beta}\underline{\beta})$, and $(\underline{\alpha}\overline{\alpha}\,\underline{\beta}\overline{\beta})$, of equipartitioning $\alpha\alpha\beta\beta$, are in the NCCS of equipartitioning $\alpha\alpha\beta\beta$, and $(\overline{\alpha\alpha}\,\underline{\beta\beta})$, $(\overline{\alpha}\underline{\alpha}\,\underline{\beta\beta})$, $(\underline{\alpha}\overline{\alpha}\,\underline{\beta\beta})$, $(\underline{\alpha\alpha}\,\overline{\beta\beta})$, $(\underline{\alpha\alpha}\,\overline{\beta}\underline{\beta})$, $(\underline{\alpha\alpha}\,\underline{\beta}\overline{\beta})$, and $(\underline{\alpha\alpha}\,\underline{\beta\beta})$ are in the CCS of equipartitioning $\alpha\alpha\beta\beta$. OMA-MA based algorithm converges to one of the following six sets: either the NCCS of equipartitioning $\alpha\alpha\beta\beta$, or the CCS of equipartitioning $\alpha\alpha\beta\beta$, or the NCCS of equipartitioning $\alpha\beta\alpha\beta$, or the CCS of equipartitioning $\alpha\beta\alpha\beta$, or the NCCS of equipartitioning $\alpha\beta\beta\alpha$, or the CCS of equipartitioning $\alpha\beta\beta\alpha$. Let *p1*, *p2*, *p3*, *p4*, *p5*, and *p6* be the current chromosome's probabilities of being in one of the sets mentioned earlier, respectively. Furthermore, assume that eighty percent of the queries accessed from the pool of queries are (A_1, A_2) or (A_3, A_4). For OMA-MA to produce the correct solution, it must generate a population containing one of the following chromosomes: $(\overline{\alpha\alpha}\,\overline{\beta\beta})$, $(\overline{\alpha\alpha}\,\overline{\beta}\underline{\beta})$, $(\overline{\alpha\alpha}\,\underline{\beta}\overline{\beta})$, $(\overline{\alpha\alpha}\,\underline{\beta\beta})$, $(\overline{\alpha}\underline{\alpha}\,\overline{\beta\beta})$, $(\overline{\alpha}\underline{\alpha}\,\overline{\beta}\underline{\beta})$, $(\overline{\alpha}\underline{\alpha}\,\underline{\beta}\overline{\beta})$, $(\overline{\alpha}\underline{\alpha}\,\underline{\beta\beta})$, $(\underline{\alpha}\overline{\alpha}\,\overline{\beta\beta})$, $(\underline{\alpha}\overline{\alpha}\,\overline{\beta}\underline{\beta})$, $(\underline{\alpha}\overline{\alpha}\,\underline{\beta}\overline{\beta})$, $(\underline{\alpha}\overline{\alpha}\,\underline{\beta\beta})$, $(\underline{\alpha\alpha}\,\overline{\beta\beta})$, $(\underline{\alpha\alpha}\,\overline{\beta}\underline{\beta})$, $(\underline{\alpha\alpha}\,\underline{\beta}\overline{\beta})$, and $(\underline{\alpha\alpha}\,\underline{\beta\beta})$. So, OMA-MA must converge to one of the sets of NCCS of equipartitioning, or one of the sets of CCS of equipartitioning $\alpha\alpha\beta\beta$. The algorithm is more accurate if the probability of converging to a set of NCCS of equipartitioning or to a set of CCS of equipartitioning αaββ (that is $p_1 + p_2$) is higher than 0.8. Next, we undertook the experimentation phase of the study. Input queries were generated in such a way that eighty percent of the queries accessed from the pool of queries were (A_1, A_2) or (A_3, A_4). Note that the initial population is considered to be of size one, that the only chromosome in the initial population is created randomly, and that it has its migratory objects at the boundary state of the actions. That is chromosome $(\overline{\alpha\alpha}\,\overline{\beta\beta})$, which belongs to the NCCS of equipartitioning $\alpha\alpha\beta\beta$, or $(\overline{\alpha}\,\overline{\beta}\,\overline{\alpha}\,\overline{\beta})$, which belongs to the NCCS of equipartitioning $\alpha\beta\alpha\beta$, or $(\overline{\alpha}\,\overline{\beta\beta}\,\overline{\alpha})$, which belongs to the NCCS of equipartitioning $\alpha\beta\beta\alpha$ are the only options. Therefore, initial values for p_1, p_3, and p_5 are considered zero, and initial values for p_2, p_4, and p_6 are considered to be 1/3. Figure 6.4(a), (b), and (c) show the evolution of p_1, p_2, p_3, p_4, p_5, and p_6 for the three different mutation operators SS-Mutation, XS-Mutation, and LS-Mutation.

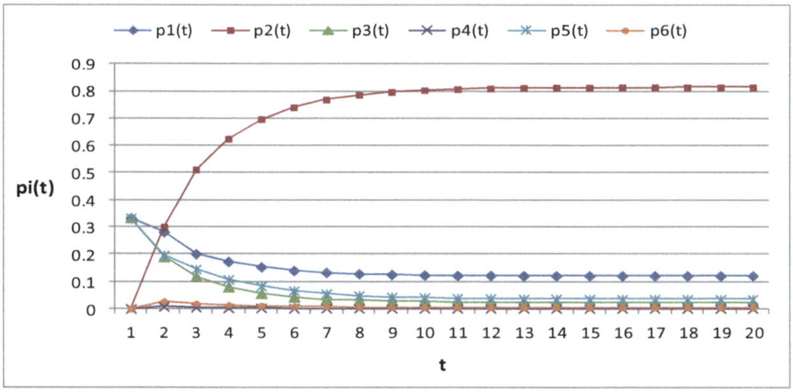

(a)

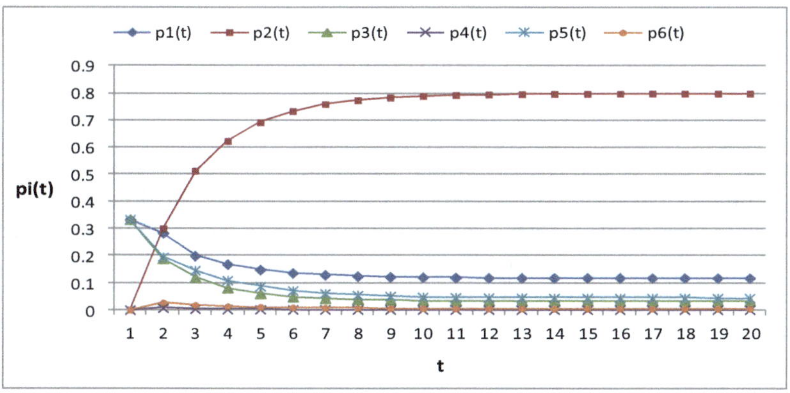

(b)

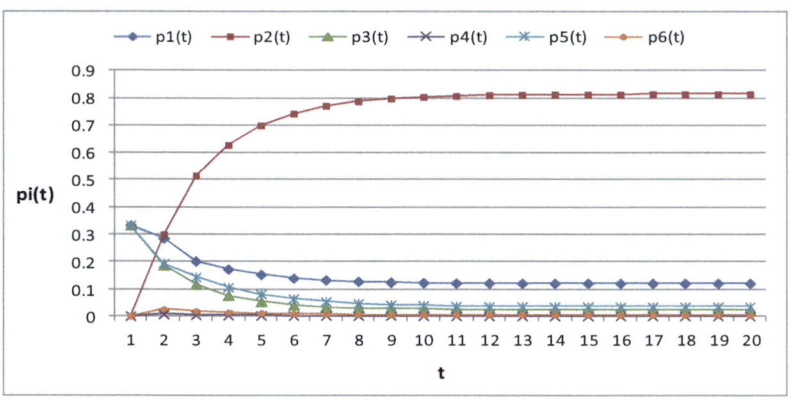

(c)

Fig. 6.4 The evolution of p_1, p_2, p_3, p_4, p_5, and p_6 in OMA-MA based algorithm with $W = 4$, $R = 2$, and $N = 2$, for the SS-Mutation**(a)**, XS-Mutation **(b)**, and LS-Mutation **(c)** operator for the EPP

Table 6.3 Accuracy of OMA-MA concerning the percentage of queries (A_1, A_2) or (A_3, A_4) and mutation operators

The percentage of queries (A_1, A_2) or (A_3, A_4)	SS-Mutation		XS-Mutation		LS-Mutation	
	Avg.	Std.	Avg.	Std.	Avg.	Std.
40	0.41	5.10E−02	0.42	4.40E−02	0.42	4.10E−02
50	0.54	7.20E−02	0.56	3.30E−02	0.56	4.20E−02
60	0.69	7.70E−02	0.69	6.50E−02	0.7	5.50E−02
70	0.8	7.80E−02	0.8	6.90E−02	0.81	6.70E−02
80	0.9	6.50E−02	0.9	7.80E−02	0.91	6.30E−02
90	0.92	6.60E−02	0.92	8.50E−02	0.94	7.20E−02
100	0.95	7.20E−02	0.94	5.60E−02	0.96	6.50E−02

These figures show that $p_1 + p_2$ approaches a value close to 0.9 for all mutation operators. That is when eighty percent of the queries accessed from the pool of queries are A_1, A_2 or A_3, A_4, OMA-MA based algorithm converges to the correct equipartitioning $(\alpha\alpha\beta\beta)$ with a probability close to 0.9. The mutation rate was set to 0.05 for this experiment. Table 6.3 presents the OMA-MA based algorithm results for different mutation operators and different query percentages (accessed from the pool of queries) being (A_1, A_2) or (A_3, A_4), with respect to the accuracy of the solution and its standard deviation. The percentage varies from 40 to 100 by increments of 10 in Table 6.3. From these results, we conclude the following:

- For all mutation operators, the accuracy of the solution generated by OMA-MA based algorithms increases as the percentage of queries (A_1, A_2) or (A_3, A_4) in the input increases.
- OMA-MA-based algorithms have the highest accuracy when the LS-Mutation operator is used.

Table 6.4 shows the results of statistical tests. From the results reported in Table 6.4, we report the following:

- For all three kinds of statistical tests (Wilcoxon, permutation, and T-test), the difference between the performance of the OMA-MA based algorithm when it uses the LS-Mutation operator and the performance of the OMA-MA based algorithm when it uses other mutation operators is not statistically significant (p-value > 0.05).

6.2.2.3 Experiment 3

This experiment aimed to study the impact of the parameter N (depth of memory) on the number of iterations required by the OMA-MA-based algorithm to find optimal

Table 6.4 The results of statistical tests for OMA-MA based algorithm with LS-Mutation operator vs. OMA-MA based algorithm with SS-Mutation and XS-Mutation operators concerning the percentage of queries (A_1, A_2) or (A_3, A_4)

The percentage of queries (A_1, A_2) or (A_3, A_4)	SS-Mutation			XS-Mutation		
	P-value			P-value		
	T-Test	Permutation	Wilcoxon	T-Test	Permutation	Wilcoxon
40	4.09E−01	5.19E−01	4.46E−01	1.00E+00	7.35E−01	8.65E−01
50	1.99E−01	3.44E−01	6.90E−01	1.00E+00	6.23E−01	5.69E−01
60	5.67E−01	1.31E−01	1.91E−01	5.25E−01	7.50E−02	9.05E−02
70	5.98E−01	7.15E−01	8.13E−01	5.73E−01	1.88E−01	3.67E−01
80	5.50E−01	2.99E−01	3.75E−01	5.89E−01	4.60E−02	6.79E−02
90	2.71E−01	3.49E−01	2.01E−01	3.34E−01	2.69E−01	3.55E−01
100	5.77E−01	7.28E−01	4.87E−01	2.12E−01	3.29E−01	2.52E−01

equipartitioning. The depth of memory was varied from 2 to 10 by increments of 2. The OMA-MA-based algorithm was tested with three different mutation operators.

Tables 6.5, 6.6 and 6.7 show the results obtained for the OMA-MA-based algorithm with the SS-Mutation, XS-Mutation, and LS-Mutation for 14 different cases, the average number of iterations required, and their standard deviation.

- For lower values of *R*, OMA-MA based algorithm requires a higher number of iterations to converge to optimal equipartitioning.
- For higher values of *W/R*, OMA-MA based algorithm requires a higher number of iterations to converge to optimal equipartitioning.
- For higher values of *W/R*, OMA-MA based algorithm requires a higher depth of memory to converge to optimal equipartitioning.
- The OMA-MA-based algorithm with a depth of memory $N = 2$ performs better than OMA-MA based algorithm with a depth of memory of $N \neq 2$.

These conclusions are because higher values of *W/R* mean a higher number of objects are placed in each class. This leads to a situation where more actions have the same class number (migratory objects) and decreases the probability that a mutation operator swaps two objects.

Tables 6.8, 6.9 and 6.10 show the p-values of the two-tailed t-test, the p-values of the two-tailed Wilcoxon rank-sum test, and the p-values of the two-tailed permutation test for OMA-MA based algorithm with SS-Mutation, XS-Mutation, and LS-Mutation, respectively. From the results reported in these tables, we report the following:

- For all three kinds of statistical tests (Wilcoxon, permutation, and T-test), the difference between the performance of the OMA-MA based algorithm with the depth of memory $N = 2$ and the performance of the OMA-MA based algorithm

Table 6.5 Number of iterations required by OMA-MA with SS-Mutation operator for different cases and different depths of memory

Algorithm	N		W = 4	W = 6		W = 9	W = 12				W = 15		W = 18			
			R = 2	R = 3	R = 2	R = 3	R = 6	R = 4	R = 3	R = 2	R = 5	R = 3	R = 9	R = 6	R = 3	R = 2
OMA-MA	2	Avg.	10	18	85	186	34	179	486	1869	188	1892	58	201	3896	9865
		Std.	2.10E+00	1.80E+00	6.50E+00	1.26E+01	6.70E+00	9.50E+00	2.56E+01	8.56E+01	1.53E+01	8.86E+01	9.50E+00	1.25E+01	9.64E+01	1.22E+02
	4	Avg.	14	20	90	178	45	175	512	1835	365	1770	86	216	3684	9635
		Std.	2.30E+00	1.80E+00	7.10E+00	8.20E+00	3.50E+00	1.84E+01	5.62E+01	3.82E+01	4.43E+01	8.52E+01	1.53E+01	2.43E+01	1.02E+02	1.56E+02
	6	Avg.	18	29	98	205	58	189	423	1663	458	1756	97	189	3356	8569
		Std.	3.50E+00	6.80E+00	1.53E+01	2.42E+01	1.53E+01	1.53E+01	8.83E+01	1.28E+02	1.42E+02	2.35E+02	1.57E+01	1.08E+01	1.16E+02	1.86E+02
	8	Avg.	28	44	102	196	79	190	486	1721	380	1656	102	225	3015	4563
		Std.	1.20E+00	5.60E+00	1.27E+01	1.42E+01	8.40E+00	2.42E+01	4.43E+01	5.78E+01	6.54E+01	1.41E+02	1.41E+01	2.81E+01	3.25E+02	1.85E+02
	10	Avg.	59	49	125	224	96	215	459	1858	412	1669	136	304	3388	5698
		Std.	6.50E+00	7.30E+00	1.42E+01	2.18E+01	9.10E+00	2.51E+01	1.84E+01	5.64E+01	2.84E+01	8.64E+01	1.28E+01	3.64E+01	2.63E+02	2.39E+02

Table 6.6 Number of iterations required by OMA-MA with XS-Mutation operator for different cases and different depths of memory

Algorithm	N		W = 4	W = 6		W = 9	W = 12				W = 15		W = 18			
			R = 2	R = 3	R = 2	R = 3	R = 6	R = 4	R = 3	R = 2	R = 5	R = 3	R = 9	R = 6	R = 3	R = 2
OMA-MA	2	Avg.	10	18	98	194	36	181	500	1884	198	1902	62	211	2443	9898
		Std.	1.50E+00	2.20E+00	6.30E+00	2.16E+01	6.50E+00	1.23E+01	2.53E+01	1.52E+01	1.95E+01	1.25E+02	6.20E+00	1.52E+01	4.52E+01	6.85E+01
	4	Avg.	12	23	112	215	55	198	523	1698	255	2015	89	226	3698	10125
		Std.	1.80E+00	3.50E+00	1.29E+01	2.54E+01	7.80E+00	1.95E+01	3.78E+01	1.25E+02	2.43E+01	2.02E+02	2.55E+01	3.89E+01	1.25E+02	2.70E+02
	6	Avg.	14	32	108	223	53	205	534	1365	273	2115	87	199	4569	9632
		Std.	2.50E+00	8.60E+00	1.29E+01	2.79E+01	9.50E+00	2.37E+01	5.83E+01	1.25E+02	6.94E+01	1.46E+02	1.62E+01	3.21E+01	2.57E+02	3.27E+02
	8	Avg.	21	39	123	245	69	218	496	1236	289	1896	112	269	4896	9986
		Std.	3.20E+00	8.60E+00	1.96E+01	3.26E+01	1.56E+01	3.53E+01	4.11E+01	6.53E+01	7.51E+01	1.05E+02	1.96E+01	3.54E+01	2.40E+02	3.56E+02
	10	Avg.	48	51	145	219	85	226	509	1456	256	2365	163	278	2369	6896
		Std.	6.50E+00	8.20E+00	1.26E+01	3.42E+01	1.51E+01	4.23E+01	5.64E+01	1.13E+02	4.75E+01	4.23E+02	2.94E+01	3.59E+01	2.05E+02	4.02E+02

Table 6.7 Number of iterations required by OMA-MA with LS-Mutation operator for different cases and different depths of memory

Algorithm	N		W = 4	W = 6		W = 9	W = 12				W = 15		W = 18			
			R = 2	R = 3	R = 2	R = 3	R = 6	R = 4	R = 3	R = 2	R = 5	R = 3	R = 9	R = 6	R = 3	R = 2
OMA-MA	2	Avg.	12	21	102	180	41	190	512	1823	212	1911	62	227	4563	9755
		Std.	2.00E+00	3.50E+00	9.80E+00	1.72E+01	1.29E+01	2.18E+01	3.13E+01	9.84E+01	6.84E+01	1.09E+02	1.67E+01	5.23E+01	1.73E+02	1.32E+02
	4	Avg.	16	22	88	195	89	165	446	1823	238	1698	78	156	3568	8856
		Std.	3.20E+00	3.10E+00	1.27E+01	3.11E+01	1.22E+01	1.83E+01	3.94E+01	1.25E+02	4.52E+01	1.25E+02	1.35E+01	2.15E+01	3.25E+02	4.53E+02
	6	Avg.	18	35	85	176	125	154	485	1912	256	1554	125	263	3998	7805
		Std.	3.50E+00	6.50E+00	1.23E+01	1.32E+01	1.94E+01	2.56E+01	3.22E+01	1.28E+02	5.52E+01	1.24E+02	1.35E+01	3.21E+01	4.01E+02	5.23E+02
	8	Avg.	32	48	95	201	147	168	456	1996	325	1396	179	325	3368	5639
		Std.	2.90E+00	3.50E+00	1.05E+01	2.54E+01	1.85E+01	2.32E+01	5.24E+01	1.55E+02	3.62E+01	1.89E+02	2.53E+01	4.52E+01	3.60E+02	4.68E+02
	10	Avg.	64	68	114	223	212	186	412	1785	378	1298	245	369	2856	3568
		Std.	8.60E+00	5.40E+00	1.43E+01	3.26E+01	1.94E+01	1.43E+01	3.54E+01	1.21E+02	4.39E+01	1.39E+02	2.64E+01	4.89E+01	2.35E+02	4.13E+02

Table 6.8 The results of statistical tests for OMA-MA with SS-Mutation and depth of memory 2 vs. OMa-MA algorithm with SS-Mutation and other depths of memory

Algorithm	N	P-value	W = 4	W = 6		W = 9	W = 12				W = 15		W = 18			
			R = 2	R = 3	R = 2	R = 3	R = 6	R = 4	R = 3	R = 2	R = 5	R = 3	R = 9	R = 6	R = 3	R = 2
OMA-MA	4	T-Test	9.77E−08	1.75E−04	8.06E−03	6.80E−03	8.63E−09	1.90E−20	2.85E−02	5.65E−02	6.58E−19	7.55E−06	2.21E−09	5.41E−03	4.16E−09	6.07E−07
		Permutation	0.00E+00	0.00E+00	5.00E−03	0.00E+00	0.00E+00	0.00E+00	4.20E−02	2.40E−02	3.12E−01	0.00E+00	3.00E−03	4.00E−03	1.00E−03	0.00E+00
		Wilcoxon	9.06E−08	1.09E−03	3.94E−03	3.76E−03	9.67E−09	1.81E−01	7.24E−02	1.44E−02	2.87E−11	9.53E−07	3.06E−09	1.56E−02	9.67E−09	2.68E−07
	6	T-Test	1.29E−11	1.96E−09	1.84E−04	6.61E−04	1.11E−08	4.96E−03	7.78E−04	4.71E−08	3.12E−11	6.00E−03	1.89E−12	4.24E−04	2.65E−18	4.02E−24
		Permutation	0.00E+00	0.00E+00	1.00E−03	0.00E+00	4.00E−03	0.00E+00	0.00E+00	0.00E+00	8.00E−03	0.00E+00	0.00E+00	0.00E+00	0.00E+00	0.00E+00
		Wilcoxon	1.62E−09	4.37E−09	7.20E−05	3.11E−03	1.06E−08	1.91E−02	2.53E−04	3.08E−08	2.87E−11	3.71E−02	5.49E−11	2.92E−04	2.87E−11	2.87E−11
	8	T-Test	3.71E−27	8.74E−21	3.80E−07	7.31E−03	3.87E−20	2.77E−02	1.00E+00	1.18E−08	1.11E−15	1.50E−08	1.43E−14	1.89E−04	1.28E−14	9.45E−42
		Permutation	0.00E+00	0.00E+00	0.00E+00	0.00E+00	0.00E+00	0.00E+00	0.00E+00	9.95E−01	2.70E−02	0.00E+00	1.00E−02	0.00E+00	0.00E+00	0.00E+00
		Wilcoxon	2.87E−11	2.87E−11	6.26E−08	1.12E−02	2.87E−11	1.68E−03	8.59E−01	2.10E−08	3.02E−11	5.09E−08	5.23E−11	3.18E−04	2.49E−10	2.87E−11
	10	T-Test	1.06E−26	5.96E−20	1.86E−14	4.11E−09	2.10E−23	4.29E−08	5.99E−05	5.61E−01	2.67E−26	8.88E−11	5.17E−22	6.08E−15	7.92E−11	2.41E−36
		Permutation	0.00E+00	0.00E+00	0.00E+00	0.00E+00	0.00E+00	0.00E+00	5.27E−01	0.00E+00	0.00E+00	0.00E+00	0.00E+00	0.00E+00	0.00E+00	0.00E+00
		Wilcoxon	2.87E−11	2.87E−11	8.15E−11	7.77E−09	2.87E−11	1.93E−08	5.97E−05	4.51E−01	2.87E−11	5.32E-10	2.87E−11	3.18E−11	1.62E−09	2.87E−11

Table 6.9 The results of statistical tests for OMA-MA with XS-Mutation and depth of memory 2 vs. OMA-MA algorithm with XS-Mutation and other depths of memory

Algorithm	N	P-value	W = 4	W = 6		W = 9	W = 12				W = 15		W = 18			
			R = 2	R = 3	R = 2	R = 3	R = 6	R = 4	R = 3	R = 2	R = 5	R = 3	R = 9	R = 6	R = 3	R = 2
OMA-MA	4	T-Test	6.25E−05	2.92E−07	9.82E−06	1.74E−03	3.76E−11	3.60E−04	9.69E−03	5.98E−09	6.30E−11	1.42E−02	4.35E−06	5.88E−02	3.77E−30	1.11E−04
		Permutation	0.00E+00	0.00E+00	3.80E−02	0.00E+00	1.00E−02	0.00E+00	0.00E+00	9.00E−03	0.00E+00	0.00E+00	2.00E−03	0.00E+00	0.00E+00	2.00E−03
		Wilcoxon	3.26E−03	1.19E−06	1.29E−05	2.21E−03	1.69E−10	1.95E−04	9.88E−03	1.49E−08	4.20E−10	2.66E−02	8.29E−06	2.46E−01	2.87E−11	5.41E−04
	6	T-Test	2.78E−08	1.64E−09	6.59E−04	1.01E−04	6.40E−09	3.14E−05	6.54E−03	5.89E−20	3.65E−06	1.28E−06	1.05E−08	7.44E−02	2.73E−28	1.46E−04
		Permutation	0.00E+00	1.00E−02	7.40E−02	0.00E+00	0.00E+00	0.00E+00	0.00E+00	4.00E−03	0.00E+00	0.00E+00	0.00E+00	1.00E−03	0.00E+00	0.00E+00
		Wilcoxon	5.53E−08	3.50E−09	2.89E−03	1.49E−04	1.37E−08	1.81E−05	9.47E−03	2.87E−11	3.70E−06	6.28E−07	7.77E−09	6.46E−02	1.48E−03	2.92E−04
	8	T-Test	1.18E−16	1.37E−13	2.72E−07	7.33E−08	1.41E−11	7.87E−06	6.53E−01	2.09E−30	5.01E−07	8.42E−01	6.87E−14	4.32E−09	7.00E−31	1.94E−01
		Permutation	2.00E−01	0.00E+00	0.00E+00	0.00E+00	8.20E−01	0.00E+00	0.00E+00	6.10E−01	0.00E+00	0.00E+00	0.00E+00	0.00E+00	0.00E+00	0.00E+00
		Wilcoxon	3.34E−11	2.87E−11	3.52E−07	2.20E−07	6.07E−11	3.97E−06	7.17E−01	2.87E−11	2.95E−08	5.06E−01	3.64E−10	2.33E−09	5.32E−10	5.44E−01
	10	T-Test	7.28E−24	3.00E−19	1.87E−17	2.06E−03	3.71E−16	4.86E−06	4.32E−01	7.39E−19	9.55E−07	3.20E−06	1.53E−17	2.56E−10	6.38E−02	5.16E−27
		Permutation	0.00E+00	0.00E+00	0.00E+00	0.00E+00	0.00E+00	0.00E+00	0.00E+00	4.08E−01	0.00E+00	0.00E+00	2.00E−03	0.00E+00	0.00E+00	0.00E+00
		Wilcoxon	2.87E−11	2.87E−11	3.69E−11	1.72E−03	3.18E−11	5.60E−07	7.45E−01	2.87E−11	2.47E−07	2.58E−06	2.87E−11	2.13E−09	2.87E−11	2.87E−11

Table 6.10 The results of statistical test for OMA-MA algorithm with LS-Mutation and depth of memory 2 vs. OMA-MA algorithm with LS-Mutation and other depths of memory

Algorithm	N	P value	W = 4	W = 6		W = 9	W = 12				W = 15		W = 18			
			R = 2	R = 3	R = 2	R = 3	R = 6	R = 4	R = 3	R = 2	R = 5	R = 3	R = 9	R = 6	R = 3	R = 2
OMA-MA	4	T-Test	2.72E−06	2.51E−01	4.67E−05	2.81E−02	4.69E−15	4.29E−05	6.58E−08	1.00E+00	9.30E−02	1.02E−07	3.21E−04	1.48E−07	4.88E−15	2.48E−11
		Permutation	0.00E+00	0.00E+00	0.00E+00	0.00E+00	0.00E+00	9.40E−02	9.88E−01	0.00E+00	0.00E+00	0.00E+00	2.40E−02	0.00E+00	1.31E−01	0.00E+00
		Wilcoxon	8.51E−07	1.60E−01	3.59E−05	1.10E−02	2.87E−11	3.16E−05	1.11E−07	8.53E−01	6.57E−02	2.28E−07	3.09E−04	4.44E−07	2.87E−11	1.15E−10
	6	T-Test	5.46E−09	2.77E−11	1.98E−06	3.21E−01	2.31E−18	2.31E−06	2.61E−03	5.28E−03	1.04E−02	1.31E−12	5.62E−16	3.21E−03	8.65E−08	2.19E−18
		Permutation	0.00E+00	0.00E+00	2.00E−03	0.00E+00	0.00E+00	1.30E−02	7.00E−03	2.00E−03	0.00E+00	0.00E+00	3.77E−01	0.00E+00	0.00E+00	0.00E+00
		Wilcoxon	5.22E−09	8.56E−11	4.58E−06	5.06E−01	2.87E−11	4.74E−06	3.03E−03	9.27E−03	6.67E−03	3.88E−11	2.87E−11	6.52E−03	1.10E−08	2.87E−11
	8	T-Test	8.00E−24	2.47E−23	1.23E−02	7.86E−04	1.59E−21	7.15E−04	2.36E−05	1.58E−05	8.06E−09	1.50E−13	3.64E−19	1.46E−08	3.44E−16	9.50E−29
		Permutation	0.00E+00	0.00E+00	0.00E+00	0.00E+00	0.00E+00	0.00E+00	0.00E+00	0.00E+00	4.00E−03	0.00E+00	1.00E−03	1.30E−02	0.00E+00	0.00E+00
		Wilcoxon	2.87E−11	2.87E−11	1.12E−02	1.52E−03	2.87E−11	1.48E−03	1.29E−05	8.58E−06	3.80E−08	3.51E−11	2.87E−11	1.49E−08	3.51E−11	2.87E−11
	10	T-Test	2.84E−24	6.33E−27	7.03E−04	5.50E−07	5.50E−27	4.08E−01	2.09E−12	1.93E−01	4.88E−12	6.94E−18	3.31E−24	9.77E−12	3.54E−24	2.74E−35
		Permutation	0.00E+00	0.00E+00	0.00E+00	0.00E+00	0.00E+00	0.00E+00	1.95E−01	0.00E+00	4.31E−01	0.00E+00	0.00E+00	0.00E+00	0.00E+00	0.00E+00
		Wilcoxon	2.87E−11	2.87E−11	1.49E−04	1.86E−06	2.87E−11	4.69E−01	2.26E−10	1.41E−01	5.32E−10	2.87E−11	2.87E−11	1.04E−10	2.87E−11	2.87E−11

with a depth of memory of $N \neq 2$ is statistically significant (p-value < 0.05) in most cases for all three types of operators.

6.2.2.4 Experiment 4

This experiment aimed to study the impact of the parameter N (depth of memory) and type of OMA connections on the number of iterations required by the OMA-MA-based algorithm to find optimal equipartitioning. The depth of memory was varied from 2 to 10 by increments of 2. The OMA-MA-based algorithm was tested with three different OMA connections; Tsetline, Krylov, and Krinsky automata. Tables 6.11, 6.12 and 6.13 show the results obtained for the OMA-MA-based algorithm with the Tsetline, Krylov, and Krinsky connections for ten different cases, concerning the average number of iterations required, their standard deviation, and the p-values of the two-tailed t-test. From the results reported in these tables, we report the following:

- For all three kinds of OMA connections (Tsetline, Krylov, and Krinsky), the OMA-MA based algorithm with a depth of memory $N = 2$ performs better than OMA-MA based algorithm with a depth of memory of $N \neq 2$ and the difference between the performance of the OMA-MA based algorithm with the depth of memory $N = 2$ and the performance of the OMA-MA based algorithm with a depth of memory of $N \neq 2$ is statistically significant (p-value < 0.05) in all.

Table 6.14 presents the results of OMA-MA based algorithm with Tsetline, Krylov, and Krinsky connections for ten different cases concerning the average number of iterations, their standard deviation, and the p-values of the two-tailed t-test. From the results reported in Table 6.14, we may conclude that OMA-MA based algorithm with Tsetline connections performs the best.

6.3 The Graph Isomorphism Problem

6.3.1 The Local Search in the Graph Isomorphism Problem

If two graphs are isomorphic to each other, then the weight and the number of input and output edges of isomorphic vertices must be equal. This is considered in the design of the local search procedure (Wang et al. 1997). Pseudocode for our local search method is given in Fig. 6.5. This local search method consists of the following steps:

1. The vertices are partitioned into several subsets of equal weight.
2. The worst gene of the current chromosome is selected (line 2 of Fig. 6.5).
3. The selected gene's value is swapped with the value of a random gene selected from the same subset (lines 3 and 4 of Fig. 6.5).

Table 6.11 Number of iterations required by MGALA based on Tsetline OMA with SS-Mutation operator for different cases and different depths of memory

Algorithm	N		W = 265	W = 795		W = 1060			W = 1410			
			R = 5	R = 5	R = 3	R = 5	R = 4	R = 2	R = 10	R = 6	R = 5	R = 4
MGALA	2	Avg.	2.57E+04	1.32E+05	2.45E+05	2.58E+05	3.01E+05	5.97E+05	2.15E+05	3.56E+05	4.25E+05	5.24E+05
		Std.	3.65E+03	8.45E+03	1.12E+04	1.45E+04	1.69E+04	2.54E+04	8.56E+03	1.75E+04	1.85E+04	1.94E+04
	4	Avg.	3.66E+04	1.75E+05	2.90E+05	3.69E+05	3.99E+05	6.35E+05	2.79E+05	4.26E+05	5.10E+05	6.01E+05
		Std.	4.52E+03	9.86E+03	1.42E+04	1.86E+04	1.86E+04	2.16E+04	1.12E+04	1.96E+04	2.24E+04	2.54E+04
		P-Value	3.68E−11	2.42E−17	6.28E−14	1.76E−21	3.41E−19	5.35E−07	5.26E−21	6.78E−15	6.04E−16	6.86E−14
	6	Avg.	4.56E+04	2.18E+05	4.24E+05	4.70E+05	5.12E+05	7.85E+05	3.46E+05	4.87E+05	5.70E+05	6.52E+05
		Std.	4.69E+03	1.13E+04	1.46E+04	1.96E+04	2.26E+04	2.86E+04	1.56E+04	2.46E+04	2.75E+04	2.90E+04
		P-Value	1.65E−17	1.01E−24	2.10E−30	4.58E−29	3.18E−27	4.05E−22	6.18E−27	1.69E−20	1.32E−20	1.22E−18
	8	Avg.	6.83E+04	3.45E+05	5.12E+05	5.36E+05	5.89E+05	8.25E+05	4.12E+05	5.63E+05	6.52E+05	7.56E+05
		Std.	5.37E+03	1.80E+04	2.37E+04	2.46E+04	2.60E+04	3.13E+04	1.98E+04	2.60E+04	3.17E+04	3.56E+04
		P-Value	1.82E−19	8.33E−26	5.09E−30	1.43E−29	1.84E−30	9.41E−36	8.28E−31	2.05E−32	1.12E−32	4.99E−34
	10	Avg.	8.65E+04	5.24E+05	6.90E+05	7.42E+05	7.85E+05	9.86E+05	5.98E+05	6.90E+05	8.02E+05	9.13E+05
		Std.	6.33E+03	2.46E+04	3.12E+04	3.40E+04	3.50E+04	4.37E+04	2.46E+04	3.21E+04	3.86E+04	4.62E+04
		P-Value	1.50E−28	5.80E−36	1.74E−34	3.33E−34	1.43E−33	1.32E−27	1.13E−35	1.10E−29	3.16E−29	1.12E−27

Table 6.12 Number of iterations required by MGALA based on Krylov OMA with SS-Mutation operator for different cases and different depths of memory

Algorithm	N		W = 265	W = 795		W = 1060			W = 1410			
			R = 5	R = 5	R = 3	R = 5	R = 4	R = 2	R = 10	R = 6	R = 5	R = 4
MGALA	2	Avg.	4.65E+04	2.37E+05	4.44E+05	4.66E+05	5.36E+05	1.05E+06	3.88E+05	6.42E+05	7.63E+05	9.62E+05
		Std.	6.49E+03	1.49E+04	1.97E+04	2.58E+04	3.07E+04	4.55E+04	1.52E+04	3.17E+04	3.37E+04	3.41E+04
	4	Avg.	6.75E+04	3.21E+05	5.09E+05	6.49E+05	7.21E+05	1.14E+06	5.10E+05	7.60E+05	9.00E+05	1.06E+06
		Std.	8.22E+03	1.77E+04	2.62E+04	3.44E+04	3.43E+04	3.78E+04	2.07E+04	3.51E+04	3.98E+04	4.46E+04
		P-Value	7.40E−12	2.19E−18	9.32E−12	2.60E−20	1.18E−19	4.42E−09	1.14E−21	3.49E−14	1.11E−14	2.80E−10
	6	Avg.	8.27E+04	3.85E+05	7.74E+05	8.46E+05	9.38E+05	1.43E+06	6.10E+05	8.70E+05	1.01E+06	1.17E+06
		Std.	8.23E+03	2.05E+04	2.60E+04	3.48E+04	3.97E+04	5.19E+04	2.83E+04	4.49E+04	4.83E+04	5.11E+04
		P-Value	7.58E−18	3.65E−24	5.54E−31	3.46E−29	4.45E−28	1.21E−23	2.90E−26	5.11E−20	4.56E−20	1.01E−17
	8	Avg.	1.24E+05	6.14E+05	9.27E+05	9.52E+05	1.08E+06	1.50E+06	7.58E+05	1.04E+06	1.15E+06	1.40E+06
		Std.	9.86E+03	3.18E+04	4.15E+04	4.40E+04	4.79E+04	5.65E+04	3.53E+04	4.60E+04	5.80E+04	6.56E+04
		P-Value	2.02E−19	5.91E−26	2.17E−30	1.03E−29	4.55E−30	1.96E−35	6.36E−31	1.41E−32	1.95E−32	4.06E−34
	10	Avg.	1.59E+05	9.46E+05	1.21E+06	1.35E+06	1.44E+06	1.78E+06	1.05E+06	1.23E+06	1.47E+06	1.60E+06
		Std.	1.16E+04	4.44E+04	5.70E+04	6.28E+04	6.39E+04	7.84E+04	4.41E+04	5.81E+04	6.97E+04	8.31E+04
		P-Value	9.70E−29	5.03E−36	8.80E−34	4.43E−34	6.54E−34	3.07E−28	2.71E−35	2.52E−29	1.20E−29	1.23E−26

Table 6.13 Number of iterations required by MGALA based on Krylov OMA with SS-Mutation operator for different cases and different depths of memory

Algorithm	N		W = 265	W = 795		W = 1060			W = 1410			
			R = 5	R = 5	R = 3	R = 5	R = 4	R = 2	R = 10	R = 6	R = 5	R = 4
MGALA	2	Avg.	5.03E+04	2.54E+05	4.70E+05	5.07E+05	5.90E+05	1.19E+06	4.22E+05	7.12E+05	8.16E+05	1.02E+06
		Std.	7.01E+03	1.68E+04	2.24E+04	2.80E+04	3.25E+04	5.00E+04	1.70E+04	3.33E+04	3.70E+04	3.70E+04
	4	Avg.	7.15E+04	3.36E+05	5.68E+05	7.30E+05	7.62E+05	1.26E+06	5.52E+05	8.24E+05	1.00E+06	1.18E+06
		Std.	8.86E+03	1.88E+04	2.75E+04	3.58E+04	3.60E+04	4.16E+04	2.24E+04	3.87E+04	4.44E+04	4.83E+04
		P-Value	3.63E−11	3.02E−17	3.00E−15	4.99E−22	3.58E−18	3.04E−06	2.53E−21	8.61E−13	5.09E−17	1.10E−14
	6	Avg.	9.08E+04	4.25E+05	8.11E+05	9.06E+05	1.01E+06	1.50E+06	6.64E+05	9.27E+05	1.13E+06	1.27E+06
		Std.	9.37E+03	2.18E+04	2.85E+04	3.74E+04	4.47E+04	5.58E+04	3.06E+04	4.90E+04	5.47E+04	5.72E+04
		P-Value	6.87E−18	5.56E−25	4.47E−30	7.10E−29	2.67E−27	5.72E−20	3.00E−26	1.88E−18	1.44E−21	2.22E−18
	8	Avg.	1.31E+05	6.69E+05	1.00E+06	1.04E+06	1.15E+06	1.62E+06	8.21E+05	1.09E+06	1.27E+06	1.44E+06
		Std.	1.06E+04	3.56E+04	4.67E+04	4.78E+04	5.00E+04	6.08E+04	3.77E+04	5.10E+04	6.12E+04	6.89E+04
		P-Value	1.90E−19	2.23E−25	1.30E−29	1.01E−29	1.09E−30	5.52E−36	4.21E−31	8.96E−33	1.79E−32	3.43E−34
	10	Avg.	1.71E+05	1.04E+06	1.31E+06	1.45E+06	1.56E+06	1.88E+06	1.14E+06	1.31E+06	1.57E+06	1.81E+06
		Std.	1.26E+04	4.74E+04	6.14E+04	6.67E+04	6.97E+04	8.33E+04	4.76E+04	6.34E+04	7.50E+04	9.01E+04
		P-Value	1.44E−28	2.15E−36	5.28E−34	3.53E−34	9.58E−34	1.55E−26	2.60E−35	1.14E−28	1.67E−29	4.00E−28

Table 6.14 Number of iterations required by MGALA based on Tsetline, Krylov, and Krinsky OMA with SS-Mutation operator for different cases and different depths of memory

Algorithm	Type of OMA Connections		W = 265	W = 795		W = 1060			W = 1410			
			R = 5	R = 5	R = 3	R = 5	R = 4	R = 2	R = 10	R = 6	R = 5	R = 4
MGALA	Tsetline	Avg.	2.57E+04	1.32E+05	2.45E+05	2.58E+05	3.01E+05	5.97E+05	2.15E+05	3.56E+05	4.25E+05	5.24E+05
		Std.	3.65E+03	8.45E+03	1.12E+04	1.45E+04	1.69E+04	2.54E+04	8.56E+03	1.75E+04	1.85E+04	1.94E+04
	Krylov	Avg.	4.65E+04	2.37E+05	4.44E+05	4.66E+05	5.36E+05	1.05E+06	3.88E+05	6.42E+05	7.63E+05	9.62E+05
		Std.	6.49E+03	1.49E+04	1.97E+04	2.58E+04	3.07E+04	4.55E+04	1.52E+04	3.17E+04	3.37E+04	3.41E+04
		P-Value	1.92E−15	9.82E−25	3.88E−29	1.84E−26	8.05E−26	4.77E−29	1.14E−30	6.79E−28	3.23E−29	3.21E−32
	Krinsky	Avg.	5.03E+04	2.54E+05	4.70E+05	5.07E+05	5.90E+05	1.19E+06	4.22E+05	7.12E+05	8.16E+05	1.02E+06
		Std.	7.01E+03	1.68E+04	2.24E+04	2.80E+04	3.25E+04	5.00E+04	1.70E+04	3.33E+04	3.70E+04	3.70E+04
		P-Value	1.20E−16	2.06E−25	1.85E−29	7.28E−28	7.51E−28	1.66E−31	7.13E−32	3.99E−30	4.15E−30	4.67E−33

Fig. 6.5 Pseudocode for the local search in the GIP

Algorithm 6-2. Local Search for GIP

```
Procedure LocalSearch( CR )
    g1 = Worst Gene(CR);
    g2 = Select a Random Gene From Same Subset(g1);
    Swap(CR(g1), CR(g2));
End LocalSearch;
```

6.3.2 Experimental Results

In this section, several experiments are described that studied the effect of different OMA-MA parameters on the GIP performance. For this purpose, we used a database with 10,000 coupled pairs of isomorphic graphs with different sizes (Foggia et al. 2001). We classified these graphs into three groups: small graphs ($n < 50$), medium graphs ($50 \leq n < 100$), and large graphs ($100 \leq n < 200$). OMA-MA-based algorithm results are compared with the results obtained from an algorithm based on a GA (Wang et al. 2011), an algorithm reported by Ullmann (1976), and the VF and VF2 algorithms (Cordella et al. 2004). The source codes for these algorithms are available at https://amalfi.dis.unina.it/graph. Every result reported is an average of 30 runs. For all experiments, an initial population of size 100 was created randomly, the chromosome size was set equal to the size of the graph, the mutation rate and crossover rate were both set to 0.05, and the selection mechanism was ($\mu + \lambda$). Each algorithm terminates when either solution finds or the number of generations exceeds 10,000. A Tsetline-based OMA is used for all experiments to represent chromosomes. We use RT to refer to running time, FE to refer to the number of fitness evaluations for the runs converged to the solution, and NR to refer to the number of runs not converged to the solution. All experiments were performed on three classes of graphs: small graphs (SG), medium graphs (MG), and large graphs (LG). We performed a parametric test (t-test) and two non-parametric tests (Wilcoxon rank-sum test and permutation test) at the 95% significance level to provide statistical confidence. The t-tests were performed after ensuring that the data followed a normal distribution (by using the Kolmogorov–Smirnov test).

6.3.2.1 Experiment 1

Experiment 1 aimed to find the optimal memory depth for different classes of graphs for the OMA-MA-based algorithm. For this purpose, we studied the effect of parameter *N* (depth of memory) on the FE, RT, and NR. The OMA-MA-based algorithm results were then compared with the Canonical Memetic Algorithm results (CMA). Note that OMA-MA based algorithm is equivalent to the CMA when $N = 0$. For this experiment, the graph density was set to 0.5, and weights for vertices and edges were chosen from [0, 100]. Table 6.15 lists the RT, FE, NR, and the standard deviation

Table 6.15 Performance of OMA-MA concerning the depth of memory

Algorithm	Depth of memory	SG					MG					LG				
		FE		RT		NR	FE		RT		NR	FE		RT		NR
		Avg.	Std.	Avg.	Std.		Avg.	Std.	Avg.	Std.		Avg.	Std.	Avg.	Std.	
CMA	0	68	4.60E+00	1.07	5.80E−02	14	373	1.27E+02	10.09	4.20E+00	18	963	8.99E+01	36.07	5.60E+00	27
OMA-MA	2	85	5.30E+00	1.09	4.40E−02	0	902	8.93E+01	20.2	3.50E+00	7	2634	2.36E+02	86.54	9.60E+00	23
	4	86	3.50E+00	1.08	5.30E−02	0	990	7.86E+01	23.01	3.20E+00	0	3346	2.68E+02	117.16	1.23E+01	9
	6	84	4.60E+00	1.1	2.30E−02	0	969	8.65E+01	21.42	4.30E+00	0	3085	3.24E+02	101.62	1.03E+01	7
	8	81	6.20E+00	1.07	4.30E−02	0	896	7.61E+01	20.04	3.60E+00	0	3351	2.68E+02	109.87	9.50E+00	6
	10	87	5.40E+00	1.13	3.30E−02	0	957	6.84E+01	21.17	4.30E+00	0	3473	2.45E+02	114.09	1.32E+01	0
	12	87	7.10E+00	1.14	6.50E−02	0	926	5.99E+01	20.42	3.10E+00	0	3246	3.26E+02	106.58	8.60E+00	0
	14	86	5.30E+00	1.14	5.40E−02	0	933	6.53E+01	20.54	3.30E+00	0	3314	2.36E+02	108.14	7.60E+00	0
	16	89	4.20E+00	1.15	3.60E−02	0	927	7.16E+01	20.49	2.30E+00	0	3529	2.89E+02	115.5	1.03E+01	0
	18	90	4.10E+00	1.18	4.70E−02	0	903	8.03E+01	19.86	3.50E+00	0	3289	3.25E+02	107.75	1.24E+01	0
	20	84	5.20E+00	1.1	5.60E−02	0	977	6.82E+01	21.48	3.10E+00	0	3397	2.59E+02	111.11	1.13E+01	0

for the different depths of memory employed. From the results, we conclude the following:

- For all classes of graphs, the minimum value for RT and FE are obtained when N = 0,
- For all classes of graphs, the maximum value for NR is obtained when N = 0, and
- For all classes of graphs, the NR is inversely proportional to the depth of memory.

Table 6.16 shows that, according to the Wilcoxon test results, a permutation test, and a T-test, the OMA-MA-based algorithm with a memory depth of $N = 2$ performs better than the OMA-MA-based algorithm for $N \neq 2$ for large graphs (LG). Figure 6.6 shows the impact of memory depth on the FE and NR for different graphs. Changes in the FE are minor for a depth of memory greater than 4 in all classes of graphs. Also, this figure shows that for all classes of graphs, a depth of memory greater than 10 causes convergence (NR = 0) in all runs.

6.3.2.2 Experiment 2

This experiment investigated the effect of graph edge and vertex weights on OMA-MA performance. We studied the effect of weights on the FE, RT, and NR for different graphs using OMA-MA and CMA (OMA-MA when $N = 0$). For this experiment, all graphs' density was set to 0.5, and N was set to 10. The experiment was repeated for five different weight ranges:[0, 20], [0, 40], [0, 60], [0, 80], and [0, 100]. The experiment was also repeated with unweighted graphs (graphs whose edge weights are chosen from {0,1}, and whose nodes have no weights). Table 6.17 gives the RT, FE, NR, and standard deviation for the different weight parameters.

From these experimental results, we conclude the following for OMA-MA based algorithm:

- For all graphs, RT and FE are minimized when the weights for vertices and edges are chosen from [0, 100] and maximized for the unweighted graph. This is because GIP only uses the properties of the local search method. If the graph's weights have higher values, then the graph's vertices are partitioned into a higher number of subsets with a lesser number of members. This is to cause just the subset vertices, which have the same weight as the worst gene, to have a chance to exchange with the worst gene in the local search method. Consequently, the local search selects an alternative vertex accurately in weighted graphs.
- For all graphs, the NR is inversely proportional to the weights of the vertices and edges.
- For all classes of graphs, the maximum value of NR is obtained when CMA is used.

Table 6.18 shows that, for all three kinds of statistical tests (Wilcoxon, permutation, and T-test), the difference between the performance of the OMA-MA based algorithm when weights for vertices and edges are chosen from [0, 100], and the

Table 6.16 The results of statistical test for OMA-MA based algorithm with the depth of memory $N = 2$ vs. OMA-MA based algorithm with other depths of memory

Algorithm	Depth of memory	SG			MG			LG		
		T-Test	Permutation	Wilcoxon	T-Test	Permutation	Wilcoxon	T-Test	Permutation	Wilcoxon
OMA-MA	4	3.96E−01	7.11E−01	5.59E−01	3.48E−04	0.00E+00	3.77E−04	8.60E−12	0.00E+00	4.40E−10
	6	4.41E−01	3.37E−01	3.33E−01	6.20E−03	3.00E−03	6.82E−03	1.02E−06	0.00E+00	6.05E−07
	8	1.18E−02	5.00E−03	1.01E−02	7.81E−01	7.59E−01	6.05E−01	7.26E−12	0.00E+00	8.15E−11
	10	1.58E−01	1.22E−01	1.52E−01	1.21E−02	6.00E−03	2.87E−02	4.82E−14	0.00E+00	2.87E−11
	12	2.26E−01	2.16E−01	1.83E−01	2.31E−01	2.39E−01	3.99E−01	3.44E−09	0.00E+00	7.12E−09
	14	4.71E−01	6.96E−01	7.96E−01	1.36E−01	1.42E−01	2.58E−01	5.10E−12	0.00E+00	1.15E−10
	16	3.00E−03	5.00E−03	5.32E−03	2.41E−01	2.45E−01	4.38E−01	9.61E−14	0.00E+00	2.87E−11
	18	3.16E−04	0.00E+00	2.60E−04	9.64E−01	9.89E−01	9.59E−01	7.95E−10	0.00E+00	3.66E−09
	20	4.67E−01	2.12E−01	2.40E−01	1.01E−03	1.00E−03	1.11E−03	1.05E−12	0.00E+00	5.77E−11

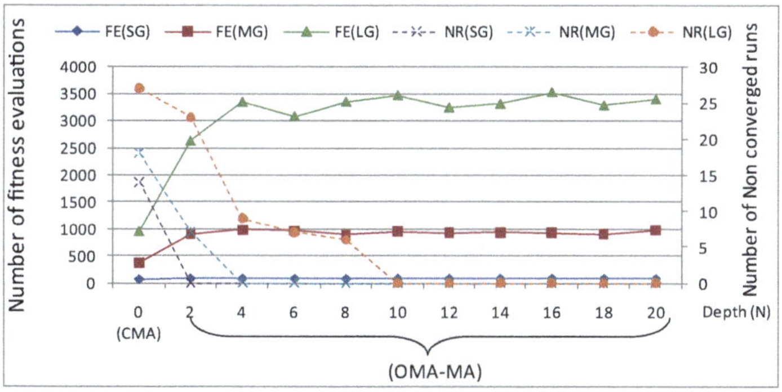

Fig. 6.6 Number of fitness evaluations (FE) and number of non-converged runs (NR) vs. depth of memory for different classes of graphs

performance of the OMA-MA based algorithm when it uses other weight parameter values are statistically significant (p-value < 0.05) for most graphs.

Figure 6.7 shows the FE for different graph classes and weights. The FE for all classes of graphs when the weights are chosen from ranges greater than [0, 20] are almost the same. Figure 6.7 also shows that with all graph classes, the NR is equal to zero when the weights are chosen from ranges greater than [0, 60]. Consequently, the FE and NR are both minimized when the weights are chosen from ranges greater than [0, 60] (ranges [0, 80] or [0, 100]).

6.3.2.3 Experiment 3

Experiment 3 studied the effect of graph density (D) on OMA-MA performance. The density of a graph is defined as $= \frac{2|E|}{|V|(|V|-1)}$ Which is the probability of the existence of an edge between any two vertices. For this experiment, the weights of vertices and edges were chosen from [0, 100], and N was set to 10. The impact of graph density on the FE, RT, NR, and the standard deviation for different classes of graphs using both OMA-MA and CMA are reported in Table 6.19. From these results, we conclude the following:

- For all classes of the graph, RT and FE are minimized when the graph density is 1.
- For all classes of graphs, NR decreases as the graph density increases.
- For all classes of graphs, a maximum value for NR is obtained when CMA is used.

Table 6.20 shows that, for all three kinds of statistical tests (Wilcoxon, permutation, and T-test), the difference between the performance of the OMA-MA based algorithm when the density is 1, and the performance of the OMA-MA based algorithm when it uses other density parameter values, is statistically significant (p-value < 0.05) for most graphs.

Table 6.17 Performance of OMA-MA concerning the weight parameter

Algorithm	Weight	SG					MG					LG				
		FE		RT		NR	FE		RT		NR	FE		RT		NR
		Avg.	Std.	Avg.	Std.		Avg.	Std.	Avg.	Std.		Avg.	Std.	Avg.	Std.	
CMA	[0, 100]	68	6.50E+00	0.99	5.30E−02	14	373	2.55E+01	10.09	3.50E+00	18	963	1.12E+02	36.07	6.50E+00	27
OMA-MA	Unweighted	102	1.52E+01	1.29	5.60E−02	1	2601	1.96E+02	59.14	8.50E+00	10	10791	6.59E+02	355.84	2.65E+01	26
	[0, 20]	99	8.20E+00	1.27	6.30E−02	0	883	1.09E+02	19.54	5.20E+00	0	3723	2.37E+02	121.33	1.43E+01	3
	[0, 40]	82	9.40E+00	1.04	2.50E−02	0	823	9.86E+01	18.22	4.60E+00	0	4262	3.25E+02	139.73	1.26E+01	1
	[0, 60]	76	6.20E+00	0.97	2.40E−02	0	892	1.15E+02	19.68	5.60E+00	0	3665	2.36E+02	119.69	1.25E+01	0
	[0, 80]	81	1.16E+01	1.03	3.60E−02	0	835	8.53E+01	18.44	3.50E+00	0	4052	3.21E+02	130.51	9.50E+00	0
	[0, 100]	74	5.20E+00	0.99	4.70E−02	0	815	7.56E+01	15.09	2.60E+00	0	3535	2.37E+02	114.94	7.50E+00	0

Table 6.18 The results of statistical tests for OMA-MA based algorithm with weight parameter [0, 100] vs. OMA-MA based algorithm with other values of the weight parameter

Algorithm	Weight of graph	SG			MG			LG		
		T-Test	Permutation	Wilcoxon	T-Test	Permutation	Wilcoxon	T-Test	Permutation	Wilcoxon
OMA-MA	Unweighted	1.87E−10	0.00E+00	1.86E−10	8.58E−29	0.00E+00	2.87E−11	2.78E−31	0.00E+00	2.87E−11
	[0, 20]	1.63E−14	0.00E+00	3.18E−11	8.68E−03	7.00E−03	3.59E−03	4.56E−03	3.00E−03	3.67E−03
	[0, 40]	3.23E−04	0.00E+00	7.10E−04	7.27E−01	7.45E−01	9.18E−01	8.48E−11	0.00E+00	3.31E−10
	[0, 60]	1.86E−01	1.29E−01	1.24E−01	4.74E−03	1.00E−03	6.24E−03	4.18E−02	2.90E−02	2.32E−02
	[0, 80]	5.28E−03	1.00E−03	8.14E−03	3.44E−01	3.60E−01	3.40E−01	8.34E−08	0.00E+00	1.11E−07

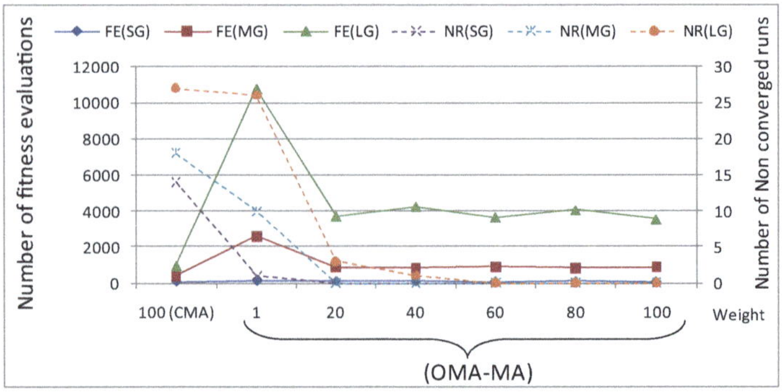

Fig. 6.7 Number of fitness evaluations (FE) and number of non-converged runs (NR) vs. the weight parameter for different classes of graphs

Figure 6.8 shows the impact of graph density on the FE for different classes of graphs. The FE remains almost fixed for graph densities greater than 0.5 for all classes of graphs. Figure 6.8 also shows that for all classes of graphs, all runs converge (NR) to zero when the graph density is greater than 0.6.

6.3.2.4 Experiment 4

The experimental goal here was to study the impact of different mutation and crossover operators on OMA-MA performance. For this experiment, all graphs' density was set to 0.5, and the depth of memory was set to 10. Weights for vertices and edges were selected from [0, 100].

Table 6.21 lists the RT, FE, NR, and the standard deviations from different mutation and crossover operators. These results lead us to conclude the following:

- A minimum NR value is obtained for all graphs when the LS-Mutation and LS-Crossover operators are used.
- For large size graphs (LG), concerning the FE and RT, OMA-MA based algorithm with the SS-Mutation and SS-Crossover operators outperforms both OMA-MA based algorithm with the XS-Mutation and XS-Crossover operators, as well as OMA-MA with the LS-Mutation and LS-Crossover operators.
- For medium-sized graphs (MG), with respect to FE and RT, OMA-MA with the LS-Mutation and LS-Crossover operators outperform both OMA-MA with the SS-Mutation SS-Crossover operators, as well as OMA-MA with the XS-Mutation and XS-Crossover operators.
- For small size graphs (SG), concerning FE and RT, OMA-MA based algorithm with the XS-Mutation and XS-Crossover operators outperforms both OMA-MA based algorithm with the SS-Mutation and SS-Crossover operators, as well as OMA-MA based algorithm with the LS-Mutation and LS-Crossover operators.

Table 6.19 Performance of OMA-MA concerning graph density

Algorithm	Density (D)	SG					MG					LG				
		FE		RT		NR	FE		RT		NR	FE		RT		NR
		Avg.	Std.	Avg.	Std.		Avg.	Std.	Avg.	Std.		Avg.	Std.	Avg.	Std.	
CMA	0.5	64	6.80E+00	0.87	6.00E−02	14	373	3.65E+01	10.09	1.62E+00	18	963	8.86E+01	36.06	3.65E+00	27
OMA-MA	0.1	85	7.30E+00	1.03	2.30E−02	0	2428	2.63E+02	56.05	3.62E+00	0	13166	1.16E+02	438.07	5.32E+00	12
	0.2	93	6.50E+00	1.01	2.50E−02	0	1114	1.36E+02	25.32	3.63E+00	0	8396	4.56E+02	276.62	2.65E+00	8
	0.3	91	8.40E+00	1.21	3.80E−02	0	1063	1.20E+02	28.35	4.56E+00	1	4470	3.56E+02	146.42	2.78E+00	5
	0.4	89	6.40E+00	1.81	1.60E−02	0	818	6.95E+01	17.65	2.37E+00	0	4116	3.66E+02	233.25	2.69E+00	5
	0.5	96	7.60E+00	1.27	2.90E−02	0	725	5.62E+01	16.35	2.76E+00	0	3357	2.36E+02	110.52	2.31E+00	4
	0.6	77	5.60E+00	1.04	3.60E−02	0	790	5.84E+01	17.66	1.69E+00	0	3369	2.38E+02	111.31	2.57E+00	0
	0.7	78	6.10E+00	1.02	3.10E−02	0	809	6.53E+01	18.18	1.67E+00	0	3530	3.13E+02	115.86	3.56E+00	1
	0.8	87	5.60E+00	1.09	3.80E−02	0	807	7.62E+01	17.81	2.36E+00	0	3284	3.45E+02	107.11	1.38E+00	0
	0.9	73	7.10E+00	0.98	1.20E−02	0	723	4.56E+01	18.17	2.49E+00	0	3259	2.89E+02	105.92	1.25E+00	0
	1	71	4.20E+00	0.95	1.50E−02	0	710	4.16E+01	19.93	1.94E+00	0	3124	2.64E+02	103.01	1.09E+00	0

Table 6.20 The results of statistical tests for OMA-MA based algorithm with density 1 vs. OMA-MA based algorithm with other values of the density parameter

Algorithm	Density of graph	SG			MG			LG		
		T-Test	Permutation	Wilcoxon	T-Test	Permutation	Wilcoxon	T-Test	Permutation	Wilcoxon
OMA-MA	0.1	5.30E−10	0.00E+00	9.76E−10	2.19E−25	0.00E+00	2.87E−11	1.69E−46	0.00E+00	2.87E−11
	0.2	1.28E−15	0.00E+00	2.87E−11	1.35E−15	0.00E+00	2.87E−11	7.77E−31	0.00E+00	2.87E−11
	0.3	1.80E−12	0.00E+00	9.92E−11	2.43E−15	0.00E+00	2.87E−11	2.26E−16	0.00E+00	2.87E−11
	0.4	1.60E−13	0.00E+00	1.04E−10	4.82E−08	0.00E+00	1.47E−07	8.13E−13	0.00E+00	1.15E−10
	0.5	9.18E−16	0.00E+00	3.02E−11	2.50E−01	2.41E−01	7.48E−02	1.16E−03	2.00E−03	1.41E−03
	0.6	5.92E−05	0.00E+00	3.83E−05	1.17E−06	0.00E+00	3.96E−07	7.35E−04	1.00E−03	7.49E−04
	0.7	1.55E−05	0.00E+00	5.66E−06	1.06E−07	0.00E+00	1.87E−07	7.52E−06	0.00E+00	3.45E−06
	0.8	3.21E−13	0.00E+00	4.73E−11	1.15E−06	0.00E+00	3.96E−07	5.30E−02	4.10E−02	5.10E−02
	0.9	1.95E−01	1.10E−01	1.33E−01	2.58E−01	2.53E−01	2.04E−01	6.90E−02	5.20E−02	7.24E−02

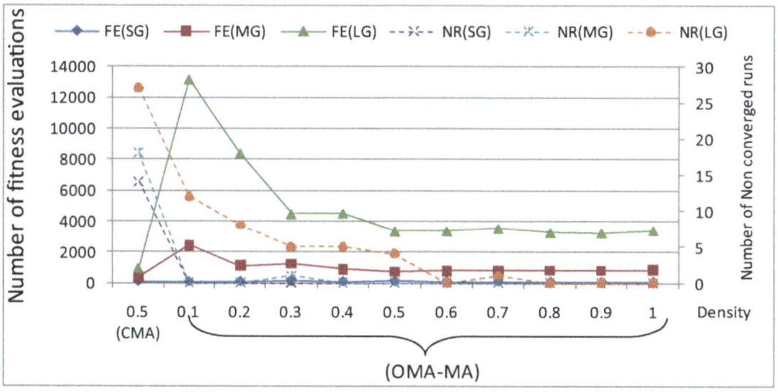

Fig. 6.8 Number of fitness evaluations (FE) and number of non-converged runs (NR) vs. density of graph for different classes of graphs

- For all classes of graphs, the maximum value for NR is obtained when CMA is used.

According to Table 6.22, the OMA-MA-based algorithm with the SS-Mutation and SS-Crossover operators performs better than the OMA-MA-based algorithm with other mutation operators for large graphs.

6.3.2.5 Experiment 5

The OMA-MA-based algorithm was compared with five other algorithms in this experiment (GA (Wang et al. 1997), Ullmann (Ullmann 1976), VF and VF2 (Cordella et al. 2004), and GALA (Rezapoor Mirsaleh and Meybodi 2006)) for the GIP, in terms of the number of fitness evaluations required. The result of this experiment is shown in Fig. 6.9. Each result was an average of 30 runs. The graph size was varied from 10 to 200 by an increment of 10. The results clearly show the superiority of OMA-MA based algorithm.

6.4 Assignment of Cells to Switches Problem (ACTSP) in Cellular Mobile Network

In a typical cellular mobile network, coverage is often geographically partitioned into hexagonal cells, as illustrated in Fig. 6.10. Each cell contains a stationary base station (BS) covering a small geographic area and an antenna for communications among users with pre-assigned frequencies. The calls first are transferred to switches by base stations, routed to another base station, or a public switched telephone network (PSTN) by switches. For various reasons and especially because of users' mobility, the signals between the mobile unit and the base station may become weaker while

Table 6.21 Performance of OMA-MA for different mutation and crossover operators

Algorithm	Operators type	SG					MG					LG				
		FE		RT		NR	FE		RT		NR	FE		RT		NR
		Avg.	Std.	Avg.	Std.		Avg.	Std.	Avg.	Std.		Avg.	Std.	Avg.	Std.	
CMA	Canonical operators	68	5.63E+00	1.1	5.60E−02	14	373	2.56E+01	10.09	1.36E+00	18	963	9.86E+01	36.06	3.69E+00	27
OMA-MA	SS-Mutation, SS-Crossover	90	9.45E+00	1.15	2.40E−02	0	951	5.63E+01	21	1.98E+00	0	3375	1.25E+02	109.16	8.96E+00	1
	XS-Mutation, XS-Crossover	86	8.35E+00	1.11	3.50E−02	0	927	4.85E+01	20.25	3.25E+00	0	3752	2.36E+02	120.95	1.37E+01	4
	LS-Mutation, LS-Crossover	92	7.36E+00	1.13	3.70E−02	0	921	7.54E+01	20.13	3.45E+00	0	4043	2.40E+02	128.65	9.54E+00	0

Table 6.22 The results of statistical tests for OMA-MA based algorithm with SS-Mutation, SS-Crossover vs. OMA-MA based algorithm with other operators

Algorithm	Operators type	SG			MG			LG		
		T-Test	Permutation	Wilcoxon	T-Test	Permutation	Wilcoxon	T-Test	Permutation	Wilcoxon
OMA-MA	XS-Mutation, XS-Crossover	9.29E−02	1.22E−01	1.98E−01	8.74E−02	8.30E−02	8.63E−02	1.64E−08	0.00E+00	1.26E−08
	LS-Mutation, LS-Crossover	3.68E−01	3.52E−01	3.18E−01	9.13E−02	9.00E−02	9.78E−02	4.64E−14	0.00E+00	3.88E−11

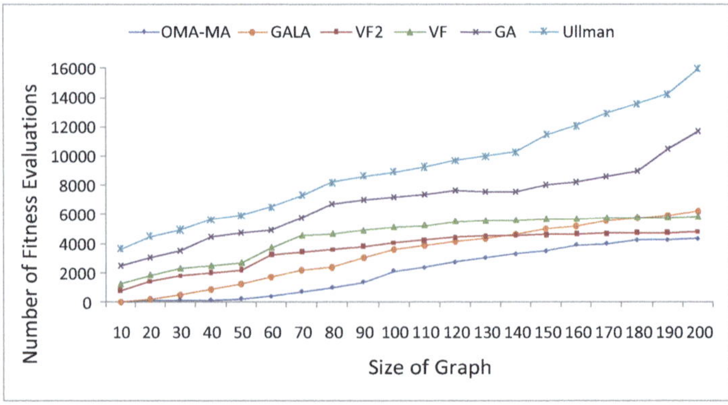

Fig. 6.9 Number of fitness evaluations vs. graph size for different algorithms

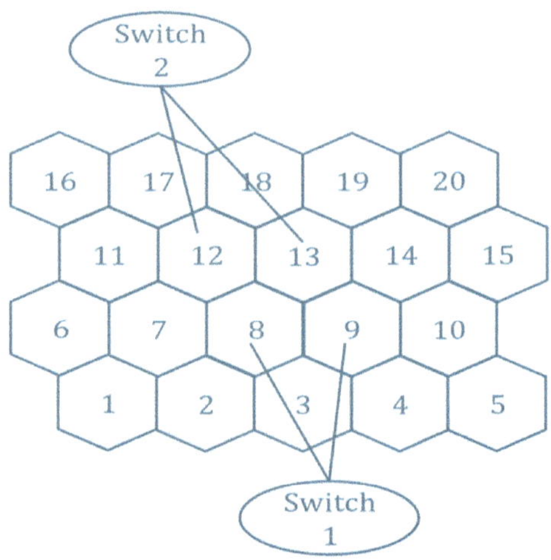

Fig. 6.10 An example of simple handoff (from cell 12 to cell 13) and complex handoff (from cell 12 to cell 8)

interference from adjacent cells increases. When a user in communication crosses the line between adjacent cells, the new cell's base station can relay this communication by assigning a new radio channel to the user. This operation is called a handoff, and it occurs when the mobile network automatically supports the transferring of communication from one cell to another adjacent cell. When a handoff occurred between two cells associated with the same switch, it is called a simple handoff; because few necessary updates are necessary. This type of handoff is relatively easy to perform and does not involve any location update in the databases that record the user's position. For example, in Fig. 6.10, a simple handoff causes by moving a user from cell 12 to cell 13. In this type of handoff, the network's database that records users' positions does not need an update. Figure 6.10 illustrates another type of handoff, complex handoff caused by moving a user from cell 12 to cell 8. A complex handoff refers to a handoff between two cells associated with different switches. In this type of handoff, handoff involves executing a complicated protocol between switch 1 and switch 2. It involves the updating of the location of the user in the network's database.

Furthermore, if the original switch (switch 1 in this case) is in charge of the billing, the handoff cannot simply replace switch 1 with switch 2. Note that the complex handoff process consumes much more network resources than the simple handoff process. Consequently, it is usually advantageous to connect cells with a high frequency of handoffs between them to the same switch.

In this section, a new algorithm is proposed for ACTSP. The new algorithm is obtained from a combination of learning automata (LA) and local search method in which the local search method searches for high-quality solutions with the minimum possible cost in terms of handoff and cabling cost, and the LA keeps the history of the local search method and manages the problem's constraints.

We compare the results of the proposed algorithm with the best-known assignment of cells to switches algorithms such as GA (Khuri and Chiu 1997), MA (Quintero and Pierre 2003a), HI (Salcedo-Sanz and Yao 2008), and HII (Salcedo-Sanz and Yao 2008). The obtained results show the new method's superiority over the other algorithms in terms of the solution's quality.

6.4.1 Background and Related Work

Let n be the number of cells and m be the number of switches in a cellular mobile network where cells and switches are predetermined. Suppose that the cell i is assigned to switch k with the cost of cabling c_{ik}. Let h_{ij} be the handoff cost per unit of time between cell i and cell j, if cell i and cell j are assigned to different switches, and $h_{ij} = 0$ when they are assigned to the same switch. The ACTSP can be described as finding matrix X that minimizes the function $Z(X)$ defined as Eq. (6.2). X is an $n \times m$ binary matrix, in which the element $x_{ik} = 1$ if cell i is assigned to switch k, and $x_{ik} = 0$ otherwise.

$$Z(X) = \sum_{i=1}^{n}\sum_{k=1}^{m} c_{ik}x_{ik} + \sum_{i=1}^{n}\sum_{j=1}^{n} h_{ij}(1 - y_{ij}) \tag{6.2}$$

Subject to:

$$\sum_{k=1}^{m} x_{ik} = 1 \qquad \forall i, \tag{6.3}$$

$$\sum_{i=1}^{n} \lambda_i x_{ik} \le M_k \qquad \forall k, \tag{6.4}$$

$$z_{ijk} = x_{ik}x_{jk} \qquad \forall i, j, k, \tag{6.5}$$

$$y_{ij} = \sum_{k=1}^{m} z_{ijk} \qquad \forall i, j. \tag{6.6}$$

The first term of Eq. (6.2) represents the cost of cabling between the cells and the switches, and the second term represents the handoff cost corresponding to handoffs between cells assigned to different switches. Every cell must be assigned to one and only one switch. This limitation imposes the constraint (6.3). The base station in cell i manages λ_i calls per unit time whereas the call handling capacity of switch k is M_k. These limitations impose the constraint (6.4). The nonlinear constraint (6.5) capture the appropriate values for z_{ijk} ($z_{ijk} = 1$ if and only if $x_{ik} = x_{jk} = 1$), whereas constraint (6.6) capture the appropriate values of variable y_{ij} ($y_{ij} = 1$ if $z_{ijk} = 1$ for any switch k).

A wide variety of approaches has been reported for ACTSP in the literature. These approaches can be classified as exact or approximation approaches. Exact approaches are practically inappropriate to solve large-sized instances of this problem. Exact approaches exhaustively examine the entire search space to find the optimal solution, and for this reason, they are only efficient for small search spaces corresponding to small-sized instances of the problem. For example, for a network with n cells and m switches, m^n solutions should be examined. Hence, different approximation algorithms have been reported in the literature for finding the near-optimal solution for ACTSP. Approximation algorithms for solving ACTSP by itself can be classified into two classes: hybrid algorithms obtained by combining evolutionary algorithms with a local search such as tabu search or simulated annealing and PSO-based algorithms. In the rest of this section, we briefly review some well-known algorithms from these two classes.

Hybrid Algorithms: In Fournier and Pierre (2005), two algorithms are reported, an ant colony-based algorithm (ACO) and a hybrid algorithm based on the ACO and the k-opt local search. The hybrid algorithm works well for large-size problems and performs the same as the ACO algorithm for small-size problems. In Quintero and Pierre (2003a), two multi-populations memetic algorithms with migrations and two sequential memetic algorithms are proposed for ACTSP. In all of these algorithms, the local search used is tabu search or simulated annealing. In Quintero and Pierre (2003b), Quintero and Samuel Pierre propose a hybrid evolutionary algorithm for ACTSP. This algorithm uses two local searches in order to combine the strengths of global and local search aspects. The local searches are tabu search and simulated annealing. In Quintero and Pierre (2003c), a comparative study of three algorithms; parallel genetic algorithms with migrations (PGAM), a hybrid algorithm based on PGAM and simulated annealing (PGAM-SA) and a hybrid algorithm based on PGAM and tabu search (PGAM-TS) for solving ACTSP are conducted. The results show that the PGAM-SA and PGAM-TS improve the solution provided by PGAM. In Menon and Gupta (2004), another hybrid algorithm based on pricing mechanism and simulated annealing is reported. The pricing mechanism provides a direction to proceed when trying to identify new solutions, which can be a powerful tool in speeding up simulated annealing, which can otherwise be extremely slow. In Fournier and Pierre (2005), a hybrid method based on the API and algorithm and tabu search is introduced. A key factor in this hybrid method's success is using an intensified search on several search space areas simultaneously. In Salcedo-Sanz and Yao (2008), a hybrid algorithm based on Hopfield network and genetic algorithm is introduced. In this algorithm a genetic algorithm searches for high quality solutions, and Hopfield network manages the problem's constraints.

PSO-Based Algorithms: In Goudos et al. (2010), two algorithms based on the barebones (BB) and the exploiting barebones (BBExp); variants of particle swarm optimization (PSO are reported for solving ACTSP. In BB-PSO, the standard PSO velocity equation is removed and replaced with samples from a normal distribution. Whereas in BBExp-PSO, approximately half of the time velocity is based on samples from a normal distribution, and the rest of the time, velocity is derived from the particle's personal best position. In Wang et al. (2011), a discrete particle swarm optimization (PSO) based on estimation of distribution algorithms (EDA) is introduced for solving ACTSP. This algorithm incorporates the global statistical information collected from personal best solutions of all particles into the PSO, and therefore each particle has comprehensive learning and searchability.

6.4.2 *The OMA-MA for Assignment of Cells to Switches Problem*

In this section, we introduce an algorithm based on the OMA-MA for solving the Assignment of cells to switches problem. The details of our algorithm are presented below. First, the solution representation for ACTSP is introduced. Then, a described description of the framework of the proposed algorithm is presented.

6.4.2.1 Solution Representation

In the proposed algorithm, the solution is represented by an object migration automaton (OMA) whose states keep information about the local search process's history. In the OMA-based representation, m actions are corresponding to m switches. Furthermore, for each action, there are a fixed number of states N. Each state in an OMA has two attributes: the value of the assigned object and the degree of association with its value. As migratory objects in the automata, the cells are assigned to states of corresponding action (switch). Figure 6.11(a) shows a representation of assigning ten cells to 6 switches using the Tsetline automaton-based OMA with the depth of memory of 5. In Fig. 6.11(a), there are six actions (switches), denoted by

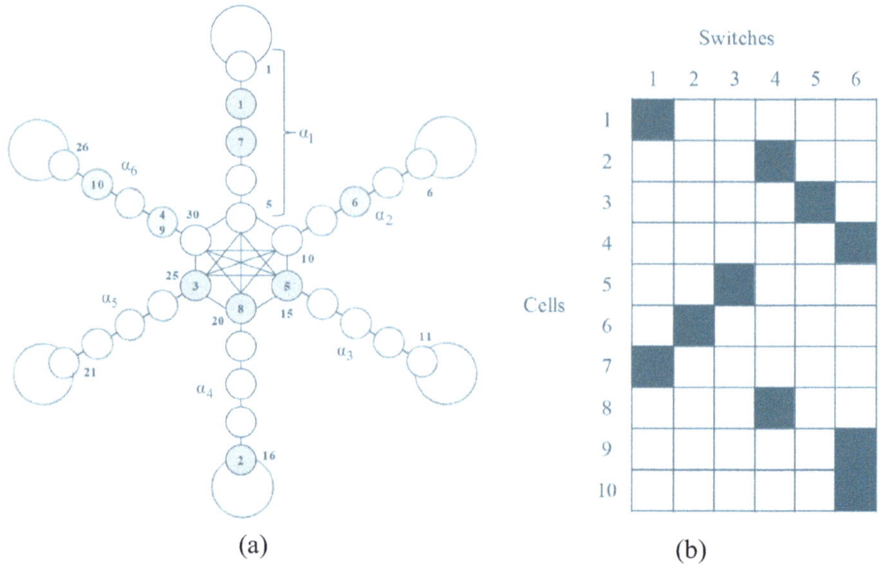

$S = [(1,2),(4,16),(5,25),(6,29),(3,15),(2,8),(1,3),(4,20),(6,29),(6,27)]$

(c)

Fig. 6.11 Example of the state transition graph of a Tsetline-based OMA **(a)**, the binary matrix X in six switches and ten cells instance (The black squares represent 1 s and the white squares represent 0 s) **(b)** and the solution vector S **(c)**

α_1, α_2, α_3, α_4, α_5, and α_6. Cells 1, 4, 6, 7, 9 and 10 are assigned to actions α_1, α_6, α_2, α_1, α_6 and α_6 (switches 1, 6, 2, 1, 6 and 6) and located at internal states 2, 4, 3, 3, 4 and 2 of their actions, respectively. Cells 3, 5, and 8 are assigned to actions α_5, α_3 and α_4 (switches 5, 3 and 4) and located at boundary states of their actions, respectively. Consequently, there is a minimum degree of association between these actions and their corresponding object. The remaining cell, cell 2, is assigned to actions α_4 (switch 4) and located at the most internal state of its action. That is, it has the maximum degree of association with action 4.

After applying a local search, if the assignment of a cell to a switch is promising, then the automaton is rewarded, and the assigned cell moves toward the most internal state of its switch; otherwise, the automaton is penalized, and the assigned cell moves toward the boundary state of its switch. The rewarding and penalizing of automaton changes the degree of association between cells and their switches. The initial assignment of cells to switches is created randomly, and cells are located in their switches' boundary state. We can transfer an OMA representation into the solution vector $S = [(s_1, t_1), (s_2, t_2), \ldots, (s_n, t_n)]$ to show the obtained assignment, where s_i is the switch assigned to cell i and t_i is its state, specifying $1 \leq s_i \leq m$ and $(s_i - 1) * N + 1 \leq t_i \leq s_i * N$. If we have a solution vector S, the corresponding binary matrix X can be calculated as:

$$x_{ij} = \begin{cases} 1, & if\ S.s_i = j, \\ 0, & otherwise \end{cases} \tag{6.7}$$

Figure 6.11(b) and (c) represent the corresponding binary matrix X and the solution vector S of the sample shown in Fig. 6.11(a), respectively. Representation of solutions based on other fixed structure learning automata is also possible. In a Krinsky-based OMA representation, the object will be associated with the most internal state (i.e., it gets the highest degree of association with the corresponding action) when rewarded and moves according to the Tsetline automaton-based OMA when it is penalized (Fig. 6.12). In the representation of a Krylov OMA, the object moves either toward the most internal state, or toward the boundary state, with a probability of 0.5 toward penalty, and moves according to the Tsetline automaton-based OMA upon reward (Fig. 6.13).

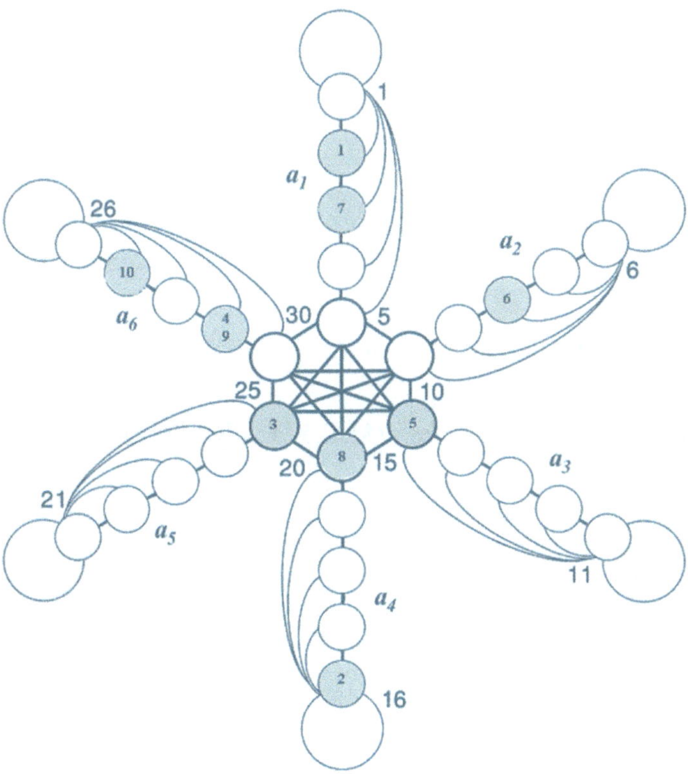

Fig. 6.12 Example of the state transition graph of a Krinsky-based OMA

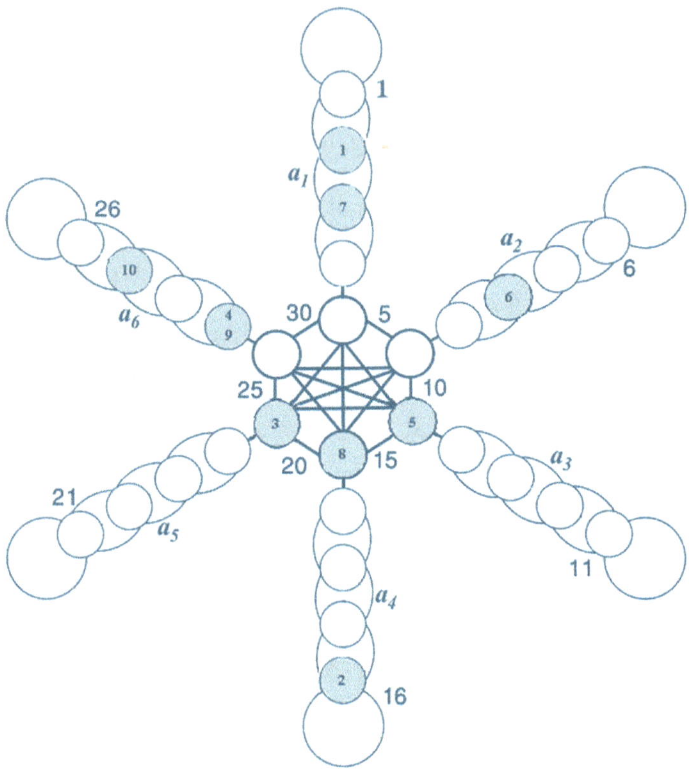

Fig. 6.13 Example of the state transition graph of a Krylov-based OMA

6.4.3 The Framework of the OMA-MA Algorithm

The proposed algorithm includes a local search which is applied to all cells in the current solution sequentially. Specifically, for a certain cell k which is assigned to switch s in current solution X, a temporary solution $X^{'}$ is created based on the current solution X in which the switch assigned to cell k (switch s) is replaced with another switch $s^{'}$ which is selected randomly. The temporary solution $X^{'}$ can be created as follows.

$$x'_{ij} = \begin{cases} x_{ij}, \ \forall i, j, \ \ 1 \le i \le n, \ \ \ i \ne k, \ \ \ i \le j \le m \\ 1, \ \ \ i = k, \ j = s' \ where \ s' \ is \ a \ random \ number \ between \ 1 \ and \ m, \ s' \ne s \\ 0, \ \ \ i = k, \ j \ne s' \end{cases} \quad (6.8)$$

If the assignment of cell k to the new switch $s^{'}$ does not exceed the capacity of the switch $s^{'}$ and the total cost of a temporary solution $(Z(X^{'}))$ is less than the total cost of the current solution ($Z(X)$), the assignment of cell k to switch s in the current solution is penalized. Otherwise, it is rewarded. By rewarding a cell, the state of that cell will move toward the most internal state according to the OMA connection. This causes the degree of association between a cell, and its corresponding switch, to be increased. Figure 6.14 illustrates an example of rewarding the assignment of cell 4 to switch 6.

The state of a cell remains unchanged if the cell is located at its most internal state, such as the assignment of cell 2 to switch 4, shown in Fig. 6.15.

Penalizing a cell causes the degree of association between the cell and its corresponding switch to be decreased. If a cell is not in the boundary state of its action, it is then penalizing causes the cell assigned to the switch to move toward the boundary state. This means that the degree of association between the cell and the corresponding switch will be decreased (Fig. 6.17).

Figure 6.16 provides a pseudo-code for rewarding the cell i in solution vector S.

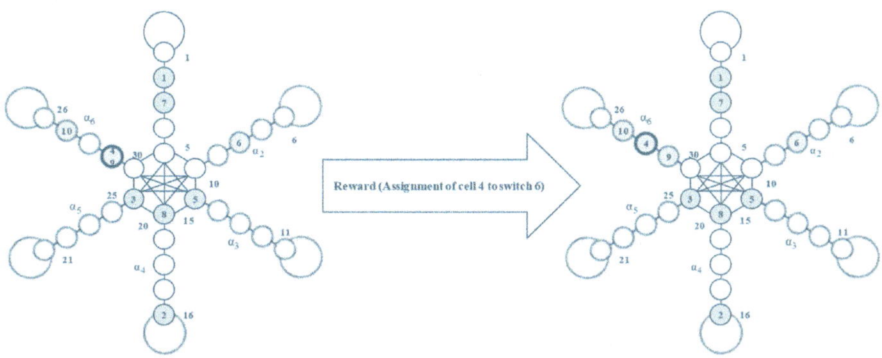

Fig. 6.14 An example of a reward function for assignment of cell 4 to switch 6

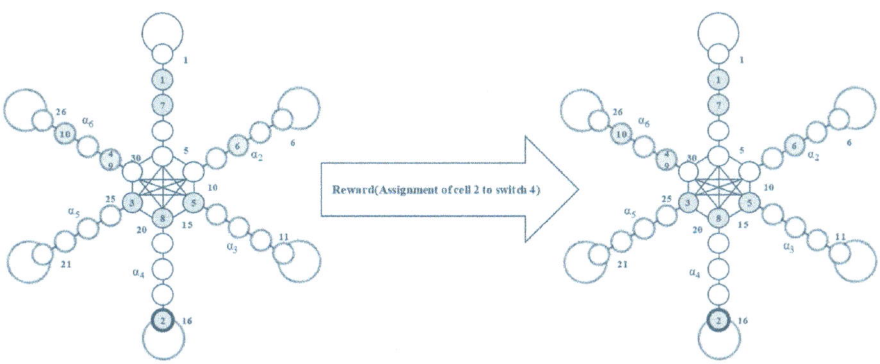

Fig. 6.15 An example of a reward function for assignment of cell 2 to switch 4

Fig. 6.16 Pseudocode for a reward function

Algorithm 6-3. Reward

```
Procedure Reward( S, i )
    If (S.t_i − 1) mod N <> 0 then
        Dec(S.t_i);
    End If
End Reward
```

Fig. 6.17 An example of a penalty function for assignment of cell 7 to switch 1

If a cell is in the boundary state of its switch, then penalizing causes the cell to be assigned to a new switch and creates a new solution. Figure 6.18 shows the effect of the penalty function on the assignment of cell 8 to switch 4.

The pseudo-code for the penalty function is shown in Fig. 6.19.

The pseudo-code for the proposed algorithm is shown in Fig. 6.20.

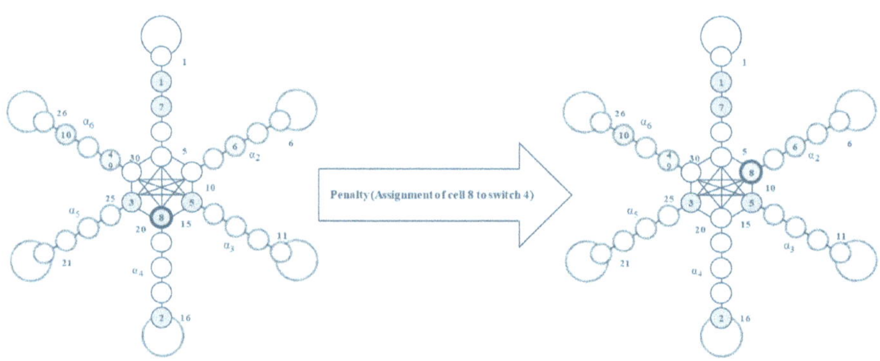

Fig. 6.18 An example of a penalty function for assignment of cell 8 to switch 4

Fig. 6.19 Pseudocode for the penalty function

```
Algorithm 6-4. Penalize
Procedure Penalize( S, i )
  If (S.t_i) mod N <> 0 then
    Inc(S.t_i);
  Else
    Select switch j randomly where j ≠ S.s_i and 1 ≤ j ≤ m.
    S.s_i = j;
    S.t_i = S.s_i * N;
  End If
End Penalize
```

```
Algorithm 6-5. Learning automata-based algorithm for assignment of cells to switches problem (LA-ACTSP)
Input: cells 1 to n with their coordinates, switch 1 to m with their coordinates, the number of states (N), the rate
of calls handled by cell i (λ_i) and the call handling capacity of switch k (M_k)
Output: The solution vector S, the binary matrix X, and the corresponding assignment of cells to switches
Begin Algorithm
    For i=1 to n do
        S.s_i = random(1..m); // Assign a switch to cell i randomly.
        S.t_i = S.s_i * N;
    End For
    Repeat
        S → X; //Transfer solution vector S into binary matrix X according to Eq. (6-7);
        For i=1 to n do
            X' = localsearch(X); //Create a temporary binary matrix X' by exchanging the switch of cell i with a
new switch s' in binary matrix X according to Eq. (6-8).
            If (Z(X') ≤ Z(X) and Σ_{i=1}^{n} λ_i x_{is'} ≤ M_{s'})
                Penalize(S, i);
            Else
                Reward(S, i);
            End If
        End For
    Until all cells located in the most internal state of their switches;
End Algorithm
```

Fig. 6.20 Pseudocode for LA-ACTSP

6.4.4 Experimental Result

In this section, several experiments are described that studied the efficiency of the proposed algorithm. Test problems were generated following the information in Merchant and Sengupta (1995), Quintero and Pierre (2003b). We consider a hexagonal grid, such as in Fig. 6.10. We assume that each cell's antenna is located in the center of the cell, and switches are uniformly distributed randomly outside the grid. The cable cost between a switch and a cell antenna is proportional to the two's geometric distance. The rate of calls (λ_i) originated in each cell was generated from a Gamma distribution with mean one and coefficient of variation 0.25. If a cell has k neighbors, we divide the range [0, 1] into $k + 1$ intervals by selecting k random numbers from a uniform distribution between 0 and 1. At the end of the service period in a cell *j*, the call could be either transferred to the i_{th} neighbor ($i = 1, \ldots, k$) with a handoff probability r_{ij} equal to the length of i_{th} interval, or ended with a probability equal to the length of the $(k+1)_{th}$ interval. The handoff rate h_{ij} is defined as:

$$h_{ij} = \lambda_i * r_{ij} \tag{6.9}$$

All the switches have the same capacity M_k calculated as:

$$M_k = \frac{1}{m}\left(1 + \frac{K}{m}\right)\sum_{i=1}^{n} \lambda_i, \tag{6.10}$$

K is uniformly chosen between 10 and 50, which ensures a global excess of 10–50% of the switches' capacity than the cells' volume of calls (Quintero and Pierre 2003b). The algorithm terminates when all the objects in only chromosomes are located in their actions' most internal state. For all experiments, a Tsetline-based OMA was used for chromosome representation. Each reported result was averaged over 50 runs. We performed a parametric test (t-test) and a non-parametric test (Wilcoxon rank-sum test) at the 95% significance level to provide statistical confidence. The t-tests were performed after ensuring that the data followed a normal distribution (by using the Kolmogorov–Smirnov test).

6.4.4.1 Experiment 1

We studied the impact of the parameter *N* (depth of memory) on the proposed algorithm's total cost to find the optimal solution. The depth of memory was varied from 2 to 10 by increments of 2. Table 6.23 shows the results obtained for the proposed algorithm for 15 different cases, concerning the average total cost (Avg.) and their standard deviation (Std.). In this table, symbols P#, CN, and SN stand for problem number, number of cells, and switches. We conclude that the proposed algorithm with different depths of memory fine feasible solution with a total cost close to the optimum solution, but the minimum total cost is obtained when $N = 4$.

Table 6.24 shows that for all two kinds of statistical tests (Wilcoxon and T-test), the difference between the performance of the proposed algorithm with the depth of memory four and the performance of the proposed algorithm with other depths of memory is statistically significant (p-value < 0.05) in all cases.

6.4.4.2 Experiment 2

In this experiment, we studied the impact of the memory depth (*N*) on the amount of reward received by cells' assignments to switches during a generation when the

Table 6.23 Comparison of the average total costs (Avg.) and their standard deviation (Std.) for different cases and different depths of memory

Type of OMA	P#	CN	SN	Proposed algorithm									
				N = 2		N = 4		N = 6		N = 8		N = 10	
				Avg.	Std.	Avg.	Std.	Avg.	Std.	Avg.	Std.	Avg.	Std.
Tsetline	1	15	3	186.12	2.40E−01	185.08	1.20E−01	186.12	2.50E−01	186.32	5.40E−01	187.25	1.23E+00
	2	15	4	372.36	4.50E−01	369.52	8.00E−02	370.63	6.50E−01	370.32	1.65E+00	372.25	2.24E+00
	3	15	5	376.14	3.25E+00	371.51	1.50E−01	374.25	1.23E+00	373.58	2.35E+00	376.14	3.45E+00
	4	30	3	808.98	2.63E+00	805.65	1.23E+00	809.65	3.60E−01	812.36	1.36E+00	810.12	4.12E+00
	5	30	4	728.34	2.35E+00	724.25	2.36E+00	732.25	1.85E+00	736.45	2.35E+00	742.25	3.45E+00
	6	30	5	698.14	4.36E+00	681.54	5.63E+00	702.54	3.25E+00	698.35	2.75E+00	689.52	2.01E+00
	7	50	3	1431.25	3.56E+00	1412.81	5.68E+00	1415.36	4.65E+00	1423.58	3.45E+00	1435.25	4.63E+00
	8	50	4	1168.32	4.68E+00	1137.64	6.12E+00	1141.25	8.24E+00	1142.36	8.14E+00	1163.25	1.26E+01
	9	50	5	956.32	4.69E+00	945.33	5.24E+00	946.79	2.36E+00	952.32	7.14E+00	962.14	1.43E+01
	10	75	3	2465.24	6.24E+00	2423.51	8.24E+00	2431.54	9.14E+00	2436.24	1.22E+01	2463.25	2.13E+01
	11	75	4	1764.32	5.14E+00	1758.39	6.89E+00	1777.33	4.35E+00	1768.47	8.14E+00	1768.36	1.21E+01
	12	75	5	2954.12	9.65E+00	2875.36	1.56E+01	2865.48	1.24E+01	2898.54	9.47E+00	2883.24	1.55E+01
	13	100	3	2963.15	1.00E+01	2879.36	1.28E+01	2886.32	1.44E+01	2912.32	1.24E+01	2924.15	1.63E+01
	14	100	4	3245.12	7.25E+00	3224.57	8.42E+00	3265.35	8.14E+00	3245.14	7.63E+00	3229.32	7.25E+00
	15	100	5	2778.23	1.24E+01	2719.49	7.52E+00	2735.89	9.12E+00	2778.65	1.44E+01	2776.36	1.24E+01

(continued)

Table 6.23 (continued)

Type of OMA	P#	CN	SN	Proposed algorithm									
				N = 2		N = 4		N = 6		N = 8		N = 10	
				Avg.	Std.	Avg.	Std.	Avg.	Std.	Avg.	Std.	Avg.	Std.
Krylov	1	15	3	187.65	1.65E+00	186.34	1.63E+00	191.63	5.62E+00	191.47	1.36E+00	192.32	5.65E+00
	2	15	4	384.18	3.25E+00	372.64	1.52E+00	375.57	2.65E+00	378.62	2.58E+00	379.51	4.23E+00
	3	15	5	396.32	4.35E+00	380.91	2.68E+00	384.58	6.84E+00	381.58	5.25E+00	381.45	7.58E+00
	4	30	3	853.64	6.35E+00	812.57	5.63E+00	856.55	3.56E+00	823.50	3.56E+00	821.57	9.45E+00
	5	30	4	736.51	7.52E+00	728.65	7.52E+00	789.57	9.12E+00	746.83	4.25E+00	753.47	1.25E+01
	6	30	5	723.37	1.06E+01	690.49	8.56E+00	775.56	8.45E+00	705.76	8.25E+00	704.64	9.54E+00
	7	50	3	1431.25	1.35E+01	1463.55	1.26E+01	1563.23	9.54E+00	1463.53	1.12E+01	1455.15	1.16E+01
	8	50	4	1536.48	9.47E+00	1169.53	1.43E+01	1196.51	1.26E+01	1168.43	1.25E+01	1178.49	1.65E+01
	9	50	5	996.45	7.52E+00	968.47	9.56E+00	989.59	1.85E+01	965.43	9.45E+00	976.42	1.85E+01
	10	75	3	2469.53	1.16E+01	2563.48	1.56E+01	2456.95	1.53E+01	2456.78	1.54E+01	2489.49	2.54E+01
	11	75	4	1965.48	1.06E+01	1804.63	2.15E+01	1896.54	2.26E+01	1789.49	1.24E+01	1789.45	1.94E+01
	12	75	5	3012.45	1.65E+01	2935.70	2.35E+01	2996.54	2.41E+01	2912.22	1.43E+01	2901.60	2.42E+01
	13	100	3	3125.38	1.27E+01	2968.45	2.50E+01	3015.55	1.95E+01	2989.21	1.97E+01	3014.57	2.56E+01
	14	100	4	3356.43	8.96E+00	3321.48	2.86E+01	3365.17	1.95E+01	3312.48	8.17E+00	3256.38	1.46E+01
	15	100	5	2874.31	1.52E+01	2896.53	1.98E+01	3012.49	2.42E+01	2796.17	1.95E+01	2789.51	1.95E+01

(continued)

Table 6.23 (continued)

Type of OMA	P#	CN	SN	Proposed algorithm									
				N = 2		N = 4		N = 6		N = 8		N = 10	
				Avg.	Std.	Avg.	Std.	Avg.	Std.	Avg.	Std.	Avg.	Std.
Krinsky	1	15	3	192.50	2.36E+00	191.63	2.36E+00	193.47	9.56E+00	194.26	9.56E+00	196.48	9.35E+00
	2	15	4	396.42	5.36E+00	379.52	4.25E+00	389.50	8.52E+00	384.63	8.25E+00	384.62	8.36E+00
	3	15	5	397.62	4.23E+00	386.50	5.24E+00	391.46	9.36E+00	386.38	7.25E+00	389.48	7.36E+00
	4	30	3	862.29	7.25E+00	825.62	8.25E+00	965.32	9.45E+00	835.33	6.56E+00	865.33	6.35E+00
	5	30	4	756.15	1.24E+01	765.50	9.21E+00	812.86	7.85E+00	765.40	1.14E+01	769.47	4.36E+00
	6	30	5	736.45	1.83E+01	712.34	1.03E+01	785.39	9.24E+00	716.48	1.53E+01	736.51	1.24E+01
	7	50	3	1456.49	1.93E+01	1469.49	1.53E+01	1572.44	1.16E+01	1476.55	9.54E+00	1469.55	1.83E+01
	8	50	4	1635.49	1.41E+01	1198.49	1.44E+01	1211.47	1.62E+01	1198.49	1.77E+01	1212.26	1.74E+01
	9	50	5	1025.25	8.24E+00	978.55	1.83E+01	996.49	1.63E+01	1011.57	1.97E+01	986.53	1.64E+01
	10	75	3	2463.47	1.63E+01	2578.60	2.44E+01	2456.17	2.24E+01	2498.40	2.16E+01	2514.29	1.92E+01
	11	75	4	2014.51	9.45E+00	1845.65	2.60E+01	1895.41	2.45E+01	1812.48	1.95E+01	1806.55	1.72E+01
	12	75	5	3045.61	2.13E+01	2998.36	1.98E+01	3016.35	1.95E+01	2946.49	1.78E+01	2916.28	2.13E+01
	13	100	3	3145.28	1.95E+01	3014.22	2.67E+01	3016.20	1.86E+01	2998.60	1.62E+01	3121.78	2.34E+01
	14	100	4	3368.52	1.62E+01	3345.21	2.65E+01	3359.55	2.35E+01	3325.47	2.16E+01	3269.53	1.97E+01
	15	100	5	2896.54	1.74E+01	2912.20	2.44E+01	3016.53	2.85E+01	2986.48	1.44E+01	2812.61	2.43E+01

Table 6.24 The results of statistical tests for the proposed algorithm with the depth of memory four vs. proposed algorithm with other depths of memory

Type of OMA	P#	CN	SN	Proposed algorithm with Tsetline OMA							
				N = 2		N = 6		N = 8		N = 10	
				t-test	Wilcoxon	t-test	Wilcoxon	t-test	Wilcoxon	t-test	Wilcoxon
Tsetline	1	15	3	4.15E−39	0.00E+00	5.99E−38	0.00E+00	7.36E−22	0.00E+00	6.96E−17	0.00E+00
	2	15	4	8.54E−43	0.00E+00	1.79E−16	4.44E−16	1.27E−03	1.92E−03	2.08E−11	5.79E−11
	3	15	5	1.52E−13	7.11E−15	6.15E−21	0.00E+00	1.02E−07	1.52E−07	1.07E−12	5.43E−12
	4	30	3	1.20E−11	9.69E−12	9.10E−30	0.00E+00	1.89E−45	0.00E+00	7.84E−10	4.36E−09
	5	30	4	8.45E−14	1.45E−11	1.65E−33	0.00E+00	1.30E−45	0.00E+00	4.21E−48	0.00E+00
	6	30	5	3.15E−29	0.00E+00	2.18E−36	0.00E+00	1.60E−29	0.00E+00	1.49E−13	1.24E−12
	7	50	3	1.52E−32	0.00E+00	1.58E−02	1.96E−02	1.24E−18	1.09E−14	2.40E−38	0.00E+00
	8	50	4	5.98E−47	0.00E+00	1.48E−02	5.52E−03	1.47E−03	2.34E−03	2.33E−20	1.29E−14
	9	50	5	7.52E−19	2.64E−14	7.76E−02	7.82E−02	2.51E−07	7.84E−07	9.19E−11	9.59E−11
	10	75	3	2.72E−47	0.00E+00	1.20E−05	4.35E−05	2.93E−08	2.25E−07	1.70E−18	2.58E−14
	11	75	4	4.69E−06	7.67E−06	9.27E−28	0.00E+00	1.59E−09	1.78E−08	2.71E−06	1.37E−06
	12	75	5	3.43E−46	0.00E+00	7.00E−04	1.71E−03	9.44E−14	1.98E−11	1.29E−02	2.71E−02
	13	100	3	8.76E−57	0.00E+00	1.20E−02	5.35E−03	2.80E−23	8.88E−16	4.67E−27	6.66E−16
	14	100	4	4.74E−23	2.22E−16	9.59E−44	0.00E+00	1.42E−22	2.22E−16	3.20E−03	7.48E−03
	15	100	5	3.43E−44	0.00E+00	4.34E−16	7.52E−13	9.62E−39	0.00E+00	3.56E−43	0.00E+00

(continued)

Table 6.24 (continued)

Type of OMA	P#	CN	SN	Proposed algorithm with Tsetline OMA							
				N = 2		N = 6		N = 8		N = 10	
				t-test	Wilcoxon	t-test	Wilcoxon	t-test	Wilcoxon	t-test	Wilcoxon
Krylov	1	15	3	1.28E−04	2.45E−04	3.28E−08	2.45E−08	1.08E−30	0.00E+00	1.50E−09	8.60E−09
	2	15	4	7.18E−34	0.00E+00	2.02E−09	1.23E−09	2.23E−23	2.22E−16	7.87E−16	2.46E−13
	3	15	5	4.60E−35	0.00E+00	7.75E−04	7.30E−04	4.25E−01	5.26E−01	6.34E−01	7.85E−01
	4	30	3	8.52E−56	0.00E+00	3.12E−61	0.00E+00	4.86E−19	1.35E−14	1.35E−07	5.59E−07
	5	30	4	9.61E−07	2.72E−06	1.86E−57	0.00E+00	2.03E−24	2.22E−16	1.35E−19	7.55E−15
	6	30	5	1.33E−30	0.00E+00	1.50E−71	0.00E+00	1.18E−14	3.03E−12	7.06E−12	3.53E−10
	7	50	3	9.10E−22	5.80E−16	6.81E−64	0.00E+00	9.95E−01	9.89E−01	7.93E−04	2.04E−03
	8	50	4	4.65E−105	0.00E+00	1.36E−16	5.56E−13	6.83E−01	9.48E−01	4.57E−03	4.32E−03
	9	50	5	6.23E−29	0.00E+00	4.56E−10	3.46E−09	1.13E−01	1.41E−01	8.73E−03	1.21E−02
	10	75	3	1.92E−53	6.86E−18	1.43E−56	6.86E−18	2.01E−56	6.86E−18	1.93E−29	1.11E−17
	11	75	4	8.74E−56	0.00E+00	9.36E−38	0.00E+00	4.74E−05	1.89E−04	3.57E−04	9.48E−04
	12	75	5	1.17E−32	0.00E+00	1.47E−22	6.66E−16	4.70E−08	2.71E−07	1.65E−10	3.05E−09
	13	100	3	6.00E−51	0.00E+00	1.82E−17	5.73E−14	1.26E−05	2.77E−05	1.02E−14	3.34E−12
	14	100	4	2.30E−11	3.69E−10	7.02E−14	1.98E−11	3.69E−02	6.08E−02	6.92E−23	2.55E−17
	15	100	5	1.12E−08	9.88E−08	4.27E−45	0.00E+00	4.35E−45	6.86E−18	2.05E−47	6.86E−18

(continued)

Table 6.24 (continued)

Type of OMA	P#	CN	SN	Proposed algorithm with Tsetline OMA							
				N = 2		N = 6		N = 8		N = 10	
				t-test	Wilcoxon	t-test	Wilcoxon	t-test	Wilcoxon	t-test	Wilcoxon
Krinsky	1	15	3	6.90E−02	4.13E−02	1.94E−01	1.00E+00	6.50E−02	8.36E−02	7.86E−04	5.47E−03
	2	15	4	3.79E−31	0.00E+00	1.89E−10	5.25E−10	2.16E−04	1.10E−03	2.54E−04	6.28E−04
	3	15	5	5.60E−20	7.99E−15	1.61E−03	1.11E−03	9.23E−01	8.20E−01	2.20E−02	1.54E−02
	4	30	3	7.37E−42	0.00E+00	3.28E−89	0.00E+00	3.70E−09	2.26E−08	1.85E−45	0.00E+00
	5	30	4	4.54E−05	1.87E−04	2.03E−47	0.00E+00	9.62E−01	6.67E−01	7.40E−03	3.61E−02
	6	30	5	5.47E−12	2.88E−11	3.09E−59	0.00E+00	1.15E−01	6.62E−02	8.03E−18	2.73E−14
	7	50	3	3.22E−04	5.32E−04	1.28E−57	0.00E+00	7.01E−03	2.51E−02	9.86E−01	8.55E−01
	8	50	4	1.61E−118	0.00E+00	5.28E−05	1.10E−04	9.99E−01	7.96E−01	3.96E−05	9.54E−05
	9	50	5	1.72E−25	0.00E+00	1.22E−06	3.79E−06	8.75E−14	3.99E−11	2.35E−02	3.61E−02
	10	75	3	1.42E−44	6.86E−18	7.46E−46	6.86E−18	1.35E−31	9.26E−18	6.79E−26	1.31E−16
	11	75	4	7.60E−48	0.00E+00	2.56E−16	3.99E−13	1.55E−10	1.31E−09	9.80E−14	2.74E−12
	12	75	5	7.34E−20	8.44E−15	1.39E−05	2.24E−05	1.32E−24	1.51E−16	2.80E−36	1.68E−17
	13	100	3	4.19E−46	0.00E+00	6.68E−01	4.99E−01	6.86E−04	1.71E−03	2.10E−38	0.00E+00
	14	100	4	9.42E−07	3.73E−06	5.15E−03	3.79E−03	9.14E−05	1.77E−04	1.56E−28	3.42E−17
	15	100	5	3.70E−04	4.99E−04	2.27E−35	0.00E+00	1.24E−30	0.00E+00	3.93E−37	9.83E−18

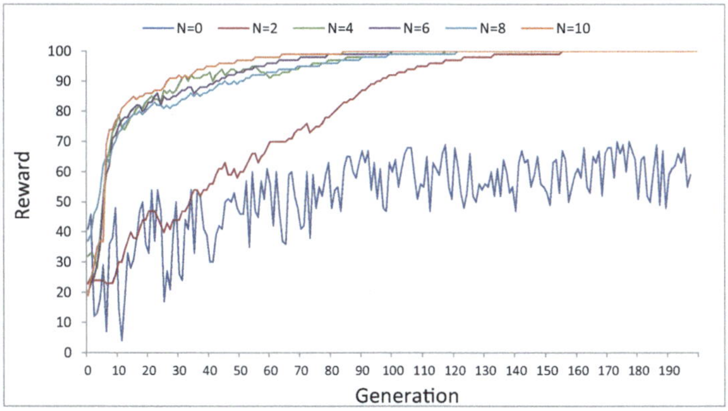

Fig. 6.21 The impact of depth of memory on the convergence rate of the proposed algorithm for test case #15

proposed algorithm is used. For this purpose, we calculated the number of assignments of cells to switches, leading to rewarding function per generation for different values of depth of memory. The depth of memory was varied from 0 to 10 by increments of 2. Note that when $N = 0$, the proposed algorithm performs without any history of local search. Figure 6.21 illustrates the result of this experiment for test case #15 (CN = 100 and SN = 5). From the result, it is evident that parameter N's value has a large impact on the proposed algorithm's performance. The best results are obtained when $N \geq 4$. The proposed algorithm is equivalent to a random algorithm when $N = 0$.

6.4.4.3 Experiment 3

In this experiment, we compared the results obtained from the proposed algorithm with Tsetline OMA with the results of four other algorithms, a genetic algorithm (GA) (Khuri and Chiu 1997), a memetic algorithm (MA) (Quintero and Pierre 2003a), and two hybrid algorithms based on Hopfield network and genetic algorithm (HI and HII), for the ACTSP, in terms of the quality of solutions. Table 6.25 presents the different algorithms for 15 different cases concerning the average total cost (Avg.) and their standard deviation (Std.). In this table, symbols P#, CN, and SN stand for problem number, number of cells, and switches. The results reported in Table 6.25 show that the proposed algorithm with the depth of memory 4 performs better than other algorithms in all cases.

Table 6.25 Comparison of the average total cost (Avg.) and their standard deviation (Std.) in proposed algorithms with Tsetline OMA and depth of memory 4 with other algorithms

P#	CN	SN	GA		MA		HI		HII		Proposed algorithm	
											With Tsetline OMA	
			Avg.	Std.	Avg.	Std.	Avg.	Std.	Avg.	Std.	Avg.	Std.
1	15	3	189.40	4.72E+00	187.41	8.46E+00	185.95	1.08E+00	185.28	4.20E−01	185.08	1.20E−01
2	15	4	379.22	1.17E+01	374.93	6.71E+00	371.28	2.15E+00	369.86	8.50E−01	369.52	8.00E−02
3	15	5	394.78	1.61E+01	376.65	5.46E+00	379.44	5.12E+00	376.66	2.63E+00	371.51	1.50E−01
4	30	3	840.28	2.32E+01	824.61	1.14E+01	832.26	2.14E+01	811.27	1.03E+01	805.65	1.23E+00
5	30	4	770.18	2.10E+01	757.24	1.62E+01	741.76	1.19E+01	744.22	1.23E+01	724.25	2.36E+00
6	30	5	740.20	1.86E+01	754.48	3.61E+01	733.45	2.52E+01	721.73	1.19E+01	681.54	5.63E+00
7	50	3	1457.17	2.91E+01	1447.81	3.07E+01	1450.63	1.73E+01	1436.19	1.45E+01	1412.81	5.68E+00
8	50	4	1205.36	2.94E+01	1182.49	1.58E+01	1221.62	4.85E+01	1178.85	2.11E+01	1137.64	6.12E+00
9	50	5	994.61	3.12E+01	980.38	3.54E+01	979.74	1.59E+01	976.73	1.32E+01	945.33	5.24E+00
10	75	3	2635.50	9.32E+01	2588.28	1.41E+02	2612.91	7.98E+01	2529.58	4.06E+01	2423.51	8.24E+00
11	75	4	1884.77	4.73E+01	1880.82	7.09E+01	1870.14	5.03E+01	1841.66	4.16E+01	1758.39	6.89E+00
12	75	5	3065.96	7.28E+01	3062.93	1.43E+02	3062.32	6.63E+01	3017.16	5.48E+01	2875.36	1.56E+01
13	100	3	3118.51	7.54E+01	3080.36	9.82E+01	3114.76	9.41E+01	3030.88	6.64E+01	2879.36	1.28E+01
14	100	4	3308.87	3.23E+01	3337.11	9.61E+01	3301.32	3.17E+01	3287.58	3.12E+01	3224.57	8.42E+00
15	100	5	2942.27	6.86E+01	2910.61	7.78E+01	2857.51	5.24E+01	2799.73	4.18E+01	2719.49	7.52E+00

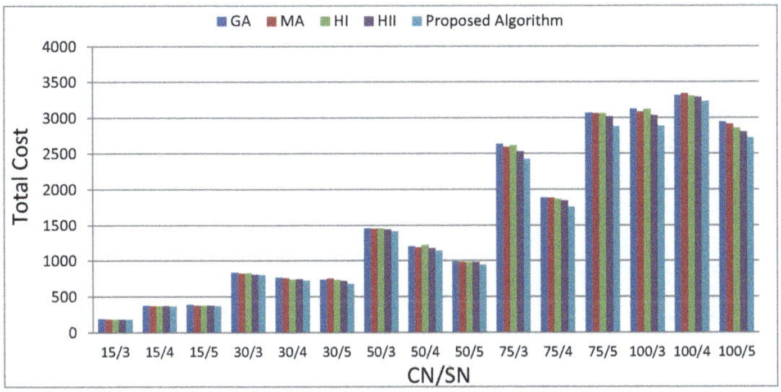

Fig. 6.22 Comparison of the total cost in proposed algorithm with Tsetline OMA and depths of memory 4 with other algorithms

Figure 6.22 shows the comparison of results reported in Table 6.25.

Table 6.26 shows the p-values of the two-tailed t-test and the p-values of the two-tailed Wilcoxon rank-sum test. The results reported in Table 6.26 shows that for all two kinds of statistical tests (Wilcoxon and T-test), the difference between the performance of the proposed algorithm with Tsetline OMA and depth of memory four and the performance of the other algorithms is statistically significant (p-value < 0.05) in all cases.

Table 6.26 The results of statistical tests for the proposed algorithm with Tsetline OMA and depth of memory 4 VS. other algorithms

P#	CN	SN	GA		MA		HI		HII	
			t-test	Wilcoxon	t-test	Wilcoxon	t-test	Wilcoxon	t-test	Wilcoxon
1	15	3	4.41E−08	5.49E−10	5.78E−02	1.20E−01	1.11E−05	6.47E−05	1.70E−03	6.60E−03
2	15	4	3.76E−07	3.53E−05	6.79E−07	1.40E−06	6.92E−07	1.44E−07	6.90E−03	1.07E−03
3	15	5	9.69E−14	4.54E−13	2.31E−08	7.43E−06	7.10E−15	3.38E−14	1.34E−18	0.00E+00
4	30	3	3.31E−14	1.78E−15	7.00E−16	7.84E−14	1.20E−11	1.27E−12	3.71E−04	5.67E−05
5	30	4	1.17E−20	2.22E−16	1.82E−19	2.22E−16	4.41E−14	3.52E−13	1.38E−15	1.42E−14
6	30	5	3.35E−29	0.00E+00	2.65E−19	0.00E+00	7.79E−20	4.44E−16	1.56E−32	0.00E+00
7	50	3	1.20E−14	6.04E−14	1.54E−10	1.07E−11	2.15E−21	2.22E−16	1.11E−15	2.44E−14
8	50	4	6.13E−22	0.00E+00	1.52E−27	0.00E+00	1.17E−16	2.89E−15	4.08E−19	4.44E−16
9	50	5	3.59E−15	2.44E−14	6.94E−09	1.58E−09	3.16E−21	0.00E+00	1.49E−23	0.00E+00
10	75	3	2.74E−21	0.00E+00	8.33E−11	1.23E−11	4.25E−22	2.22E−16	2.27E−24	0.00E+00
11	75	4	1.74E−24	2.22E−16	1.54E−16	3.38E−14	5.70E−21	2.22E−16	3.61E−19	4.44E−16
12	75	5	1.81E−24	0.00E+00	2.48E−12	1.67E−12	4.16E−26	0.00E+00	1.05E−24	0.00E+00
13	100	3	4.58E−28	0.00E+00	1.74E−19	4.44E−16	3.39E−23	0.00E+00	1.09E−21	0.00E+00
14	100	4	9.73E−25	0.00E+00	7.04E−11	8.36E−11	3.49E−23	0.00E+00	1.17E−19	2.00E−15
15	100	5	3.69E−28	0.00E+00	1.05E−22	0.00E+00	3.42E−24	0.00E+00	1.73E−18	1.78E−15

6.5 Conclusion

In this chapter, the equipartitioning problem (EPP), the graph isomorphism problem (GIP), and the assignment of cells to switches problem in cellular mobile network (ACTSP) were used to investigate the performance of OMA-MA. The OMA-MA-based algorithm was also compared with some other well-known algorithms for the EPP, GIP, and ACTSP applications. Our experimental results showed the superiority of the proposed algorithm in terms of quality of solution and convergence rate. This research line could be extended in several directions, such as applying OMA-MA for designing memetic algorithms for solving optimization problems where the environment is dynamic. Extension examples include the dynamic shortest path problem and the dynamic traveling salesman problem, improving the proposed algorithms by designing new mutation or crossover operators, and designing new object migrating automata to be used for chromosome representation. Another direction that may be pursued is the development of a mathematical framework for analyzing the proposed algorithm.

References

Cordella, L.P., Foggia, P., Sansone, C., Vento, M.: A (sub) graph isomorphism algorithm for matching large graphs. IEEE Trans. Pattern Anal. Mach. Intell. **26**, 1367–1372 (2004)

Foggia, P., Sansone, C., Vento, M.: A database of graphs for isomorphism and sub-graph isomorphism benchmarking. In: Proceedings of the 3rd IAPR TC-15 International Workshop on Graph-based Representations, pp. 176–187 (2001)

Fournier, J.R., Pierre, S.: Assigning cells to switches in mobile networks using an ant colony optimization heuristic. Comput. Commun. **28**, 65–73 (2005)

Goudos, S.K., Baltzis, K.B., Bachtsevanidis, C., Sahalos, J.N.: Cell-to-switch assignment in cellular networks using barebones particle swarm optimization. IEICE Electron. Express **7**, 254–260 (2010)

Khuri, S., Chiu, T.: Heuristic algorithms for the terminal assignment problem. In: Proceedings of the 1997 ACM Symposium on Applied Computing, pp. 247–251 (1997)

Menon, S., Gupta, R.: Assigning cells to switches in cellular networks by incorporating a pricing mechanism into simulated annealing . IEEE Trans. Syst. Man Cybern. Part B (Cybern.) **34**, 558–565 (2004)

Merchant, A., Sengupta, B.: Assignment of cells to switches in PCS networks. IEEE/ACM Trans. Netw. **3**, 521–526 (1995)

Oommen, B.J., Ma, D.C.Y.: Deterministic learning automata solutions to the equipartitioning problem. IEEE Trans. Comput. **37**, 2–13 (1988)

Quintero, A., Pierre, S.: Assigning cells to switches in cellular mobile networks: a comparative study. Comput. Commun. **26**, 950–960 (2003a)

Quintero, A., Pierre, S.: Evolutionary approach to optimize the assignment of cells to switches in personal communication networks. Comput. Commun. **26**, 927–938 (2003b)

Quintero, A., Pierre, S.: Sequential and multi-population memetic algorithms for assigning cells to switches in mobile networks. Comput. Netw. **43**, 247–261 (2003c)

Rezapoor Mirsaleh, M., Meybodi, M.R.: Improving GA+ LA algorithm for solving graph isomorphic problem. In: 11th Annual CSI Computer Conference of Iran, Tehran, Iran, pp. 474–483 (2006)

Salcedo-Sanz, S., Yao, X.: Assignment of cells to switches in a cellular mobile network using a hybrid Hopfield network-genetic algorithm approach. Appl. Soft Comput. **8**, 216–224 (2008)
Ullmann, J.R.: An algorithm for subgraph isomorphism. J. ACM (JACM) **23**, 31–42 (1976)
Wang, Y.-K., Fan, K.-C., Horng, J.-T.: Genetic-based search for error-correcting graph isomorphism. IEEE Trans. Syst. Man Cybern. Part B (Cybern.) **27**, 588–597 (1997)
Wang, J., Cai, Y., Zhou, Y., Wang, R., Li, C.: Discrete particle swarm optimization based on estimation of distribution for terminal assignment problems. Comput. Ind. Eng. **60**, 566–575 (2011)

Chapter 7
An Overview of Multi-population Methods for Dynamic Environments

Javidan Kazemi Kordestani, Mehdi Razapoor Mirsaleh, Alireza Rezvanian, and Mohammad Reza Meybodi

Abstract Dynamic optimization problems (DOPs) can be found almost everywhere, from ship navigation at sea (Michalewicz et al. 2007) to aerospace design (Mack et al. 2007). In general terms, all aspects of science and engineering include the optimization of a set of complex problems, in which the objectives of the optimization, some restrictions, or other elements may vary over time. Since exact algorithms are impractical in dynamic environments, stochastic optimization techniques have gained much popularity. Among them, evolutionary computation (EC) techniques have attracted a great deal of attention due to their potential for solving complex optimization problems. Nevertheless, EC methods should undergo certain adjustments to work well when applying on DOPs. Diversity loss is by far the most severe challenge to EC methods in DOPs. This issue appears due to the tendency of the individuals to converge to a single optimum. As a result, when the global optimum is shifted away, the number of function evaluations (FEs) required for a partially converged population to relocate the optimum is quite harmful to the performance. In this chapter, we provide an overview of the multi-population methods for dynamic environments.

J. Kazemi Kordestani
Department of Computer Engineering, Science and Research Branch, Islamic Azad University, Tehran, Iran

M. Razapoor Mirsaleh
Department of Computer Engineering and Information Technology, Payame Noor University (PNU), P.O. BOX 19395-3697, Tehran, Iran
e-mail: mrezapoorm@aut.ac.ir

A. Rezvanian (✉)
Department of Computer Engineering, University of Science and Culture, Tehran, Iran
e-mail: rezvanian@usc.ac.ir

M. R. Meybodi
Computer Engineering Department, Amirkabir University of Technology (Tehran Polytechnic), Tehran, Iran
e-mail: mmeybodi@aut.ac.ir

J. Kazemi Kordestani et al. (eds.), *Advances in Learning Automata and Intelligent Optimization*, Intelligent Systems Reference Library 208, https://doi.org/10.1007/978-3-030-76291-9_7

7.1 Introduction

Dynamic optimization problems (DOPs) can be found almost everywhere, from ship navigation at sea (Michalewicz et al. 2007) to aerospace design (Mack et al. 2007). In general terms, all aspects of science and engineering include the optimization of a set of complex problems, in which the objectives of the optimization, some restrictions, or other elements may vary over time.

A typical DOP can be formulated by a quintuple $\mathcal{F} = \{\Omega, \vec{x}, \phi, f, t\}$ where $\Omega \subseteq \mathbb{R}^D$ denotes to the search space, $\vec{x}$ is a feasible solution in the search space, ϕ represents the system control parameters that determine the distribution of the solutions in the fitness landscape, f: $\Omega \rightarrow \mathbb{R}$ is a static objective function, and $t \in \mathbb{N}$ is the time. With these definitions, the problem $\mathcal{F}$ is then defined as follows (Novoa-Hernández et al. 2011):

$$\begin{aligned} &optimize_x \; f^t(\vec{x}), \vec{x} = (x_1, x_2, \ldots, x_D) \\ &subject\ to: \; \vec{x} \in \Omega \end{aligned} \tag{7.1}$$

The goal of optimizing the DOP $\mathcal{F}$ is to find the set of global optima $X(t)$, in every time t, such as $X(t) = \left\{x^* \in \Omega \middle| \forall x \in \Omega \Rightarrow f^t(x^*) \succsim f^t(x)\right\}$. Here, $\succsim$ is a comparison relation which means *is better or equal than*, hence $\succsim \in \{\leq, \geq\}$. Regarding the above equation, DOP $\mathcal{F}$ is composed of a series of static instances $f^1(x), f^2(x), \ldots, f^{end}(x)$. Hence, the optimization goal in such problems is no longer just locating the optimal solution(s) but instead tracking the shifting optima over time.

Since exact algorithms are impractical in dynamic environments, stochastic optimization techniques have gained much popularity. Among them, evolutionary computation (EC) techniques have attracted a great deal of attention due to their potential for solving complex optimization problems. Nevertheless, EC methods should undergo certain adjustments to work well when applying on DOPs. Diversity loss is by far the most severe challenge to EC methods in DOPs. This issue appears due to the tendency of the individuals to converge to a single optimum. As a result, when the global optimum is shifted away, the number of function evaluations (FEs) required for a partially converged population to relocate the optimum is quite harmful to the performance.

Over the years, researchers have proposed various techniques to improve traditional EC methods for solving DOPs. According to Nguyen et al. (2012), the existing proposals can be categorized into the following six approaches:

i. Increasing the diversity after detecting a change in the environment.
ii. Maintaining diversity during the optimization process.
iii. Employing memory schemes to retrieve information about previously found solutions.
iv. Predicting the location of the next optimal solution(s) after a change is detected.
v. Making use of the self-adaptive mechanisms of ECs.

vi. Using multiple sub-populations to handle separate areas of the search space concurrently.

Among the approaches mentioned above, the multi-population (MP) approach is very effective for handling DOPs, especially for multimodal fitness landscapes. The success of MP methods can be attributed to three reasons (Nguyen and Yao 2012):

1. While different populations search in different sub-areas in the fitness landscape, the overall population diversity can be maintained at the global level.
2. It is possible to locate and track multiple changing optima simultaneously. This feature can facilitate tracking of the global optimum, given that one of the being-tracked local optima may become the new global optimum when changes occur in the environment.
3. It is easy to extend any single-population approach, e.g., diversity increasing/maintain schemes, memory schemes, adaptive schemes, etc., to the MP version.

The importance of familiarity with MP methods for following the next chapters compels us to devote a chapter exclusively to MP approaches. The primary objectives of this chapter are to (i) provide a categorization of the current MP approaches which are addressing the moving peaks problem (MPB) (the most extensively used synthetic benchmark for continuous DOPs), (ii) review each category and its corresponding methods in detail, (iii) emphasize on benefits and drawbacks of each class. We hope that the knowledge collected in this chapter can contribute toward designing new methods for DOPs by giving more insights into the advantages and disadvantages of the current techniques and assessing different strategies dealing with challenges in dynamic environments. Moreover, the present chapter can facilitate the study of existing MP methods in the literature and compare them.

Several studies have already been made in the literature with a broader scope to review the DOPs in general. For example, Jin and Branke (2005) addressed and discussed various types of uncertainties in evolutionary optimization. Another study in the literature, which also focused on methods designed for MPB, is the published book chapter titled "Dynamic Function Optimization: The Moving Peaks Benchmark" by Moser and Chiong (2010). In this work, the authors have overviewed the available approaches in the literature for solving the MPB. They have examined various methods for MPB in four groups as *Evolutionary Algorithms*, *Swarm Intelligence Algorithms*, *Hybrid Approaches,* and *Other Approaches.* Cruz et al. (2011) have contributed to the advancement of knowledge relevance of the field by accomplishing a twofold task: (1) they have collected a large number of relevant references about the topic over the past decade and then categorized the references based on the *type of publications*, *type of dynamism*, *methods for solving DOPs*, *performance measures*, *applications* and *publication year*, (2) afterwards, they have performed an overview of the research done on DOPs based on the collected repository. Recently, Nguyen et al. (2012) have provided an in-depth survey in evolutionary optimization in dynamic environments. The study has examined the state-of-the-art of academic research from four different points of view: (a) benchmark problems, (b) performance measures, (c) methodology, and (d) theory. At the end of the study, they have presented some discussions about the current gaps between academic research and real-world problems. Some future research directions have also been suggested. Li et al. (Nguyen and Yao 2012) analyzed the critical challenges of MP methods when applying to DOPs using experimental studies. The study attempts to find answers for critical questions like when to react to environmental changes, adapt the number of populations to changing environments, and determine each population's search area. Besides, several other issues, e.g., communication between populations, overlapping search, creating multiple populations, detecting changes, and local search operators are also discussed. Lately, Mavrovouniotis et al. (2017) presented a review on algorithms and applications of swarm intelligence methods for dynamic environments arranged by different classes of problems.

All mentioned studies are precious and beneficial in DOPs research and give classified information about different aspects of the topic. However, since the second part of this book is focused on applying learning automaton (LA) in dynamic environments, we provide the readers with enough information about different aspects of DOPs and MP methods to make them able to understand the rest of the book. The rest of this chapter is as follows: First, we examine different aspects of MPB and existing performance measures in the following two sections. Next, we review various MP methods in the literature in Sect. 7.4. Then, we provide numerical results of different MP methods on MPB. Finally, we conclude the chapter.

7.2 Moving Peaks Benchmark

MPB is a real-valued synthetic DOP with a D-dimensional landscape consisting of N peaks, where the height, the width, and the position of each peak are changed slightly every time a change occurs in the environment (Branke 1999). The definition of the landscape depends on the shape of the peaks. Two peak functions are used to define the peaks' shape: *function1* (sharp peaks) and *cone* (conical peaks). The function $f_t(x)$ for *function1* and *cone* is defined as follows:

$$f_t(x) = \max_{i=1,\dots,N} \frac{H_t(i)}{1 + W_t(i)\sum_{j=1}^{D}(x_t(j) - X_t(i,j))^2} \tag{7.2}$$

$$f_t(x) = \max_{i=1,\dots,N} H_t(i) - W_t(i)\sqrt{\sum_{j=1}^{D}(x_t(j) - X_t(i,j))^2} \tag{7.3}$$

where $H_t(i)$ and $W_t(i)$ are the height and the width of peak i at time t, respectively. The coordinates of each dimension $j \in [1, D]$ related to the location of peak i at time t, is expressed by $X_t(i, j)$, and D is the problem dimensionality. A typical change of a single peak can be modeled as follows:

$$H_{t+1}(i) = H_t(i) + height_{severity}.\sigma_h \tag{7.4}$$

$$W_{t+1}(i) = W_t(i) + width_{severity}.\sigma_w \tag{7.5}$$

$$\vec{X}_{t+1}(i) = \vec{X}_t(i) + \vec{v}_{t+1}(i) \tag{7.6}$$

$$\vec{v}_{t+1}(i) = \frac{s}{|\vec{r} + \vec{v}_t(i)|}((1-\lambda)\vec{r} + \lambda\vec{v}_t(i)) \tag{7.7}$$

where σ_h and σ_w are two random Gaussian numbers with zero mean and standard deviation one. Moreover, the shift vector $\vec{v}_{t+1}(i)$ is a combination of a random vector $\vec{r}$, which is created by drawing random numbers in $[-0.5, 0.5]$ for each dimension, and the current shift vector $\vec{v}_t(i)$, and normalized to the length s. Parameter $\lambda \in [0.0, 1.0]$ specifies the correlation of each peak's changes to the previous one. This parameter determines the trajectory of changes, where $\lambda = 0$ means that the peaks are shifted in completely random directions and $\lambda = 1$ means that the peaks always follow the same direction until they hit the boundaries where they bounce off.

The goal of optimization in MPB at each time is to locate the set of optimums ξ_t, that:

$$\xi_t = \{x^* \in \Omega | f_t(x^*) \geq f_t(x), \forall x \in \Omega\} \tag{7.8}$$

As can be seen from the above equation, MPB is a real-valued dynamic maximization problem.

MPB has three sets of standard configurations to provide a unified testbed for the researchers to test their approaches in the same condition, referred to as *Scenario 1, Scenario 2,* and *Scenario 3.* Table 7.1 presents the default values for MPB parameters corresponding to each scenario. Moreover, Fig. 7.1 illustrates an instance of the fitness landscape of each scenario.

7.2.1 Extended Versions of MPB

Recently, other types of changes have also been introduced for MPB. For instance, du Plessis and Engelbrecht proposed a new kind of change in MPB that allows the number of peaks to fluctuate when a change occurs in the environment (du Plessis and Engelbrecht 2013). The number of peaks is then determined as follows:

$$m_{t+1} = \begin{cases} max\{1, m_t - M * \sigma * p\} & if\ \sigma < 0.5 \\ min\{M, m_t + M * \sigma * p\} & otherwise \end{cases} \tag{7.9}$$

where M is the maximum number of peaks in the landscape, and $p \in [0.0, 1.0]$ is the maximum fraction of the M that can be added or removed from the current peaks after a change in the environment. Therefore, p is used to control the severity of the change in the number of peaks. For example, $p = 0.0$ means no change in the number

Table 7.1 Parameter settings for each scenario of the moving peaks benchmark

Parameter	Scenario 1	Scenario 2	Scenario 3
Number of peaks (m)	5	10–200	50
Height severity	7.0	7.0	1.0
Width severity	0.01	1.0	0.5
Peak function	function1	cone	cone
Number of dimensions (D)	5	5	5
Height range (H)	[30.0 70.0]	[30.0 70.0]	[30.0 70.0]
Width range (W)	[0.0001 0.2]	[1.0 12.0]	[1.0 12.0]
Standard height (I)	50.0	50.0	0.0
Standard width (K)	0.1	0.0	0.0
Search space range (A)	[0.0 100.0]	[0.0 100.0]	[0.0 100.0]
Frequency of change (f)	5000 FEs	1000–5000 FEs	1000 FEs
Shift severity (s)	[0.0 2.0]	[0.0 3.0]	1.0
Correlation coefficient (λ)	0.0	[0.0,1.0]	0.5
Change step size	Constant	Constant	Constant
Basis function	False	False	True

of peaks, $p = 0.5$ will change up to 50% in M, and p = 1.0 will cause up to M peaks being added or removed. M and p are included as parameters of the MPB.

Fernandez-Marquez and Arcos incorporated noise to the MPB as follows (Fernandez-Marquez and Arcos 2009):

$$f_t(x) = f_t(x) + \frac{2 * \upsilon - 1}{2}\gamma, \tag{7.10}$$

where υ generates a uniform random number between [0.0, 1.0] and γ control the percentage of noise in the problem, the noise in the fitness function brings another challenge to the problem, which causes the algorithms to confuse the noise with the changes in the environment.

Another group of authors (Novoa-Hernández et al. 2011) modified MPB to use other peak functions, e.g., *Ackley*, *Sphere*, *Schwefel*, *Quadric*, etc., and study the performance of their proposed methods.

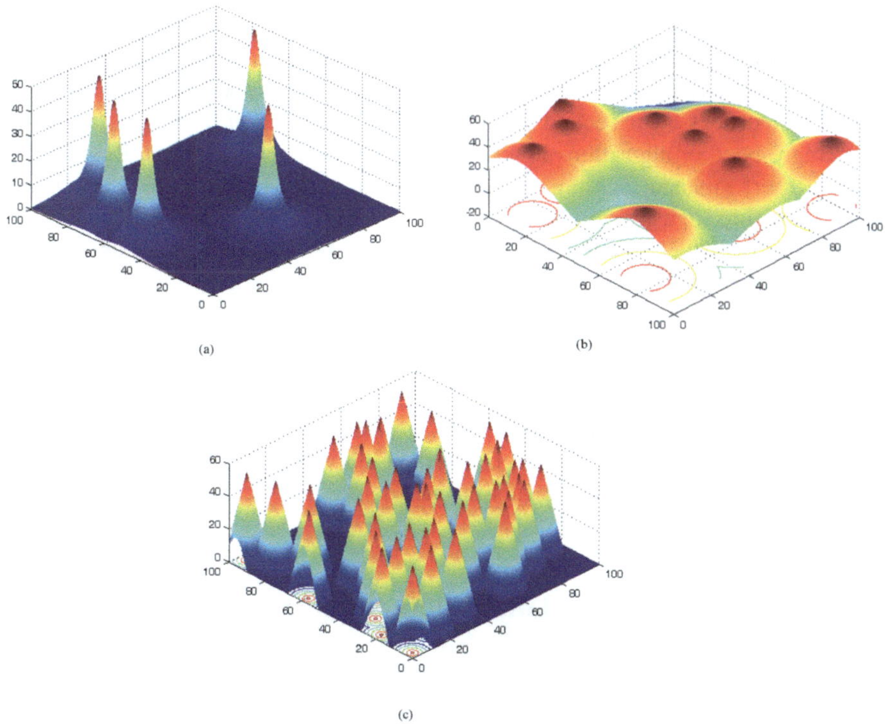

Fig. 7.1 Different landscapes generated by MPB related to (a) scenario 1, (b) scenario 2, and (c) scenario 3

Recently, Nasiri et al. proposed (Nasiri et al. 2016) added a pendulum motion characteristic to the MPB to allow the previous environments to re-appear in the future. This feature is useful when evaluating the performance of memory-enhanced approaches.

7.3 Performance Measurement

Several measures were proposed to assess the efficiency of the optimization algorithms and evaluate different aspects of the methods for DOPs. For example, Weicker (2002) introduced *accuracy*, *stability,* and *reactivity* as key characteristics of optimization in dynamic environments. Alba and Sarasola (2010) proposed $\beta_{degradation}$ to measure the degradation of the optimization algorithms in dynamic environments. A more general approach for measuring an attribute of DOPs was proposed in Sarasola and Alba (2013), which calculates an attribute of a population as the area below the curve. Ayvaz et al. (2012) proposed the *dissimilarity factor*, which evaluates an optimization algorithm's ability to recover after a change. Recently, Kordestani et al.

(2019c) proposed a set of two measures as a framework for comparing algorithms in dynamic environments. The first measure, called *alpha-accuracy,* calculates the degree of closeness of the optimization error to a certain expected level during the optimization process. The second measure is *fitness adaptation speed* that evaluates the average speed of algorithms in adapting to environmental changes. Despite the numerous performance measures in the literature, most authors have used *offline error* to measure their proposed algorithms and compare their methods with other peer algorithms. which is defined as the average of the smallest error found by the algorithm in every time step:

$$E_{off} = \frac{1}{T} \sum\nolimits_{t=1}^{T} e_t^* \tag{7.11}$$

where T is the maximum number of FEs so far and e_t^* is the minimum error obtained by the algorithm at the time step t. It is assumed that a function evaluation made by the algorithm corresponds to a single time step.

7.4 Types of Multi-population Methods

Different criteria can be taken into account when categorizing MP methods. Here, we focus on the way the populations are created. The MP approach's main idea is to divide the task of optimization between several sub-populations and execute the sub-populations concurrently. First, we can simply study MP methods in three general categories: (a) MP with a fixed number of populations, (b) MP with varying numbers of populations, and (c) MP with an adaptive number of populations.

7.4.1 Methods with a Fixed Number of Populations

A very straightforward way to create an MP scheme is to define the number of populations before the actual optimization process.

7.4.1.1 Methods Based on Atom Analogy

The main idea of these methods is to establish a mutual repulsion among several fixed-size populations. Thus, they are placed over different promising areas of the search space. The pioneering work in this context was done by Blackwell and Branke (2004). They proposed two multi-swarm algorithms based on particle swarm optimization (PSO), namely mCPSO and mQSO. In mCPSO, each swarm is composed of *neutral* and *charged* particles. Neutral particles update their velocity and position according to the principles of pure PSO.

Algorithm 7-1. Methods inspired by atom analogy

```
Setting the parameters r_excl, r_conv, M, pop_size; /*Set the exclusion radius, convergence radius, multi-population cardinality and population size*/
Randomly initialize M population of individuals in the search space; /*Initialize populations into the search space*/
while termination criteria are not met, do
    Apply anti-convergence operator; /*if (r_conv=0) then anti-convergence is off */
    Apply exclusion operator; /*if (r_excl=0) then exclusion is off */
    Detect change and react to it;
    for each population m do
        Evolve individuals based on their type (neutral, charged, quantum, etc.), and their basic algorithm (e.g., PSO, DE, etc.);
    end-for
end
```

Fig. 7.2 Template for methods based on atom analogy

On the other hand, charged particles move in the same way as neutral particles, but they are also mutually repelled from other charged particles residing in their swarm. Therefore, charged particles help to maintain the diversity inside the swarm. In mQSO, instead of having charged particles, each swarm contains *quantum* particles. Quantum particles change their positions around the center of the swarm's best particle according to a random uniform distribution with radius r_{cloud}.

Consequently, they never converge and provide a suitable level of diversity to swarm to follow the shifting optimum. The authors also introduced the *exclusion* operator, which prevents populations from settling on the same peak. In another work (Blackwell and Branke 2006), the same authors proposed a second operator referred to as *anti-convergence*, triggered after all swarms converge and reinitialize the worst swarm in the search space.

The template for methods based on atom analogy is shown in Fig. 7.2.

Several attempts have been made to elevate the work done in Blackwell and Branke (2006) by modifying different aspects of mQSO that can be examined into four groups discussed in the rest of this sub-section.

7.4.1.1.1 Implementing the Essence of mQSO or Its Variants with Other Algorithms

Some researchers have adopted the general idea of Blackwell and Branke (2006) to capture several peaks in the search space and implement it with other methods. The main purpose of these approaches is to benefit from other algorithms' intrinsic positive characteristics to increase the performance of Nguyen et al. (2012). For instance, Mendes and Mohais (2005) adopted the essence of Blackwell and Branke (2006) and introduced a multi-population differential evolution (DE) algorithm for dynamic environments called DynDE. In DynDE, several genomes populations are initialized into the search space and explore for multiple peaks in the landscape incorporating *exclusion*, which prevents populations from converging to the same optima, and *increasing diversity*, enabling the partially converged population to track the shifting optimum. They used three methods to introduce diversity to each population: Quantum individuals, Brownian individuals, and *Entropic differential evolution*. Moreover, DynDE uses random values for F and CR parameters taken from the uniform distribution, eliminating the need for exhausting the task of parameter fine-tuning. Several DE schemes are also investigated, indicated that greedy schemes are more preferable. Xiao and Zuo (Xiao and Zuo 2012; Zuo and Xiao 2014) proposed an MP algorithm based on a hybrid operator, i.e., *DEPSO* operator from DE and PSO. Their method, called Multi-DEPSO, starts with several populations equal to the number of peaks and tries to place each population on a distinct peak. The operator *DEPSO* carries out DE and PSO's operations for each population in a sequential manner. For each individual x, two individuals are generated by *DEPSO*: (1) an individual p is generated using DE with the greedy scheme DE/best/1/bin, (2) an individual q is generated by applying PSO candidate solution p. Finally, the best individual among x, p, and q is used as an updated individual in the next generation.

7.4.1.1.2 Changing the Number and the Distribution of the Quantum Particles

A group of studies analyzed the effect of changing the number and the distribution of quantum particles on the performance of mQSO. For instance, Trojanowski (2008a) proposed a new class of limited area distribution for quantum particles in which the uniformly distributed candidate solutions within a hyper-sphere with radius r_{cloud} are wrapped using von Neumann's acceptance-rejection method. In another work (Trojanowski 2008b), the same author introduced a two-phase method for generating the cloud of quantum particles in the search space's entire area based on a direction vector θ and the distance d from the original position using an α-stable random distribution. This approach allows particles to be distributed equally in all directions. The findings of both studies revealed that changing the distribution of quantum particles has a significant effect on the performance of mQSO. del Amo et al. (2009) investigated the effect of changing the number of quantum and neutral particles on the performance of mQSO. The three major conclusions of their study can be summarized as follow: (a) an equal number of quantum and neutral particles is not the best configuration for mQSO, (b) configurations in which the number of neutral particles is higher than the number of quantum particles usually performs better, and (c) quantum particles are most helpful immediately after a change in the environment.

7.4.1.1.3 Modifying the Exclusion Operator

Since the initial introduction of the exclusion operator, researchers have made several attempts to alleviate its disadvantages. As stated in Kordestani et al. (2019a), exclusion variations in the literature can be classified and discussed in four main categories. They are approaches for setting the exclusion radius, approaches for detecting situations where collided populations stand on distinct optima, approaches for preventing the reinitialized population from redundant search, and approaches for saving the weaker population's search history. The rest of this section describes each variation in detail.

- **Setting the exclusion radius**

The first category of studies on exclusion is focused on setting the parameter exclusion radius (r_{excl}) to determine the search area of each population in the fitness landscape. These methods can be further categorized into two groups: (*a*) methods that tune the parameter r_{excl} and (*b*) methods that control the parameter r_{excl}.

A. Tuning the parameter r_{excl}

Most existing MP methods try to tune the parameter r_{excl}. Finding a good value for it before the algorithm's actual run and then running the algorithm using this value remains fixed during the run. This is done by experimenting with different values of r_{excl} and it is then selecting the one that gives the best results on the tested DOPs at hand.

Although hand-tuning the parameters based on experimentation is a common practice in the EC community, it has some inherent drawbacks when applying on r_{excl} that can be summarized as follows:

- The process of parameter tuning for param r_{excl} is time-consuming, especially if the search domain is continuous.
- A good value for the parameter r_{excl} depends on different criteria like the number of peaks, search space range, search space dimensionality, the shape of the peaks, and the distribution of the fitness landscape peaks. Moreover, the suitable value for r_{excl} can be varied in different methods. For example, the exclusion radius of all populations is set to 30 in mQSO (Blackwell and Branke 2006) and HmSO (Kamosi et al. 2010b), but 25 in FMSO (Li and Yang 2008).
- For a given DOP, the selected value for the parameter r_{excl} is not necessarily optimal, even if the attempt made for setting the r_{excl} was significant.
- Since the value of r_{excl} is fixed during the run, tuning the value of r_{excl} is not a practical approach for DOPs with a varying number of optima.

B. Controlling the parameter r_{excl}

A few studies have been conducted to control the parameter r_{excl} during the run. In the preliminary work done by Blackwell and Branke, the linear diameter of the basin of attraction of any peak was used to predict the optimal value for r_{excl}, based on the assumption that all p peaks are evenly distributed in search space X^d as follows (Blackwell and Branke 2006):

$$r_{excl} = \frac{X}{2p^{1/d}} \tag{7.12}$$

Where X is the search space range of the d dimensions, and p is the number of existing peaks in the landscape. The disadvantage of this adaptation is that it requires information about the environment, e.g., the number of optima that may not be available in real-world applications.

Several adaptations have been proposed for situations when the information on the number of optima is not available. For example, Blackwell (2007) suggests estimating the exclusion radius based on the number of current populations in the search space by modifying Eq. (7.12) as follows:

$$r_{excl} = \frac{X}{2M^{1/d}} \quad (7.13)$$

where M is the current number of populations in the search space.

Rezazadeh et al. (2011) used the fuzzy c-means (FCM) clustering method for adaptive determination of the exclusion radius with respect to the position of the particles in the search space. In this approach, in each iteration, all particles are first clustered using FCM. Then the distance between centers of the clusters are used to adjust the exclusion radius as follows:

$$mindis_i = \text{minimum}\left(dist\left(center_i, center_j\right)\right), where\ 1 \leq i < j \leq n, i \neq j \quad (7.14)$$

$$r_{excl} = \frac{mean\left(\sum_{i=1}^{n} mindist_i\right)}{2^{-n} + 2} \quad (7.15)$$

In above equations, $dist$ is a function which calculates the Euclidian distance between centers of the clusters i and j in the D-dimensional search space using Eq. (7.15), n is the total number of clusters determined by FCM, and $mindist_i$ denotes the distance from ith cluster to the closest cluster in the search space.

$$dist(i, j) = \sqrt{\sum_{d=1}^{D} \left(x_i^d - x_j^d\right)^2} \quad (7.16)$$

This adaptation also eliminated the need to have information about the search domain for calculating r_{excl}.

- Detecting situations where collided populations are standing on distinct optima

One of the technical drawbacks of exclusion is that it ignores the situations when two populations stand on two distinct but highly close peaks (i.e., within the exclusion radius of each other). In these cases, the exclusion operator removes the worst population and leaves one of the peaks unpopulated. A group of studies has tried to address this issue by adding an extra routine to the exclusion, which is executed when a collision between two populations is detected. For example, du Plessis and Engelbrecht (2012) proposed a reinitialization midpoint check (RMC) to see whether different peaks are located within each other's exclusion radius. In RMC, a simpler version of hill-valley detection (Ursem 2000), once a collision occurs between two populations, the midpoint's fitness on the line between the best individuals in each population is evaluated. Suppose the midpoint fitness has a lower value than the fitness value of both populations' best individuals. In that case, it implies that the two populations reside on distinct peaks and that neither should be reinitialized. Otherwise, the worst-performing population will be reinitialized in the search space. Although RMC is effective, it cannot detect all extremely close peaks correctly, as pointed by the authors.

Zuo and Xiao (2014) applied hill-valley detection with three checkpoints to determine whether two colliding populations are located on different peaks. In their approach, three points between the best solutions of the collided populations x and y are examined. After that, if there exists a point $z = c \cdot x + (1 - c) \cdot y$ for which $f(z) < min\{f(x), f(y)\}$, where $c \in \{0.05, 0.5, 0.95\}$, then two populations are on different peaks and they remain unchanged. Otherwise, they reside at the same peak.

- **Preventing the reinitialized population from redundant search**

Exclusion in its native form simply restarts the worst-performing population when an overlapping between two populations is detected. However, there is a great chance that the reinitialized population is again moving toward the other populations or populated peaks, triggering another exclusion. This can waste the precious computing resources that could be otherwise used to explore the search space's non-visited areas or further exploit the previously found peaks. Therefore, some studies devised some strategies to overcome the mentioned limitation. For instance, in Zuo and Xiao (2014), for each population marked to be reinitialized by the exclusion, a part of its individuals is generated by an opposition-based initialization operator (Rahnamayan et al. 2007). This operator can improve the capability of the algorithm in searching unexplored areas.

In du Plessis and Engelbrecht (2013), du Plessis and Engelbrecht used a mechanism to remove populations that the exclusion operator frequently reinitializes to stop the unproductive search.

Sharifi et al. (2015) proposed a hybrid method based on PSO and local search for DOPs. This method used the exclusion operator to control the number of local search agents in different search space areas. In this regard, when the number of local search agents in an area, which is determined by r_{excl}, exceeds a predefined threshold, the fittest local search agent is kept and the other ones will be removed from the search space.

- **Saving the information obtained the weaker population**

Another disadvantage of basic exclusion is that the weaker population's information will be lost due to restart. This can waste precious computing resources and reduce the performance of the MP algorithms. Some authors proposed modifications to alleviate the information loss caused by reinitializing the worst-performing population upon collision. For instance, in Xiao and Zuo (2012), when the overlapping among two populations is detected, before restarting the worst population, the best individual of the worst population replaces the worst individual of the best population if it is fitter. In Yang and Li (2010), if the degree of overlapping among collided populations becomes greater than a pre-defined threshold, two populations will be merged. This way, it is possible that the information obtained by the weaker population is maintained in the next generation.

7.4.1.1.4 Managing Function Evaluations

Various studies in the literature considered the allocation of FEs to populations from the resource management perspective. These studies aim to manage the FEs to be mostly allocated to the search space's most promising areas. These strategies can be roughly categorized into two classes.

- **Strategies that aim to cut down the allocation of FEs to unproductive populations**

The first group of strategies tries to identify unproductive populations and cut down the allocation of FEs to those populations. This way, more FEs will be available for productive populations to improve the algorithm's overall performance. Hibernation (Kamosi et al. 2010b) is among the earliest function evaluation management schemes for detecting and terminating the allocation of FEs to unproductive populations. In this method, when a population converges, it is deactivated until a change occurs in the environment. Later, Novoa-Hernández et al. (2011) incorporated a new mechanism, named *swarm control mechanism*, into mQSO to enhance its performance. The swarm control mechanism uses a fuzzy membership function to identify so-called bad behavior swarms as those with low diversity and bad fitness and stop them from consuming FEs. du Plessis and Engelbrecht (2008) proposed an extension to DynDE, referred to as favored populations DE. In this method, FEs between two successive changes in the environment are scheduled into three phases: (1) all populations are evolved according to normal DynDE for ζ_1 generations to locate peaks, (2) the weaker populations are frozen and stronger populations are executed for another ζ_2 generation, and (3) frozen populations are back to search process, and all populations are evolved for ζ_3 generations in a usual DynDE manner. This strategy adds three parameters ζ_1, ζ_2, and ζ_3 to DynDE, which must be tuned manually. In another work (du Plessis and Engelbrecht 2012), they equipped DynDE with a new mechanism, namely *Competitive Population Evaluation* (CPE). The primary motivation behind CPE is to manage resources (i.e., FEs) to enable the optimization algorithm to reach the lowest error faster. In CPE, populations compete to take the FEs, and the FEs are allocated to populations based on their performance. Each population's performance is measured based on the current fitness of the best individual in the population and the best individual's improvement during the previous evaluation of the population as defined in du Plessis and Engelbrecht (2012). Hence, the best-performing population will thus be the population with the highest fitness and improvement. At each iteration, the best-performing population evolves until its performance drops below that of another population. Then another population takes the resource, and this process continues during the run. This adaptation scheme allows FEs to mostly spent on higher peaks. Kordestani et al. (2019b) proposed two FE management methods to suitably exploit the FEs assigned to each sub-population. The first method combines the *success rate* and the quality of the best solution found by the sub-populations in a single criterion called *performance index*. The performance index is then used to distribute the FEs among populations. The second method uses a variable-structure leaning automaton for choosing the sub-population that should be executed at each time step. Kordestani and Meybodi (Kordestani et al. 2020) suggested combining the concept of function evaluation management with a mechanism to control the operation performed by each population.

7.4.1.1.5 Methods that Are Explicitly Allocating More FEs to the Best-Found Individual

Another group of methods explicitly allocate more FEs to the best-performing population/individual at each iteration. For instance, Sepas-Moghaddam et al. (2012) incorporated a novel local search into the multi-swarm PSO in which one quantum particle is used to improve the best-performing population at each iteration. This quantum particle can perform a certain number of local-searches, determined with the parameter *try_number*, around the global best position.

Sharifi et al. (2015) proposed a competitive version of hybrid PSO and local search. In this adaptive algorithm, an extra local search operation is allocated to the global best individual at the end of each iteration. The experimental results show that the proposed approach is a very suitable tool for optimizing DOPs.

7.4.1.2 Collaborative Methods

Apart from the methods based on atom analogy, another group of methods uses collaborative mechanisms between populations to share information about promising peaks in the landscape. For example, Lung and Dumitrescu (2007) introduced a hybrid collaborative approach for dynamic environments called Collaborative Evolutionary Swarm Optimization (CESO). CESO has two equal-size populations: the main population to maintain a set of local and global optima during the search process using crowding-based differential evolution (CDE), and a PSO population acting as a local search operator around solutions provided by the first population. During the search process, information is transmitted between both populations via collaboration mechanisms. In another work (Lung and Dumitrescu 2010), the same authors added a third population to the CESO. The extra population acts as a memory to recall some useful information from past generations of the algorithm. Inspired by CESO, Kordestani et al. (2014) proposed a bi-population hybrid algorithm. The first population, called QUEST, is evolved by CDE principles to locate the optima in the fitness landscape. The second population, called TARGET, uses PSO to exploit useful information in the vicinity of the QUEST's best position. When the TARGET population's search process around the best-found position of the QUEST becomes unproductive, the TARGET population is stopped, and all genomes of the QUEST population are allowed to perform an extra operation using hill-climbing local search. The authors also used several mechanisms to conquer the existing challenges in the DOPs.

7.4.1.3 Strength, Weaknesses, and Challenges

Methods falling in this category can provide several advantages as follows:

- They can capture multiple promising areas of the search space in parallel. This property is especially useful in multimodal DOPs where a previously good optimal solution may become bad or vice versa due to a change in the peaks' height.
- They can recall some information from the past. The MP nature of these methods allows them to retain old solutions. This feature is specifically advantageous in recurrent dynamic environments where the optima return to their previous locations or the previous states of the environment are repeated.
- They are simple and easy to implement. Compared to other MP approaches, implementing these methods is easier.

However, this approach in its generic form also has some shortcomings and must conquer some difficulties to produce promising results as follows:

- The proper number of populations is difficult to determine. In methods falling in this category, the number of populations is specified before the optimization process and according to the optima number. It should be noted that the information about the environment may not be in hand in real-world problems. On the other hand, too many populations may deteriorate the optimization's performance, and too few populations may slow down the search progress.
- Specifying the search area of each population is difficult. As was mentioned earlier, the search areas of the populations are separated using an exclusion operator. However, the exclusion also has different intrinsic deficiencies. Therefore, defining the search radius of each population is a challenging task.
- The radius r_{cloud} in quantum particles requires having information about the shift severity. This problem can be partially redeemed by adding an extra procedure to the optimization algorithm to learn the environment's shift severity.
- The proper number of individuals in each population is hard to determine. The number of individuals in each population has a critical effect on the algorithms' performance in dynamic environments. Most of the existing literature works try to find the optimal population size empirically. In contrast, the task of hand-tuning the size of each population is time-consuming, and the value of this parameter is usually problem-dependent. Therefore, methods which adaptively adjust the number of individuals for each population would be preferable.
- FEs are equally distributed between populations. Since the available FEs are limited, they should be spent efficiently in promising search space areas. In general, more FEs should be devoted to areas with higher quality. However, this approach allocates an equal number of FEs to different parts of the search space.

Table 7.2 provides summarized information of different MP methods with a fixed number of populations in dealing with existing challenges of DOPs.

Table 7.2 A summary of MP methods with a fixed number of populations for MPB

Method	Algorithm	Upon detecting change	Redundant search control	Diversity increasing/maintenance	Managing FEs
mQSO (Blackwell and Branke 2006)	PSO	The memory of all particles is reevaluated	If the distance between two swarms is less than the *exclusion radius*, then the worse swarm is reinitialized	Each swarm has some quantum particles which maintain a certain level of diversity	–
mCPSO (Blackwell and Branke 2006)	PSO	The memory of all particles is reevaluated	If the distance between two swarms is less than the *exclusion radius*, then the worse swarm is reinitialized	An inter-particle Coulomb repulsion maintains particle diversity between charged particles	–
DynDE (Mendes and Mohais 2005)	DE	All genomes are reevaluated	If the distance between two populations is less than the *exclusion radius*, then the worse population is reinitialized	Three diversity-increasing mechanisms are introduced: Brownian individuals, Quantum individuals, and Entropy Differential Evolution	–
DynDE + LA (Kordestani et al. 2019b)	DE	All genomes are reevaluated	If the distance between two populations is less than the *exclusion radius*, then the worse population is reinitialized	Three diversity-increasing mechanisms are introduced: Brownian individuals, Quantum individuals, and Entropy Differential Evolution	A variable-structure learning automaton is used to distribute FEs among populations

(continued)

Table 7.2 (continued)

Method	Algorithm	Upon detecting change	Redundant search control	Diversity increasing/maintenance	Managing FEs
mQSOE (Novoa-Hernández et al. 2011)	PSO	The memory of all particles is reevaluated	If the distance between two swarms is less than the *exclusion radius*, then the worse swarm is reinitialized	Each swarm has some quantum particles which maintain a certain level of diversity	Swarms with *low* diversity and *bad* fitness are stopped from consuming FEs
RMCCPE (du Plessis and Engelbrecht 2012)	DE	All genomes are reevaluated	*Exclusion* is performed between collided populations subject to reinitialization Midpoint Check	Brownian individuals are used for introducing diversity to the populations	At each generation, FEs are devoted to the best performing population
Multi-DEPSO (Zuo and Xiao 2014)	DE + PSO	A random normal vector is added to the position of the fraction of the individuals in each population. Then, all individuals are reevaluated	*Exclusion* is performed between collided populations based on a hill-valley detection scheme. In case two populations are residing on the same peak, then the worst population is re-initialized based on the *opposition-based method*	Diversity is increased after detecting a change in the environment	–

7.4.2 Methods with a Variable Number of Populations

A critical drawback of the MP method proposed in Blackwell and Branke (2006) is that the number of swarms should be defined before the optimization process. In contrast, in real-world problems, the information about the environment may not be in hand. Blackwell and Branke (2006) showed that the best performance of the algorithm is obtained when the number of swarms (m) is equal to the number of peaks (p) in the landscape. They pointed out that when $p < m$, there will be $(m - p)$ swarms that will perpetually be reinitialized as they become removed from previously captured peaks. These swarms consume many FEs and cannot contribute much to the optimization process. Thus, they deteriorate the performance of the method.

On the other hand, if $p > m$, there are more peaks than swarms, and the anti-convergence operator will alleviate performance deterioration due to low swarm diversity. Their simulation results also support the mentioned claims. Following previous remarks, it is desirable to adjust the number of swarms in the search space according to the number of existing peaks in the landscape.

7.4.2.1 Methods with a Parent Population and Variable Number of Child Populations

Methods falling in this category consist of a parent population and a varying number of child populations. The parent population is responsible for continuously exploring the search space and maintaining the diversity, and child populations are used to exploit promising sub-regions found by the parent population. Figure 7.3 illustrates the procedure in which different parts of the search space are assigned to various child populations by the parent population.

The very first attempt to develop such a method was made by Branke et al. (2000). Borrowing the forking genetic algorithm (Tsutsui et al. 1997), they proposed an MP method called Self-Organizing Scouts (SOS). In SOS, the optimization process begins with a large parent population exploring through the entire search space to find promising areas. When such areas are located by the parent, a child population (i.e., scout population) is split off from the parent population and independently explores the respective sub-space. In contrast, the parent population continues to search in the remaining search space for locating new optima. The number of individuals in each population is determined using a quality measure calculated as defined in Branke et al. (2000). The higher the value of the quality measure for a population, the more individuals are assigned to the population. The maximum and minimum population sizes of the parent population and child populations are also specified as the algorithm's parameters. Finally, the search area of the populations is separated using the exclusion.

Algorithm 2 in Fig. 7.4 shows the high-level template for methods based on a parent population and some child populations.

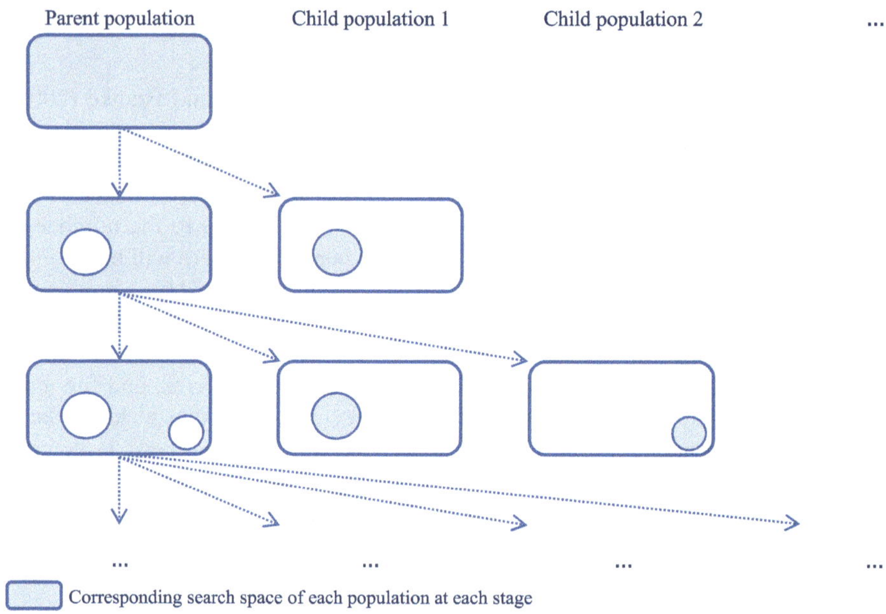

Fig. 7.3 Division of the search space in methods based on a parent population and several child populations (Branke et al. 2000)

Algorithm 7-2. Template of the algorithms based on a parent population and some child populations

```
Setting the parameters r_excl, parent_size, and child_size; /*set the search radius for child populations, number
    of individuals in parent population, and number of individuals in child populations*/
Randomly initialize the parent population in the search space;
while termination criteria are not met, do
  | Detect change and react to it;
  | Evolve child populations;
  | Evolve parent population;
  | Apply exclusion operator; /*if (r_excl=0) then exclusion is off */
  | if the parent population locates a promising area, then
  |   | Create a new child population in the corresponding sub-region of the space;
  | end
end
```

Fig. 7.4 Template for methods based on a parent population and variable number of child populations

Later, other researchers implemented the general idea of SOS using different optimization algorithms. For instance, Li and Yang (2008) proposed a fast multi-swarm (FMSO) algorithm for locating and tracking optima in DOPs. FMSO starts with a large parent swarm exploring through the search space to discover promising regions. Whenever the best particle in the parent swarm improves, a child swarm is split off from the parent swarm to exploit its corresponding sub-space. They also used exclusion to separate the search area of the child swarm. Similar ideas can also be found in mPSO (Kamosi et al. 2010b) and HmSO (Kamosi et al. 2010b).

Recently, Yazdani et al. (2013) proposed a multi-swarm algorithm based on a finder and various tracker sub-populations (FTMPSO). In this algorithm, the finder population is exploring the search space to find the peaks. In this case, a tracker swarm will be generated by transferring the fittest particles from the finder swarm into the newly created tracker swarm. The corresponding tracker swarm is then responsible for covering the peak summit and tracking its movements after the change. Meanwhile, the finder swarm is then reinitialized into the search space to capture other uncovered peaks. The exclusion is also applied between every pair of tracker swarms to avoid exploiting the same peak. Besides, if the finder swarm converges to a populated area, it is reinitialized into the search space without generating any tracker swarm.

7.4.2.1.1 Strength, Weaknesses, and Challenges

Methods with a parent population and some child populations can bring different advantages. They can:

- The number of populations is dynamically changed during the run. Compared to MP with a fixed number of populations, this approach's main advantage is that there is no need to set the number of populations before the run.
- Like other MP methods, they can capture multiple promising areas of the search space in parallel.
- Like other MP methods, they can recall some information from the past.

However, this approach also needs to overcome some difficulties to gain better performance:

- The number of individuals in the parent population and child populations is hard to determine. Contrary to the SOS in which individuals are dynamically allocated to different populations based on a quality measure, the other methods following this approach have a fixed number of individuals in each of their populations. Therefore, the proper population size for the parent population and child populations should be determined manually.
- Specifying the search area of each population is difficult. These methods are also used exclusion to separate the search area of different populations.
- Specifying the maximum number of child populations is a challenging task. Since an excessive number of child populations can be quite harmful to the methods' performance, these methods should control the number of current child populations in the fitness landscape.

7.4.2.2 Methods Based on Space Partitioning

The main idea of these methods is to partition the search space into some sub-spaces and guide the search process in promising sub-spaces. Hashemi and Meybodi were the first to use the space partitioning idea in MPB (Hashemi and Meybodi 2009c). They incorporated PSO, and cellular automata into an algorithm referred to as Cellular PSO. In Cellular PSO, the search space is partitioned into some equally sized cells using cellular automata. Then, particles of the swarm are allocated to different cells according to their search space distribution. At any time, particles residing in each cell use their best personal experience and the best solution found in the cell's neighborhood for searching for an optimum.

Moreover, whenever the number of particles within a cell exceeds a predefined limit, randomly selected particles from the overcrowded cell are allocated to randomly chosen cells within the search space. Besides, each cell has a memory used to keep track of the best position found within its boundary and the best position among its neighbors. In Hashemi and Meybodi (2009a), the same authors changed some particles' role in each cell, from standard particles to quantum particles, for a few iterations upon detecting a change in the environment. Figure 7.5 shows the application of a 2D cellular automaton for space partitioning.

Several authors have followed the same approach to solve MPB. For example, Noroozi et al. adopted the concept of Cellular PSO and proposed an algorithm based on DE called CellularDE (Noroozi et al. 2011). The CellularDE employs DE/rand-to-best/1/bin scheme to provide local exploration capability for individuals residing in each cell. After a change is detected in the environment, the whole population performs a random local search for several upcoming iterations. In another work (Noroozi et al. 2012), they equipped CellularDE with a mechanism to estimate the presence of a peak within each cell's boundary or its neighbors. In this method, called Alpinist CellularDE, a type is assigned to each cell from set $T \in \{PEAK, SLOPE, PLATUE, UNKNOWN\}$. Each type is representing the nature of the fitness landscape within the corresponding cell. Initially, since there is no information about the fitness landscape, all cells are labeled as UNKNOWN. During the search process, the cells' type is changed using the fitness value received from different landscape positions. PEAK cells are those fittest cells among the neighborhood where the best individual located within the cell improves less than a threshold Ø during the last t updates. All adjacent cells to a PEAK cell are considered as a SLOPE. Finally, PLATUE cells are those for which the cell's local best memory is located on a neighboring SLOPE cell. To exploit FEs more efficiently, the maximum allowed number of individuals in SLOPE and PLATUE cells is decreased to half. This strategy allocates more FEs to the PEAK cells, which are more likely to contain local optimum. Besides, the search process in PEAK cells is conducted by a hill-climbing search around the cell's best position. Moreover, the authors also used a local search to improve the efficiency of tracking local optima after detecting a change in the environment.

Sharifi et al. (2012) proposed a two-phased collaborative algorithm, named Two Phased Cellular PSO (TP-CPSO), to improve Cellular PSO's performance. In TP-CPSO, each cell can function in two different phases: the exploration phase, in which the search is performed using a modified PSO algorithm, and the exploitation phase, in which the search process is conducted by a local search with a high capability of exploitation. Initially, all cells are in the exploration phase, but they go to the

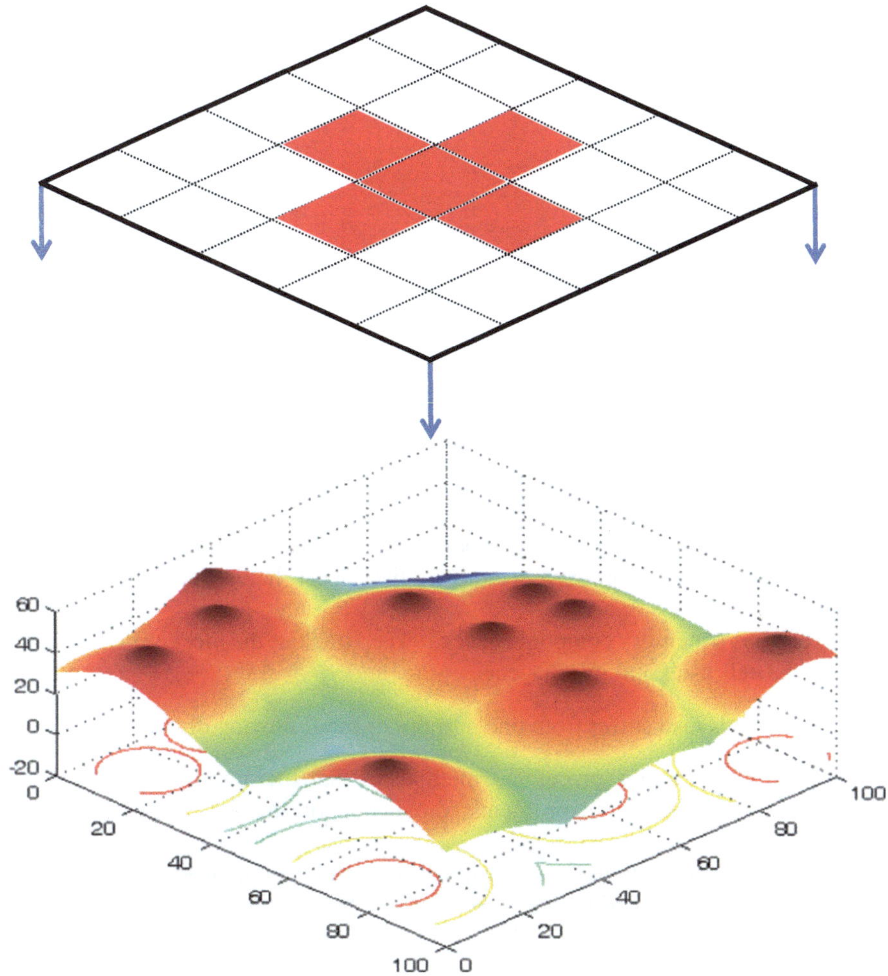

Fig. 7.5 Embedding a cellular automaton in a two-dimensional search space. The 1st-order von Neumann neighborhood is shown for the central cell

exploitation phase upon converging to local optima. Algorithm 3 in Fig. 7.6 shows the general template for methods based on space partitioning.

Strength, Weaknesses, and Challenges
Methods following the space partitioning approach offer the following advantages:

- Like other MP methods, they can capture multiple promising areas of the search space in parallel.
- These methods eliminate the need for using the exclusion to separate the search area of the populations. Since the number of individuals in each cell is kept below a

Algorithm 7-3. Template of the algorithms based on space partitioning

```
Initialize a cellular automaton with C^D equal-sized cells; /*partitions the search space into C^D cells*/
Randomly initialize a population of individuals in the cellular automaton; /*Initialize the population into the
partitions*/
while termination criteria are not met, do
    Detect change and react to it;
    for each cell_i in the cellular automaton, do
        Update memories of cell_i;
        Evolve active individuals in cell_i based on their type (neutral, charged, quantum,
        etc.), and their basic algorithm (e.g., PSO, DE, etc.);
        Check the number of individuals residing in cell_i and adjust them according to
        threshold θ; /* θ=∞ means that there is no constraint on the number of individuals
        within the corresponding cell*/
    end-for
end
```

Fig. 7.6 Template for methods based on space partitioning

predefined threshold, space partitioning methods do not use exclusion to separate the search areas of different populations.

- Finding local optima in these methods is faster since the populations' search process is conducted in limited areas of the fitness landscape from the very beginning.
- Each cell has a memory component used to keep track of the most promising area inside the cell itself and its adjacents.

These methods also suffer from one or more of the following drawbacks:

- The main disadvantage of cellular methods is that the number of cells exponentially increases as the problem's dimension or the number of partitions increase.
- The size of the cells should be determined before the commencement of the run. Therefore, the process of finding a suitable size for the cells is exhaustive.
- These methods seem to work well only for problems where the solution space is bounded.

7.4.3 Methods Based on Population Clustering

This approach's main objective is to use a clustering method to split the main population into several sub-populations based on the individuals' distribution in the fitness landscape. Several attempts have been made to employ the population clustering in the context of DOPs, and specifically MPB. For instance, Parrott and Li (2006) proposed a speciation-based multi-swarm PSO (SPSO), which dynamically distributes particles of the main swarm over a variable number of so-called *species*. In SPSO, a population of randomly generated particles is initialized into the search space. Afterward, at each iteration of the algorithm, different species are first

Algorithm 7-4. Template of the algorithms based on population clustering

1. Randomly initialize a population of individuals in the search space; /*Initialize the population into the partitions*/
2. **while** *termination criteria are not met,* **do**
3. Detect change and react to it;
4. Cluster the main population into several sub-populations using a clustering method; /* K-means, hierarchical clustering, competitive clustering, etc.*/
5. Evolve all sub-populations using the inner optimizer (e.g., *PSO*, *DE*, etc.);
6. Check the status of the sub-populations; /*In terms of overlapping, redundancy and convergence*/
7. **end**

Fig. 7.7 Template for methods based on population clustering

constructed based on *species seed* as defined in Parrott and Li (2006) and *species radius* r_s.

On the other hand, a number of p_{max} particles that fall within the r_s distance from a species seed are classified as the same species. Each species is then used to explore the corresponding sub-region of the search space. In another work (Bird and Li 2007), the performance of the SPSO was improved by estimating the location of the peaks using a least-squares regression method. Similarly, Yang and Li (2010) applied a hierarchical clustering method to divide an initial cradle swarm into various sub-swarms, each of which covers a different sub-area of the landscape. Each sub-swarm is evolved using a modified PSO. Afterward, each sub-swarm status (i.e., overlapping, overcrowding, and convergence) is checked, and an operator is executed on each sub-swarm accordingly. In another work (Li and Yang 2012), the fundamental idea of CPSO was extended by introducing a novel framework for covering undetectable DOPs.

Similar ideas have also been proposed using competitive clustering (Nickabadi et al. 2012), k-means clustering (Halder et al. 2013), and affinity propagation clustering (Liu et al. 2020). The template for MP methods based on population clustering is illustrated in Fig. 7.7.

7.4.3.1 Strength, Weaknesses, and Challenges

The main advantage of the cluster-based MP algorithms is that:

- The algorithm automatically adjusts the number of subpopulations and each subpopulation size based on individuals' distribution in the fitness landscape.

  Apart from that:

- The computational cost of the clustering operation can be too high, especially when the DOP is large-scale.

Table 7.3 summarizes different MP methods with a variable number of populations in dealing with the existing challenges of DOPs.

Table 7.3 A summary of MP methods with a variable number of populations for MPB

Method	Algorithm	Upon detecting change	Redundant search control	Diversity increasing/maintenance	Managing FEs
HmSO (Kamosi et al. 2010b)	PSO	The memory of all particles in the parent swarm is reevaluated. Hibernated child swarms are awakened. All particles in each child swarm are sample around the best-found position of their corresponding child swarm	If the distance between two child swarms is less than a *radius*, then the worse child swarm is removed		Hibernation mechanism is used to deactivate unproductive populations
CellularPSO (Hashemi and Meybodi 2009b)	PSO	Reset the memory of cells and the particles' history	A threshold is specified for the number of particles in each cell	Some randomly selected particles of each saturated cell are reinitialized to a random cell	–
Alpinist CellularDE (Noroozi et al. 2012)	DE	All genomes are reevaluated, and the memory of all cells will be cleared. A directed local search is performed on cell best memory	A threshold is specified for the number of particles in each cell	The worst genomes of each saturated cell are reinitialized to a random cell. Noise is added to individuals	The maximum allowed number of individuals in SLOPE and PLATUE cells is decreased to half

(continued)

Table 7.3 (continued)

Method	Algorithm	Upon detecting change	Redundant search control	Diversity increasing/maintenance	Managing FEs
SPSO (Parrott and Li 2006)	PSO	All particles' personal bests are reset to their associated current positions	Only a limited number of candidate members (p_{max}) with the highest fitness values will be assigned as members of a species	Lower fitness candidate members that would exceed species members are reinitialized at random positions in the solution space and are not allocated to a new species until the following iteration	
CPSO (Yang and Li 2010)	PSO	Reevaluate the global best particle of all sub-swarms. Save the best position of each sub-swarm, then all particles reinitialized to create a new cradle swarm. Worst particles are replaced by saved positions			
CCPSO (Nickabadi et al. 2012)	PSO	The memories of all particles are refreshed		A group of free particles	

7.4.4 *Self-adapting the Number of Populations*

The last group of studies includes those methods that monitor the algorithm's search progress and adapt the number of populations based on the number of current promising peaks in the fitness landscape using one or more feedback parameters. To this end, Blackwell (2007) introduced a self-adaptive version of mQSO, which dynamically adjusts the number of swarms by spawning new swarms into the search space or destroying redundant ones. The algorithm, i.e., AmQSO, has two types of populations: *free* swarms, whose expansion is larger than radius r_{conv}, and *converged* swarms. Once the expansion of a *free* swarm becomes smaller than a radius r_{conv}, it is converted to the *converged* swarm. Whenever the number of *free* swarms that have not converged to an optimum (M_{free}) has dropped to zero, a *free* swarm is initialized to the search space for capturing undetected peaks. On the other hand, *free* swarms are removed if M_{free} is higher than n_{excess}, a parameter of the algorithm. This adaptation can be expressed using the following formula:

$$M_0 = 1$$
$$M_{t+1} = \begin{cases} M_t + 1, \; if \; M_{free} = 0 \\ M_t - 1, \; if \; M_{free} > n_{excess} \end{cases} \tag{7.17}$$

Therefore, the number of *free* swarms acts as a feedback parameter for the algorithm. Some researchers adapted the idea of AmQSO and used it in conjunction with other ECs. The DMAFSA (Yazdani et al. 2012) and MMCSA (Fouladgar and Lotfi 2016) are some examples of those methods.

du Plessis and Engelbrecht (2013) proposed dynamic population DE (DynPopDE), in which the populations are adaptively spawned and removed required, based on their performance. In this approach, whenever all of the current populations fail to improve their fitness after spending their last respective FEs, DynPopDE produces a new population of random individuals in the search space. They defined a function $\Upsilon(t)$, which indicates if all existing populations in the search space have been stagnated, as follow:

$$\Upsilon(t) = \begin{cases} true \quad if(\Delta f_k(t) = 0)\forall k \in \kappa \\ false \; otherwise \end{cases} \tag{7.18}$$

where κ denotes to the set of current populations and $\Delta f_k(t)$ is the amount of improvement in the kth population at time *t*, which is calculated as follow:

$$\Delta f_k(t) = |f_k(t) - f_k(t-1)| \tag{7.19}$$

The DynPopDE starts with a single population and gradually adapts to an appropriate number of populations. If the number of populations surplus the number of peaks in the landscape, redundant populations are detected as frequently reinitialized by the exclusion operator. Therefore, a population *k* will be removed from the search space when it is marked for restart due to exclusion, and it has not improved since its last FEs ($\Delta f_k(t) \neq 0$).

7.5 Numerical Results

In this section, the numerical results obtained by some of the known MP algorithms that have been tested on the MPB are reported. The results of different MP algorithms taken from the literature are reported in Table 7.4.

Table 7.4 Comparison of the average offline error ± standard error of the recent or well-known algorithms for dynamic environments modeled with MPB (Scenario 2)

Method	*m*								
	1	5	10	20	30	40	50	100	200
mQSO (Blackwell and Branke 2006)	4.49 ± 0.14	1.72 ± 0.05	1.77 ± 0.05	2.49 ± 0.05	2.58 ± 0.04	2.58 ± 0.04	2.50 ± 0.03	2.27 ± 0.02	2.12 ± 0.02
AmQSO (Blackwell and Branke 2004)	1.27 ± 0.04	1.30 ± 0.06	1.64 ± 0.05	2.03 ± 0.04	2.11 ± 0.03	2.18 ± 0.02	2.18 ± 0.02	2.14 ± 0.01	2.12 ± 0.01
CellularPSO (Hashemi and Meybodi 2009b)	2.79 ± 0.19	1.94 ± 0.18	1.93 ± 0.08	2.73 ± 0.12	3.08 ± 0.11	3.28 ± 0.11	3.34 ± 0.07	3.48 ± 0.11	3.37 ± 0.08
mPSO (Kamosi et al. 2010b)	0.90 ± 0.05	1.21 ± 0.12	1.66 ± 0.08	2.05 ± 0.08	2.18 ± 0.06	2.24 ± 0.06	2.30 ± 0.04	2.32 ± 0.04	2.34 ± 0.03
HmSO (Kamosi et al. 2010b)	1.01 ± 0.03	1.23 ± 0.05	1.47 ± 0.04	1.67 ± 0.04	1.72 ± 0.03	1.78 ± 0.03	1.75 ± 0.02	1.78 ± 0.02	1.82 ± 0.01
APSO (Rezazadeh et al. 2011)	0.53 ± 0.01	1.05 ± 0.06	1.31 ± 0.03	1.69 ± 0.05	1.78 ± 0.02	1.86 ± 0.02	1.95 ± 0.02	1.95 ± 0.01	1.90 ± 0.01
TP-CPSO (Sharifi et al. 2012)	0.40 ± 0.01	0.75 ± 0.07	0.96 ± 0.05	1.18 ± 0.03	1.32 ± 0.03	1.36 ± 0.03	1.40 ± 0.03	1.50 ± 0.02	1.56 ± 0.02
DMAFSA (Yazdani et al. 2012)	0.55 ± 0.06	0.78 ± 0.06	1.01 ± 0.05	1.42 ± 0.06	1.63 ± 0.06	–	1.84 ± 0.07	1.95 ± 0.05	1.99 ± 0.04
FTMPSO (Yazdani et al. 2013)	0.18 ± 0.01	0.47 ± 0.05	0.67 ± 0.04	0.93 ± 0.04	1.14 ± 0.04	–	1.32 ± 0.04	1.61 ± 0.03	1.67 ± 0.03
DynPopDE (du Plessis and Engelbrecht 2013)	–	1.03 ± 0.07	1.39 ± 0.03	–	–	–	2.10 ± 0.03	2.34 ± 0.02	2.44 ± 0.02
CCPSO (Nickabadi et al. 2012)	**0.09 ± 0.00**	**0.25 ± 0.01**	0.75 ± 0.06	1.21 ± 0.08	1.40 ± 0.07	1.47 ± 0.08	1.50 ± 0.09	1.76 ± 0.09	–

(continued)

Table 7.4 (continued)

Method	*m*								
	1	5	10	20	30	40	50	100	200
CDEPSO (Kordestani et al. 2014)	0.41 ± 0.00	0.97 ± 0.01	1.22 ± 0.01	1.54 ± 0.01	2.62 ± 0.01	–	2.20 ± 0.01	1.54 ± 0.01	2.11 ± 0.01
DynDE + LA (Kordestani et al. 2019b)	3.07 ± 0.12	1.41 ± 0.08	1.32 ± 0.06	2.60 ± 0.07	3.05 ± 0.10	3.34 ± 0.07	3.56 ± 0.09	3.88 ± 0.11	3.71 ± 0.09
DynDE + PI (Kordestani et al. 2019b)	2.70 ± 0.11	1.32 ± 0.09	1.47 ± 0.08	2.46 ± 0.08	2.91 ± 0.11	3.28 ± 0.09	3.38 ± 0.10	3.78 ± 0.11	3.62 ± 0.09
DynDE + HLA (Kordestani et al. 2020)	0.92 ± 0.04	0.98 ± 0.05	1.16 ± 0.08	2.27 ± 0.07	2.79 ± 0.09	3.03 ± 0.08	3.22 ± 0.09	3.50 ± 0.08	3.41 ± 0.09
BfCS-wVN (Kordestani et al. 2018)	0.30 ± 0.06	0.38 ± 0.21	**0.51 ± 0.11**	**0.74 ± 0.12**	**1.00 ± 0.11**	**0.97 ± 0.07**	**0.84 ± 0.06**	1.11 ± 0.06	1.18 ± 0.08
CHPSO (Sharifi et al. 2015)	0.19 ± 0.00	0.44 ± 0.02	0.64 ± 0.02	0.91 ± 0.01	**0.99 ± 0.01**	1.02 ± 0.01	1.03 ± 0.01	**1.04 ± 0.01**	**1.01 ± 0.00**

From Table 7.4, we can observe no omnipotent algorithm to beat all the others on all tested DOPs. Considering Table 7.4, when the number of peaks $m = 1, 5$, CCPSO is the best-performing method. However, as the number of peaks in the landscape is increased, BfCS-wVN and CHPSO are the best-performing algorithms.

7.6 Conclusions

This chapter attempts to provide a review of some of the multi-population methods proposed for the MPB problem. To this end, multi-population methods are categorized based on their distinctive features and how they try to improve the overall performance of the optimization in dynamic environments. Hopefully, the contents of this chapter help the readers to understand the following chapters.

References

Alba, E., Sarasola, B.: Measuring fitness degradation in dynamic optimization problems. In: European Workshops on Applications of Evolutionary Computation, EvoApplications, pp. 572–581. Springer, Cham (2010)

Ayvaz, D., Topcuoglu, H.R., Gurgen, F.: Performance evaluation of evolutionary heuristics in dynamic environments. Appl. Intell. **37**, 130–144 (2012). https://doi.org/10.1007/s10489-011-0317-9

Bird, S., Li, X.: Using regression to improve local convergence. In: 2007 IEEE Congress on Evolutionary Computation, Singapore, pp. 592–599. IEEE (2007)

Blackwell, T.: Particle swarm optimization in dynamic environments. In: Yang, S., Ong, Y.-S., Jin, Y. (eds.) Evolutionary Computation in Dynamic and Uncertain Environments, pp. 29–49. Springer, Heidelberg (2007)

Blackwell, T., Branke, J.: Multiswarms, exclusion, and anti-convergence in dynamic environments. IEEE Trans. Evol. Comput. **10**, 459–472 (2006). https://doi.org/10.1109/TEVC.2005.857074

Blackwell, T., Branke, J.: Multi-swarm optimization in dynamic environments. In: Applications of Evolutionary Computing, pp. 489–500. Springer, Heidelberg (2004)

Branke, J.: Memory enhanced evolutionary algorithms for changing optimization problems. In: Proceedings of the 1999 IEEE Congress on Evolutionary Computation, pp. 1875–1882. IEEE (1999)

Branke, J., Kaussler, T., Smidt, C., Schmeck, H.: A multi-population approach to dynamic optimization problems. In: Parmee, I.C. (ed.) Evolutionary Design and Manufacture: Selected Papers from ACDM 2000, pp. 299–307. Springer, London (2000)

Cruz, C., González, J.R., Pelta, D.A.: Optimization in dynamic environments: a survey on problems, methods and measures. Soft. Comput. **15**, 1427–1448 (2011). https://doi.org/10.1007/s00500-010-0681-0

del Amo, I.G., Pelta, D.A., González, J.R., Novoa, P.: An analysis of particle properties on a multi-swarm PSO for dynamic optimization problems. In: Conference of the Spanish Association for Artificial Intelligence, pp. 32–41. Springer, Cham (2009)

du Plessis, M.C., Engelbrecht, A.P.: Differential evolution for dynamic environments with unknown numbers of optima. J. Global Optim. **55**, 73–99 (2013). https://doi.org/10.1007/s10898-012-9864-9

du Plessis, M.C., Engelbrecht, A.P.: Using competitive population evaluation in a differential evolution algorithm for dynamic environments. Eur. J. Oper. Res. **218**, 7–20 (2012). https://doi.org/10.1016/j.ejor.2011.08.031

du Plessis, M.C., Engelbrecht, A.P.: Improved differential evolution for dynamic optimization problems. In: 2008 IEEE Congress on Evolutionary Computation (IEEE World Congress on Computational Intelligence), pp. 229–234 (2008)

Fernandez-Marquez, J.L., Arcos, J.L.: An evaporation mechanism for dynamic and noisy multimodal optimization. In: ACM, pp. 17–24 (2009)

Fouladgar, N., Lotfi, S.: A novel approach for optimization in dynamic environments based on modified cuckoo search algorithm. Soft. Comput. **20**, 2889–2903 (2016). https://doi.org/10.1007/s00500-015-1951-7

Halder, U., Das, S., Maity, D.: A cluster-based differential evolution algorithm with external archive for optimization in dynamic environments. IEEE Trans. Cybern. **43**, 881–897 (2013). https://doi.org/10.1109/TSMCB.2012.2217491

Hashemi, A.B., Meybodi, M.R.: A multi-role cellular PSO for dynamic environments. In: Proceedings of the 14th International CSI Computer Conference, pp. 412–417. IEEE (2009a)

Hashemi, A.B., Meybodi, M.R.: Cellular PSO: a PSO for dynamic environments. In: Cai, Z., Li, Z., Kang, Z., Liu, Y. (eds.) Advances in Computation and Intelligence, ISICA 2009, pp. 422–433. Springer, Heidelberg (2009b)

Jin, Y., Branke, J.: Evolutionary optimization in uncertain environments—a survey. IEEE Trans. Evol. Comput. **9**, 303–317 (2005). https://doi.org/10.1109/TEVC.2005.846356

Kamosi, M., Hashemi, A.B., Meybodi, M.R.: A hibernating multi-swarm optimization algorithm for dynamic environments. In: 2010 Second World Congress on Nature and Biologically Inspired Computing (NaBIC), Fukuoka, Japan, pp. 363–369. IEEE (2010a)

Kamosi, M., Hashemi, A.B., Meybodi, M.R.: A new particle swarm optimization algorithm for dynamic environments. In: Panigrahi, B.K., Das, S., Suganthan, P.N., Dash, S.S. (eds.) Swarm, Evolutionary, and Memetic Computing, SEMCCO 2010, pp. 129–138. Springer, Heidelberg (2010b)

Kordestani, J.K., Abedi Firouzjaee, H., Meybodi, M.R.: An adaptive bi-flight cuckoo search with variable nests for continuous dynamic optimization problems. Appl. Intell. **48**, 97–117 (2018). https://doi.org/10.1007/s10489-017-0963-7

Kordestani, J.K., Meybodi, M.R., Rahmani, A.M.: A note on the exclusion operator in multi-swarm PSO algorithms for dynamic environments. Connection Sci. 1–25 (2019a). https://doi.org/10.1080/09540091.2019.1700912

Kordestani, J.K., Meybodi, M.R., Rahmani, A.M.: A two-level function evaluation management model for multi-population methods in dynamic environments: hierarchical learning automata approach. J. Exper. Theor. Artif. Intell. 1–26 (2020). https://doi.org/10.1080/0952813X.2020.1721568

Kordestani, J.K., Ranginkaman, A.E., Meybodi, M.R., Novoa-Hernández, P.: A novel framework for improving multi-population algorithms for dynamic optimization problems: a scheduling approach. Swarm Evol. Comput. **44**, 788–805 (2019b). https://doi.org/10.1016/j.swevo.2018.09.002

Kordestani, J.K., Rezvanian, A., Meybodi, M.R.: New measures for comparing optimization algorithms on dynamic optimization problems. Nat. Comput. **18**, 705–720 (2019c). https://doi.org/10.1007/s11047-016-9596-8

Kordestani, J.K., Rezvanian, A., Meybodi, M.: CDEPSO: a bi-population hybrid approach for dynamic optimization problems. Appl. Intell. (2014). https://doi.org/10.1007/s10489-013-0483-z

Li, C., Yang, S.: Fast multi-swarm optimization for dynamic optimization problems. In: 2008 Fourth International Conference on Natural Computation, Jinan, China, pp. 624–628. IEEE (2008)

Li, C., Yang, S.: A general framework of multipopulation methods with clustering in undetectable dynamic environments. IEEE Trans. Evol. Comput. **16**, 556–577 (2012). https://doi.org/10.1109/TEVC.2011.2169966

Liu, Y., Liu, J., Jin, Y., Li, F., Zheng, T.: An affinity propagation clustering based particle swarm optimizer for dynamic optimization. Knowl. Based Syst. **195**, 105711 (2020). https://doi.org/10.1016/j.knosys.2020.105711

Lung, R.I., Dumitrescu, D.: Evolutionary swarm cooperative optimization in dynamic environments. Nat. Comput. **9**, 83–94 (2010). https://doi.org/10.1007/s11047-009-9129-9

Lung, R.I., Dumitrescu, D.: A collaborative model for tracking optima in dynamic environments. In: IEEE Congress on Evolutionary Computation, pp. 564–567 (2007)

Mack, Y., Goel, T., Shyy, W., Haftka, R.: Surrogate model-based optimization framework: a case study in aerospace design. In: Yang, S., Ong, Y.-S., Jin, Y. (eds.) Evolutionary Computation in Dynamic and Uncertain Environments, pp. 323–342. Springer, Heidelberg (2007)

Mavrovouniotis, M., Li, C., Yang, S.: A survey of swarm intelligence for dynamic optimization: algorithms and applications. Swarm Evol. Comput. **33**, 1–17 (2017). https://doi.org/10.1016/j.swevo.2016.12.005

Mendes, R., Mohais, A.S.: DynDE: a differential evolution for dynamic optimization problems. In: IEEE Congress on Evolutionary Computation, vol. 3, pp. 2808–2815 (2005)

Michalewicz, Z., Schmidt, M., Michalewicz, M., Chiriac, C.: Adaptive business intelligence: three case studies. In: Yang, S., Ong, Y.-S., Jin, Y. (eds.) Evolutionary Computation in Dynamic and Uncertain Environments, pp. 179–196. Springer, Heidelberg (2007)

Moser, I., Chiong, R.: Dynamic function optimisation with hybridised extremal dynamics. Memetic Comput. **2**, 137–148 (2010). https://doi.org/10.1007/s12293-009-0027-6

Nasiri, B., Meybodi, M., Ebadzadeh, M.: History-Driven Particle Swarm Optimization in dynamic and uncertain environments. Neurocomputing **172**, 356–370 (2016). https://doi.org/10.1016/j.neucom.2015.05.115

Nguyen, T.T., Yang, S., Branke, J.: Evolutionary dynamic optimization: a survey of the state of the art. Swarm Evol. Comput. **6**, 1–24 (2012). https://doi.org/10.1016/j.swevo.2012.05.001

Nguyen, T.T., Yao, X.: Continuous dynamic constrained optimization—the challenges. IEEE Trans. Evol. Comput. **16**, 769–786 (2012). https://doi.org/10.1109/TEVC.2011.2180533

Nickabadi, A., Ebadzadeh, M.M., Safabakhsh, R.: A competitive clustering particle swarm optimizer for dynamic optimization problems. Swarm Intell. **6**, 177–206 (2012). https://doi.org/10.1007/s11721-012-0069-0

Noroozi, V., Hashemi, A.B., Meybodi, M.R.: CellularDE: a cellular based differential evolution for dynamic optimization problems. In: Dobnikar, A., Lotrič, U., Šter, B. (eds.) Adaptive and Natural Computing Algorithms, ICANNGA 2011, pp. 340–349. Springer, Heidelberg (2011)

Noroozi, V., Hashemi, A.B., Meybodi, M.R.: Alpinist CellularDE: a cellular based optimization algorithm for dynamic environments. In: Proceedings of the 14th Annual Conference Companion on Genetic and Evolutionary Computation, Philadelphia, PA, USA, pp. 1519–1520. ACM Press (2012)

Novoa-Hernández, P., Corona, C.C., Pelta, D.A.: Efficient multi-swarm PSO algorithms for dynamic environments. Memetic Comput. **3**, 163–174 (2011)

Parrott, D., Li, X.: Locating and tracking multiple dynamic optima by a particle swarm model using speciation. IEEE Trans. Evol. Comput. **10**, 440–458 (2006). https://doi.org/10.1109/TEVC.2005.859468

Rahnamayan, S., Tizhoosh, H.R., Salama, M.M.: A novel population initialization method for accelerating evolutionary algorithms. Comput. Math. Appl. **53**, 1605–1614 (2007)

Rezazadeh, I., Meybodi, M.R., Naebi, A.: Adaptive particle swarm optimization algorithm for dynamic environments. In: Tan, Y., Shi, Y., Chai, Y., Wang, G. (eds.) Advances in Swarm Intelligence, pp. 120–129. Springer, Heidelberg (2011)

Sarasola, B., Alba, E.: Quantitative performance measures for dynamic optimization problems. In: Alba, E., Nakib, A., Siarry, P. (eds.) Metaheuristics for Dynamic Optimization, pp. 17–33. Springer, Heidelberg (2013)

Sepas-Moghaddam, A., Arabshahi, A., Yazdani, D., Dehshibi, M.M.: A novel hybrid algorithm for optimization in multimodal dynamic environments. In: 2012 12th International Conference on Hybrid Intelligent Systems (HIS), pp. 143–148. IEEE (2012)

Sharifi, A., Kordestani, J.K., Mahdaviani, M., Meybodi, M.R.: A novel hybrid adaptive collaborative approach based on particle swarm optimization and local search for dynamic optimization problems. Appl. Soft Comput. **32**, 432–448 (2015). https://doi.org/10.1016/j.asoc.2015.04.001

Sharifi, A., Noroozi, V., Bashiri, M., Hashemi, A.B., Meybodi, M.R.: Two phased cellular PSO: a new collaborative cellular algorithm for optimization in dynamic environments. In: 2012 IEEE Congress on Evolutionary Computation, Brisbane, QLD, Australia, pp. 1–8. IEEE (2012)

Trojanowski, K.: Tuning quantum multi-swarm optimization for dynamic tasks. In: Rutkowski, L., Tadeusiewicz, R., Zadeh, L.A., Zurada, J.M. (eds.) Artificial Intelligence and Soft Computing – ICAISC 2008, pp. 499–510. Springer, Heidelberg (2008)

Trojanowski, K.: Adaptive non-uniform distribution of quantum particles in mQSO. In: Li, X., Kirley, M., Zhang, M., Green, D., Ciesielski, V., Abbass, H., Michalewicz, Z., Hendtlass, T., Deb, K., Tan, K.C., Branke, J., Shi, Y. (eds.) Proceedings of the Simulated Evolution and Learning: 7th International Conference, SEAL 2008, Melbourne, Australia, 7–10 December 2008, pp. 91–100. Springer, Heidelberg (2008b)

Tsutsui, S., Fujimoto, Y., Ghosh, A.: Forking genetic algorithms: GAs with search space division schemes. Evol. Comput. **5**, 61–80 (1997)

Ursem, R.K.: Multinational GAs: multimodal optimization techniques in dynamic environments, pp. 19–26. Morgan Kaufmann Publishers Inc. (2000)

Weicker, K.: Performance measures for dynamic environments. In: Guervós, J.J.M., Adamidis, P., Beyer, H.-G., Schwefel, H.-P., Fernández-Villacañas, J.-L. (eds.) Parallel Problem Solving from Nature—PPSN VII: Proceedings of the 7th International Conference Granada, Spain, 7–11 September 2002, pp. 64–73. Springer, Heidelberg (2002)

Xiao, L., Zuo, X.: Multi-DEPSO: a DE and PSO based hybrid algorithm in dynamic environments, pp. 1–7. IEEE (2012)

Yang, S., Li, C.: A clustering particle swarm optimizer for locating and tracking multiple optima in dynamic environment. IEEE Trans. Evol. Comput. **14**, 959–974 (2010). https://doi.org/10.1109/TEVC.2010.2046667

Yazdani, D., Akbarzadeh-Totonchi, M.R., Nasiri, B., Meybodi, M.R.: A new artificial fish swarm algorithm for dynamic optimization problems. In: 2012 IEEE Congress on Evolutionary Computation, Brisbane, QLD, Australia, pp. 1–8. IEEE (2012)

Yazdani, D., Nasiri, B., Sepas-Moghaddam, A., Meybodi, M.R.: A novel multi-swarm algorithm for optimization in dynamic environments based on particle swarm optimization. Appl. Soft Comput. **13**, 2144–2158 (2013). https://doi.org/10.1016/j.asoc.2012.12.020

Zuo, X., Xiao, L.: A DE and PSO based hybrid algorithm for dynamic optimization problems. Soft. Comput. **18**, 1405–1424 (2014). https://doi.org/10.1007/s00500-013-1153-0

Chapter 8
Learning Automata for Online Function Evaluation Management in Evolutionary Multi-population Methods for Dynamic Optimization Problems

Javidan Kazemi Kordestani, Mehdi Razapoor Mirsaleh, Alireza Rezvanian, and Mohammad Reza Meybodi

Abstract Multi-population (MP) approach is among the most successful methods for solving continuous dynamic optimization problems (DOPs). Nevertheless, the MP approach has to conquer several obstacles to reach its maximum performance. One of these obstacles, which is the subject of this chapter, is how the MP methods exploit function evaluations (FEs). Since the calculation of FEs is the most expensive component of the evolutionary computation (EC) methods for solving real-world DOPs, we should find a way to spend a major portion of FEs around the most promising search area space. In generic form, the MP approach as a sub-population located far away from the optimal solution(s) is assigned the same amount of FEs as that near-optimal solution(s), which in turn exert deleterious effects on the performance of the optimization process. Therefore, one major challenge is how to suitably assign the FEs to each sub-population to enhance MP methods' efficiency for DOPs. This chapter generalizes the application of variable-structure learning automaton (VSLA) and fixed-structure learning automaton (FSLA) for FE management to improve MP methods for DOPs. The present work is applied to DE-based MP methods, MP version of particle swarm optimization (PSO), firefly algorithm (FFA), and JAYA.

J. Kazemi Kordestani
Department of Computer Engineering, Science and Research Branch, Islamic Azad University, Tehran, Iran

M. Razapoor Mirsaleh
Department of Computer Engineering and Information Technology, Payame Noor University (PNU), P.O. BOX, 19395-3697 Tehran, Iran

A. Rezvanian (✉)
Department of Computer Engineering, University of Science and Culture, Tehran, Iran
e-mail: rezvanian@usc.ac.ir

M. R. Meybodi
Computer Engineering Department, Amirkabir University of Technology (Tehran Polytechnic), Tehran, Iran
e-mail: mmeybodi@aut.ac.ir

J. Kazemi Kordestani et al. (eds.), *Advances in Learning Automata and Intelligent Optimization*, Intelligent Systems Reference Library 208, https://doi.org/10.1007/978-3-030-76291-9_8

8.1 Introduction

Multi-population (MP) approach is among the most successful methods for solving continuous dynamic optimization problems (DOPs). Nevertheless, the MP approach has to conquer several obstacles to reach its maximum performance. One of these obstacles, which is the subject of this chapter, is how the MP methods exploit function evaluations (FEs). Since the calculation of FEs is the most expensive component of the evolutionary computation (EC) methods for solving real-world DOPs, we should find a way to spend a major portion of FEs around the most promising search area space. At least in its generic form, the MP approach significantly reduces the utilization of FEs and delays the process of finding the global optimum by sharing an equal portion of FEs among sub-populations. In other words, sub-populations located far away from the optimal solution(s) are assigned the same amount of FEs as those located near to optimal solution(s), which in turn exert deleterious effects on the performance of the optimization process. Therefore, one major challenge is how to suitably assign the FEs to each sub-population to enhance MP methods' efficiency for DOPs. A recent study (Kazemi Kordestani et al. 2019b) investigated the application of variable-structure learning automaton (VSLA) (Rezvanian et al. 2019b) for FE management to improve the performance of MP methods for DOPs. In this work, the VSLA-based FE management was used to enhance three MP differential evolution (DE) methods, including DynDE, jDE, and mSQDE.

Interestingly, the experimental results showed that using a FE management approach achieves a better performance. In particular, using the VSLA-based FE management is more promising than the other FE management schemes. Following the work in Kazemi Kordestani et al. (2019b), the authors extended their work in Kazemi Kordestani and Meybodi (2020) by investigating the fixed-structure learning automaton (FSLA) (Rezvanian et al. 2018c) for FE management in dynamic environments.

This chapter generalizes the works in (Kazemi Kordestani et al. 2019b, Kazemi Kordestani and Meybodi) by proposing a general model for FE management in MP methods. Besides, while the previous works only focused on DE-based MP methods, the present work is applied to other MP algorithms. Specifically, in this chapter, the MP version of particle swarm optimization (PSO), firefly algorithm (FFA), and JAYA are considered.

The rest of this chapter is structured as follows: Sect. 8.2 discusses the related works and factors that deteriorate MP methods' performance when solving DOPs. The theory of learning automaton (LA) is described in Sect. 8.3. Section 8.4 reviews the EC methods understudy in brief. Section 8.5 explains our proposed model and a framework for MP methods with FE management in detail. Section 8.6 reviews three FE-management methods based on the proposed model—the experimental study regarding the proposed model Sect. 8.7. Finally, Sect. 8.8 concludes the chapter with some directions for future works.

8.2 Preliminaries

A considerable amount of FEs in MP methods are wasted during optimization. There are various reasons for wasting FEs in MP methods, some of which can be summarized as follows:

8.2.1 *Waste of FEs Due to Change Detection*

Change detection is a vital task in solving DOPs. Although several approaches exist, the most common way for detecting change is to re-evaluate one or more "sentry" individuals of the algorithm, called sensors, at each iteration. If the changes occur globally, one sensor is enough to detect the changes. However, if the changes occur locally, choosing a good number of sensors would be challenging. In both situations, the MP method needs to consume additional FEs to detect the changes. A group of studies suggested the methods that detect changes based on algorithm behavior. Most of these methods detect changes based on monitoring the drop in value of the average of best-found solutions over a time period. This approach has the advantage of not requiring any additional FEs. Although there is no guarantee that changes are detected, this approach may cause false positives (Nguyen et al. 2012).

8.2.2 *Waste of FEs Due to the Excessive Number of Sub-populations*

As mentioned in different studies, the number of sub-populations can dramatically affect the performance of MP methods. For example, Blackwell and Branke (2006b) showed that the optimal numbesr of populations (m) is equal to the number of peaks (p) in the fitness landscape for the moving peaks benchmark (MPB) problem with a small number of peaks (i.e., less than ten peaks). They pointed out that when $m < p$, there will be $(p - m)$ swarms that will be perpetually reinitialized as they become removed from previously captured peaks. These swarms consume many FEs and cannot contribute much to the optimization process. Thus, they deteriorate the performance of the method. Recently proposed studies confirmed that the optimal number of populations should be relevant to the number of promising peaks (du Plessis and Engelbrecht 2013b). Therefore, when the number of populations is greater than the number of promising peaks, a share of FEs is wasted without contributing to the optimization process's performance.

8.2.3 Waste of FEs Due to Overcrowding of Subpopulations in the Same Area of the Search Space

Another reason for wasting FEs in MP methods is the overcrowding of two or more sub-populations in the same fitness landscape area. Several proposals exist in the literature about defining and separating each sub-population area in the search space. For example, Blackwell and Branke (2006b) proposed the exclusion operator to ensure that no more than one population surrounds a single peak by reinitializing worse populations that cover the same area. Later, many researchers borrowed the concept of exclusion in their MP algorithms (Mendes and Mohais 2005; Kamosi et al. 2010d).

Apart from the exclusion, Hashemi and Meybod (Hashemi and Meybodi 2009c) proposed cellular automata's application to split the search space into some equally sized partitions. The maximum number of individuals in each partition is then determined using a density control parameter. A similar idea can also be found in (Hashemi and Meybodi 2009a; Noroozi et al. 2011b; Noroozi et al. 2012; Nabizadeh et al. 2012b; Sharifi et al. 2012).

8.2.4 Waste of FEs Due to Exclusion Operator

The exclusion operator is a key component of MP methods to separate the populations' search areas and prevent the redundant search. However, as stated in Kordestani et al. (2019a), the exclusion operator in its generic form suffers from various drawbacks. one of the main disadvantages of the exclusion operator is that the weaker population's information will be lost due to restart. This can waste precious computing resources and reduce the performance of the MP algorithms. Since the initial introduction of exclusion, several attempts have been made by researchers to improve this operator. For example, Kazemi Kordestani modified the mQSO algorithm so that sub-populations do not move toward locations that have been already captured by the other ones (Kordestani et al. 2019a). This way, the waste of FEs due to the exclusion operator is alleviated.

8.2.5 Allocation of FEs to Unproductive Populations

Another straightforward source of FE wastage is allocating FEs to sub-populations that are not contributing to the search process (e.g., converged populations are one of those). Various studies exist in the literature that used different techniques to detect and stop unproductive populations. For example, Kamosi et al. (2010a) proposed an interesting mechanism called hibernation for stopping the execution of unproductive populations in a multi-swarm algorithm. In their method, when the radius of a child swarm, say c, becomes less than a constant parameter r_{conv} (i.e., $r_c < r_{conv}$), and the difference between fitness of the child swarm and the global best-found position is less than a threshold (i.e., $f_c < f_{gbest} - \xi$) the corresponding child swarm is deactivated. In turn, more FEs are available for productive populations. In (Novoa-Hernández et al. 2011), the authors proposed a fuzzy resource management mechanism, called swarm control mechanism, for DOPs. The swarm control mechanism is activated on the swarms with low diversity and bad fitness and stops them from consuming FEs. Sharifi et al. (2015) recently generalized the idea of hibernation to stop the search process in individuals that are not contributing to the search performance.

8.2.6 Unsuitable Parameter Configuration of the EC Methods

Finding good parameter values for EC methods is a conventional approach for improving these methods' efficiency in static environments. A group of studies tried to maximize the utilization of FEs by adjusting the parameters of EC methods to increase their convergence speed and adaptivity in DOPs. For example, Nickabadi et al. (2011) proposed an adaptive strategy to control the value of inertia weight in PSO during the optimization process. A time-varying inertia weight, called oscillating triangular inertia weight, was also proposed by Kordestani et al. (2016). Both mentioned approaches reported an improvement over the basic method on the MPB problem. Recently, Kazemi Kordestani et al. (2018) proposed a bi-flight MP cuckoo search algorithm that uses LA to adjust the search strategy of each sub-population by adapting the parameter β in Levy distribution.

8.2.7 Equal Distribution of FEs Among Sub-populations

As shown in Kazemi Kordestani et al. (2019b), the equal distribution of FEs among sub-populations is not a good choice. The reasons are:

i. The main goal in dynamic optimization is to locate the global optimum in a minimum amount of time and track its movements in the solution space. Therefore, strategies that can locate the global optimum faster would be preferable.
ii. A DOP can contain several peaks or local optima. However, as the heights of the peaks are different, they do not have the same importance from the optimality point of view. Therefore, spending an equal number of FEs on all of them postpones reaching the lower error.
iii. Many real-world optimization problems are large-scale in nature. For these problems, the equal distribution of FEs among sub-populations would be detrimental to the algorithms' performance.

Therefore, the MP methods should execute the populations in a manner that better-performing populations receive more FEs. To this end, du Plessis and Engelbrecht (du Plessis and Engelbrecht 2012b) proposed a new mechanism, namely competitive population evaluation (CPE), to manage resources to enable the optimization algorithm to reach the lowest error faster. In CPE, populations compete to obtain FEs, which are allocated to populations according to their performance. Each population's performance is measured based on a performance metric (du Plessis and Engelbrecht 2012b). According to this metric, the best-performing population will be the population with the highest fitness and improvement values. At each iteration, the best performing population takes the FEs and evolves itself until its performance drops below that of another population, where the other population takes the resource, and this process continues during the run. Kazemi Kordestani et al. (Kazemi Kordestani et al. 2019b) proposed two FE management methods to suitably exploit the FEs assigned to each sub-population. The first method combines the success rate and the quality of the best solution found by the sub-populations in a single performance index criterion. The performance index is then used to distribute the FEs among populations. The second method uses VSLA to distribute FEs among sub-populations. In Ozsoydan and Baykasoğlu (2019), the authors proposed two sub-population selecting strategies based on sequential selection and roulette wheel selection. Kazemi Kordestani et al. (Kazemi Kordestani and Meybodi) also proposed the application of FSLA for sub-population selection in MP methods for DOPs. In another work (Kordestani et al. 2020), Kazemi Kordestani et al. proposed a two-level FE management based on hierarchical LA to control further the operation performed by each sub-population. The present work also falls within this category, where the main objective is to distribute FEs among populations in MP methods to maximize their final performance.

8.3 Theory of Learning Automata

An LA (Narendra and Thathachar 1989; Rezvanian et al. 2018b; Vafashoar et al. 2021a) is an adaptive decision-making unit that improves its performance by learning the way to choose the optimal action from a finite set of allowable actions through iterative interactions with an unknown random environment. The action is chosen at random based on a probability distribution kept over the action-set, and at each instant, the given action is served as the input to the random environment. The environment responds to the taken action in turn with a reinforcement signal. The action probability vector is updated based on the reinforcement feedback from the environment. An LA's objective is to find the action sets optimal action to minimize the average penalty received from the environment.

The environment can be described by a triple $E = \langle \alpha, \beta, c \rangle$ where $\alpha = \{\alpha_1, \alpha_2, \ldots, \alpha_r\}$ represents the finite set of the inputs, $\beta = \{\beta_1, \beta_2, \ldots, \beta_m\}$ denotes the set of the values that can be taken by the reinforcement signal, and $c = \{c_1, c_2, \ldots, c_r\}$ denotes the set of the penalty probabilities, where the element c_i is associated with the given action α_i. If the penalty probabilities are constant, the random environment is said to be a stationary random environment, and if they vary with time, the environment is called a non-stationary environment. The environments depending on the nature of the reinforcement signal β can be classified into P-model, Q-model, and S-model. The environments in which the reinforcement signal can only take two binary values, 0 and 1, are referred to as P-model environments. In another class of the environments, a finite number of the interval [0, 1] can be taken by the reinforcement signal. Such an environment is referred to as a Q-model environment. In S-model environments, the reinforcement signal lies in the interval $[a, b]$. Learning automata can be classified into two main families (Narendra and Thathachar 2012): fixed structure learning automata and variable structure learning automata.

If the probabilities of the transition from one state to another and probabilities of correspondence of action and state are fixed, the automaton is said to be fixed-structure automata. Otherwise, the automaton is said to be variable-structure automata.

8.3.1 Fixed Structure Learning Automata

A fixed structure learning automaton is a quintuple $\langle \alpha, \Phi, \beta, F, G \rangle$ where:

- $\alpha = (\alpha_1, \cdots, \alpha_r)$ is the set of actions that it must choose from.
- $\Phi = (\Phi_1, \cdots, \Phi_s)$ is the set of states.
- $\beta = \{0, 1\}$ is the set of inputs where 1 represents a penalty and 0 represents a reward.
- $F : \Phi \times \beta \rightarrow \Phi$ is a map called the transition map. It defines the transition of the state of the automaton on receiving input; F may be stochastic.
- $G : \Phi \rightarrow \alpha$ is the output map and determines the action taken by the automaton if it is in state Φ_j.

The operation of a typical FSLA can be described as follows:

At the first step, the selected action $\alpha(n) = G[\Phi(n)]$ serves as the input to the environment, which in turn emits a stochastic response $\beta(n)$ at the time n. $\beta(n)$ is an element of $\beta = \{0, 1\}$ and is the feedback response of the environment to the automaton. In the second step, the environment penalizes (i.e., $\beta(n) = 1$) the automaton with the penalty probability c_i, which is action-dependent. Based on the response $\beta(n)$, the state of the automaton is updated by $\Phi(n+1) = F[\Phi(n), \beta(n)]$. This process continues until the desired result is obtained. The early models of such automata were Tsetline, Krinsky, and Krylov automata (Rezvanian et al. 2018c).

8.3.2 Variable Structure Learning Automata

A variable structure learning automaton is represented by a quadruple $\langle \beta, \alpha, p, T \rangle$ where $\beta = \{\beta_1, \beta_2, \ldots, \beta_m\}$ is the set of inputs, $\alpha = \{\alpha_1, \alpha_2, \ldots, \alpha_r\}$ is the set of actions, $p = \{p_1, p_2, \ldots, p_r\}$ is the probability vector, which determines the selection probability of each action, and T is a learning algorithm that is used to modify the action probability vector, i.e., $p(n+1) = T[(n), (n), p(n)]$. Let $\alpha(n)$ and $p(n)$ denote the action chosen at instant n and the action probability vector on which the chosen action is based, respectively. The recurrence equations shown by Eq. (8.1) and Eq. (8.2) is a linear learning algorithm by which the action probability vector p is updated as follows:

$$p_j(n+1) = \begin{cases} p_j(n) + a.\big(1 - p_j(n)\big) & if\ i = j \\ p_j(n).(1-a) & if\ i \neq j \end{cases} \tag{8.1}$$

when the taken action is rewarded by the environment (i.e., $\beta(n) = 0$), and

$$p_j(n+1)\begin{cases} p_j(n).(1-b) & if\ i = j \\ \frac{b}{r-1} + (1-b).p_j(n) & if\ i \neq j \end{cases} \tag{8.2}$$

when the taken action is penalized by the environment (i.e., $\beta(n) = 1$), in the above equations, r is the number of actions. Finally, a and b denote the reward and penalty parameters and determine the amount of increases and decreases of the action probabilities, respectively. If $a = b$, the recurrence Eqs. (8.1) and (8.2) are called linear reward—penalty (L_{R-P}) algorithm, if $a \gg b$ the given equations are called linear reward—ε penalty ($L_{R-\varepsilon P}$), and finally, if $b = 0$ they are called linear reward–inaction (L_{R-I}). In the latter case, the action probability vectors remain unchanged when the environment penalizes the taken action.

Learning automaton has been shown to perform well in parameter adjustment (Kazemi Kordestani et al. 2014; Mahdaviani et al. 2015a), networking (Rezvanian et al. 2018a), social networks (Rezvanian et al. 2018a), to name a few.

8.4 EC Techniques under Study

8.4.1 Particle Swarm Optimization

PSO is a versatile population-based stochastic optimization method that was first proposed by Kennedy and Eberhart (Blackwell 2007b) in 1995. PSO begins with a population of randomly generated particles in a D-dimensional search space. Each particle i of the swarm has three features: $\vec{x}_i$ that shows the current position of the particle i in the search space, $\vec{v}_i$ which is the velocity of the particle i and $\vec{p}_i$ which

Algorithm 8-1. Pseudocode for canonical PSO

```
Setting parameters ω, c1 and c2 ;
Generate the initial swarm of the particles with random positions and velocities;
Evaluate the fitness of each particle of the swarm;
repeat
    for each particle i in the swarm do
        update particle i according to Eq. (8-3) and Eq. (8-4);
        if (fitness(p_i)<fitness(x_i)) then
            p_i:= x_i;
            if (fitness(p_g)<fitness(x_i)) then
                p_g:= x_i;
            end-if
        end-if
    end-for
until a termination condition is met
```

Fig. 8.1 Pseudocode of canonical particle swarm optimization

denotes the best position found so far by the particle i. Each particle i updates its position in the search space, at every time step t, according to the following equations:

$$\vec{v}_i(t+1) = \omega\vec{v}_i(t) + c_1r_1\big[\vec{p}_i(t) - \vec{x}_i(t)\big] + c_2r_2\big[\vec{p}_g(t) - \vec{x}_i(t)\big] \tag{8.3}$$

$$\vec{x}_i(t+1) = \vec{x}_i(t) + \vec{v}_i(t+1) \tag{8.4}$$

where ω is an inertia weight that governs the amount of speed preserved from the previous iteration. c_1 and c_2 denote the cognitive and social learning factors used to adjust the degree of particles' movement toward their personal best position and the global best position of the swarm, respectively.r_1 and r_2 are two independent random variables drawn with uniform probability from [0, 1]. Finally, $\vec{p}_g$ is the global best position found so far by the swarm. The pseudocode of the PSO is shown in Fig. 8.1.

8.4.2 Firefly Algorithm

FFA is another nature-inspired stochastic algorithm proposed by Yang (2010), inspired by fireflies' flashing behavior. A firefly presents each candidate solution, and the light intensity of each firefly determines its fitness value. In this algorithms, for any two flashing fireflies i and j, the less bright one, say i, will move towards the brighter one, say j, according to the following equation:

$$\vec{x}_i(t+1) = \vec{x}_i(t) + \beta_0 e^{-\gamma r_{ij}^2}\big(\vec{x}_j(t) - \vec{x}_i(t)\big) + \alpha(\epsilon_i - 0.5) \tag{8.5}$$

Algorithm 8-2. Pseudocode for firefly algorithm

```
setting parameters β0, ϕ and α;
generate the initial population of the fireflies with random positions;
evaluate the fitness of each firefly of the population;
repeat
    for each firefly i in the population do
        for each firefly j in the population do
            if (fitness j is greater than the fitness i) then
                move firefly i towards firefly j in all dimensions according to Eq. (8-6);
                evaluate the fitness of the newly generated solution;
            end-if
        end-for
    end-for
until a termination condition is met
```

Fig. 8.2 Pseudocode of Firefly algorithm

where β_0 is the attractiveness at $r = 0$, γ is the light absorption coefficient, r_{ij}^2 is the Euclidean distance between the firefly i and firefly j, α is the randomization parameter and ϵ_i is a uniform random number drawn from [0, 1].

As the distance between pair of fireflies increases, the value of $\beta_0 e^{-\gamma r_{ij}^2}$ approaches to zero quickly. As a result, fireflies might be stuck in their positions throughout the following generations. To prevent this, γ can be fixed to very small values; however, obtaining an appropriate value for γ is not an easy task because it affects the search's efficiency. Therefore, in this chapter, we replace the Eq. (8.5) with the following equation as suggested in Ozsoydan and Baykasoglu (2015):

$$\vec{x}_i(t+1) = \vec{x}_i(t) + \beta_0\left(\frac{1}{\phi + r}\right)\left(\vec{x}_j(t) - \vec{x}_i(t)\right) + \alpha(\epsilon_i - 0.5), \tag{8.6}$$

where ϕ is a very small number like 0.000001 to prevent division by zero, the pseudocode of the FFA is shown in Fig. 8.2.

8.4.3 Jaya

Jaya (Venkata Rao 2016) is a relatively new population-based algorithm that has obtained very promising results for constrained and unconstrained optimization problems. This algorithm's main idea is to reach the best solution while moving away from the worst one. The initial population of the Jaya algorithm is randomly generated within the range of the search space. At each iteration, the location of the new candidate solution $\vec{x}_i$ is determined according to the following equation:

$$\vec{x}_i(t+1) = \vec{x}_i(t) + r_1\left[\vec{x}_{best}(t) - |\vec{x}_i(t)|\right] - r_2\left[\vec{x}_{worst}(t) - |\vec{x}_i(t)|\right] \tag{8.7}$$

Algorithm 8-3. Pseudocode for Jaya algorithm

```
setting parameters;
generate the initial population of the solutions with random positions;
evaluate the fitness of each solution of the population;
repeat
    for each solution i in the population do
        update solution i according to Eq. (8-7);
        evaluate the fitness of the newly generated solution;
    end-for
    accept the new solutions if better;
until a termination condition is met
```

Fig. 8.3 Pseudocode of Jaya algorithm

where $\vec{x}_{best}$ and $\vec{x}_{worst}$ are the location of the best and worst solutions found by the algorithm so far, respectively. r_1 and r_2 are two independent random variables drawn with uniform probability from [0, 1].

Regarding Eq. (8.7), the term $r_1\left[\vec{x}_{best}(t) - |\vec{x}_i(t)|\right]$ determines the tendency of the candidate solution to move toward the best-found position in the search space. Similarly, the term $-r_2\left[\vec{x}_{worst}(t) - |\vec{x}_i(t)|\right]$ indicates the tendency of the candidate solution to avoid the worst found position.

The interesting point about the Jaya is that it does not contain any algorithm-specific parameters rather than population size. Figure 8.3 summarizes the pseudocode for the Jaya algorithm.

8.5 LA-Based FE Management Model for MP Evolutionary Dynamic Optimization

Our proposed model's overall architecture is illustrated in Fig. 8.4. It consists of three major components: General manager, Action selector, and State/Action updater.

The general manager unit is a two-way interface between the MP methods and our model, which manages the whole process of communication between the resource management system and MP algorithm. The general manager closely observes the progress of the optimization process. If a subpopulation contributes to the search progress, then the general manager unit sends a positive signal to the state/action updater unit. Otherwise, a negative signal is sent to the state/action updater unit by the general manager unit.

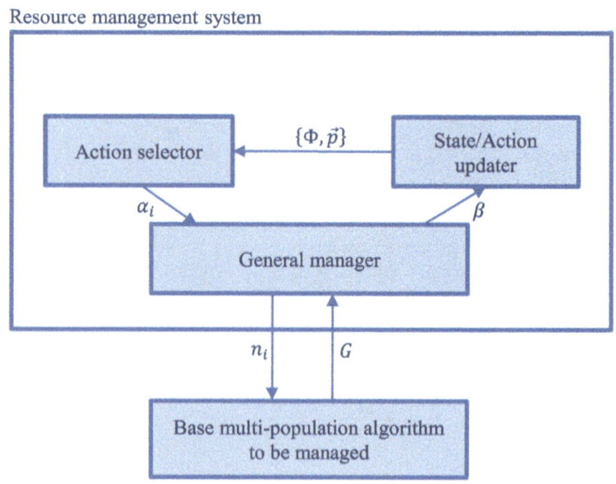

Fig. 8.4 The architecture of the proposed model

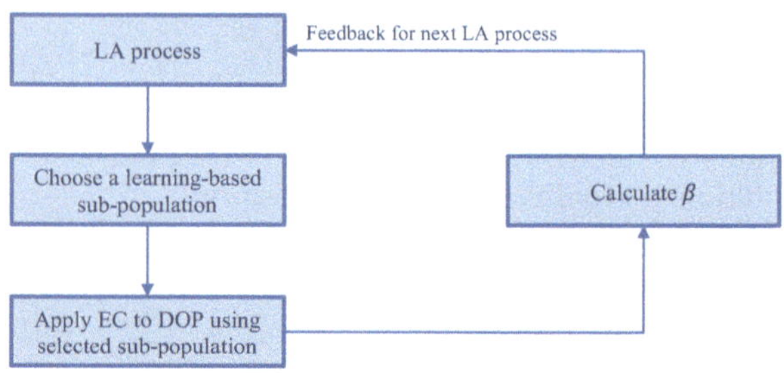

Fig. 8.5 The learning process of LA in the proposed model

The action selector unit is responsible for choosing an action for the LA from its set of allowable actions. If the LA is a VSLA, then the action selector unit randomly chooses one action according to the probability vector $\vec{p}$. On the other hand, if the LA is an FSLA, then the action selector unit chooses an action using the output mapping $G[\Phi(n)]$.

State/Action updater contains the learning procedure for updating the state (in case of FSLA) or probability vector $\vec{p}$ (in case of VSLA) of the LA-based on the reinforcement signal β received from the general manager unit.

The learning process of LA in the proposed model can be summarized in Fig. 8.5.

Regarding the above model, any EC method can be converted into the MP version with an FE management scheme using the framework shown in Fig. 8.6. In the rest of this section, different parts of the framework are studied in detail.

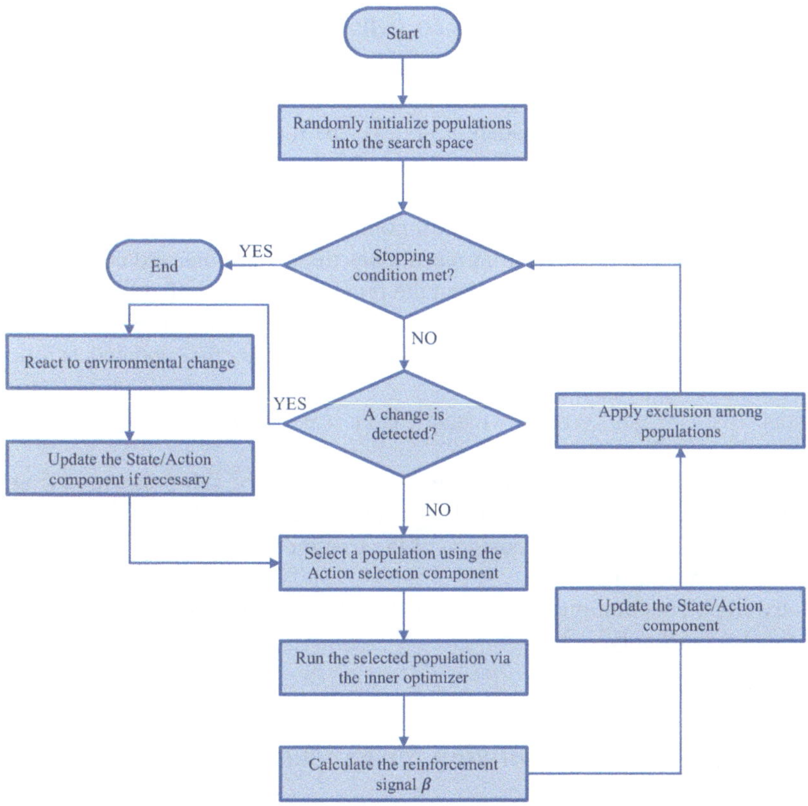

Fig. 8.6 The general framework for MP with FE management based on the proposed model

8.5.1 Initialization of Sub-populations

The proposed framework in Fig. 8.6 starts with N populations of randomly generated individuals in a D-dimensional search space. Each sub-population contains i individuals which are simply randomized into the boundary of the search space according to a uniform distribution as follows:

$$\vec{x}_{i,k} = lb_j + r[0,1] \times \left(ub_j - lb_j\right) \tag{8.8}$$

where $\vec{x}_{i,k}$ is the position of the i-th individual of sub-population k, $r[0, 1]$ is a uniformly distributed random number in [0, 1]. Finally, lb_j and ub_j are the lower and upper bounds of the search space corresponding to j-th dimension of the search space.

8.5.2 Detection and Response to Environmental Changes

Several methods have been suggested for detecting changes in dynamic environments. See, for example Nabizadeh et al. (2012b) for a good survey on the topic. The most frequently used method consists of reevaluating the algorithm's memories for detecting inconsistencies in their corresponding fitness values. In the proposed framework, the best solution for each sub-population was considered. At the beginning of each iteration, the algorithm reevaluates these solutions and compares their current fitness values against those from the previous iteration. If at least one inconsistency is found from these comparisons, one can assume that the environment has changed. Once a change is detected, all sub-populations are first updated as follows:

$$\vec{x}_{i,k}(e+1) = r[-1, 1] \cdot \vec{s} + \vec{x}_g(e) \tag{8.9}$$

where $\vec{x}_{i,k}(e+1)$ is the position of the i-th individual of sub-population k at the beginning of environment $e+1$, $r[-1, 1]$ is a random number in $[-1, 1]$, $\vec{s}$ is the change severity vector and $\vec{x}_g(e)$ is the position of the best individual of sub-population k at the end of environment e.

Afterward, all populations are reevaluated, and the state/action updater component is modified accordingly.

8.5.3 Choose a Sub-population for Execution

At this point, the action selector component selects a sub-population for execution with the underlying optimization method. Depending on the nature of the LA (i.e., VSLA or FSLA), the LA either uses a probability vector $\vec{p}$ or output mapping $G[\Phi(n)]$ for choosing one of its actions. Afterward, the selected action is sent to the general manager unit. The general manager unit then informs the EC to execute the corresponding sub-population with the inner optimizer for one iteration. Finally, the fitness of the individuals of the sub-population is evaluated using the fitness function.

8.5.4 Evaluate the Search Progress of Populations and Generate the Reinforcement Signal

A very important part of the proposed model is to measure the contribution of a sub-population in improving the algorithm's performance. In this regard, the researchers have already used various criteria to measure a sub-population's effectiveness in DOPs. For example, Kamosi et al. (Kamosi et al. 2010a) considered the combination of diversity and fitness value of each sub-swarm's best particle for measuring their performance in HmSO. The same criterion has been also used by Novoa-Hernández et al. (Novoa-Hernández et al. 2011). Kazemi Kordestani et al. (Kazemi Kordestani et al. 2019b) combined the success rate and fitness value of each subpopulation best individual in a single criterion to measure the performance of each sub-population. We use the algorithm's global best individual fitness value to measure the search progress in this work. In other words, if the execution of a sub-population improves the global best fitness value found by the algorithm, then the corresponding subpopulation has made a good contribution to the performance of the MP method.

After evolving the selected sub-population, the general manager unit checks whether the selected sub-population's execution improved the MP algorithm's overall performance. To this end, the general manager unit compares the global best-found solution's fitness value against that from before executing the selected sub-population. Then the general manager unit generates a reinforcement signal as follows:

$$Reinforcement\ signal\ \beta = \begin{cases} 0\ iff\left(\vec{X}_{G,t}\right) < f\left(\vec{X}_{G,t+1}\right) \\ 1\ otherwise \end{cases} \tag{8.10}$$

where f is the fitness function, and $\vec{X}_{G,t}$ is the global best individual of the algorithm at time step t. The above equation implies that if the MP algorithm's quality of the best-found solution improves by executing the selected sub-population, a positive signal is generated by the general manager unit. The generated reinforcement signal is then sent to the state/action unit. Based on the received reinforcement signal, the state/action unit determines whether the selected action was right or wrong.

8.5.5 Exclusion

In order to prevent different sub-populations from settling on the same peak, an exclusion operator is applied. This operator triggers when the distance between the best individuals of any two sub-populations becomes less than the threshold r_{excl}, and reinitializes the worst-performing sub-population in the search space. The parameter r_{excl} is calculated as follows:

$$r_{excl} = \frac{X}{2m^{1/d}}, \tag{8.11}$$

where X is the range of the search space for each dimension, and m is the number of sub-populations in the landscape. The pseudocode for the proposed framework is shown in Fig. 8.7.

Algorithm 8-4. The framework for MP algorithms with a resource management scheme

```
setting the parameters r_excl, N, pop_size; /*Set the exclusion radius, multi-population cardinality and the size of each population*/
randomly initialize M populations of individuals in the search space; /*Initialize populations into the search space*/
while termination condition is not met do
    if a change in the environment is detected then
        introduce diversity to each population according to Eq. (8-9);
        re-evaluate all individuals;
        modify the State/Action component;//if necessary
    end-if
    select a population using action selector component;
    execute the selected population using the inner optimizer;// PSO, FFA, JAYA, etc.
    evaluate the effectiveness of the executed sub-population using the feedback parameter
    β according to Eq. (8-10);
    update the State/Action component;
    apply exclusion operator according to Eq. (8-11);
end
```

Fig. 8.7 Pseudocode of the proposed framework

Algorithm 8-5. Pseudocode for action/selector in VSLA

```
rnd := rand();
sum := 0;
selectedPopulation := NULL;
for each action j∈[1,2,...,N] of VSLA do
    if rnd < sum + p_j then
        selectedPopulation := j;
        break;
    end-if
    sum := sum + p_j;
end
```

Fig. 8.8 Pseudocode for action selection in VSLA

8.6 FE-Management in MP Method with a Fixed Number of Populations

8.6.1 VSLA-Based FE Management Strategy

The proposed method is modeled with an N-action VSLA, which its actions (i.e., $\alpha_1, \alpha_2, \ldots, \alpha_N$) corresponds to the execution of different sub-populations (i.e., $n_1, n_2, \ldots, n_N$). In order to select a new sub-population to be executed, the VSLA selects one of its actions (e.g., α_i) according to its action probability vector $\vec{p}$. The pseudocode for the action selection procedure in VSLA is shown in Fig. 8.8.

Algorithm 8-6. FE-management strategy based on VSLA

```
setting the parameters r_excl, N, pop_size; /*Set the exclusion radius, multi-population cardinality and population size*/
randomly initialize N population of individuals in the search space; /*Initialize populations into the search space*/
initialize the VSLA with action set α = {1,2,..., N}, action probability vector p⃗ = {(1/N), (1/N),..., (1/N)}, reward parameter a = 0.15, penalty prameter b = 0.05;
while termination criterion not met do
    if a change in the environment detected then
        re-evaluate all sub-populations;
        reset the action probability vector of VSLA;
    end
    select a sub-population by VSLA, say i, and evolve it using the inner optimizer using Algorithm 8-5 as show in Figure 8-8;/*e.g. PSO, FFA, JAYA, etc.*/
    calculate the reinforcement signal β according to Eq. (8-10);
    for each action j∈[1,2, ..., r] of VSLA do
        if (β==0) then //favorable response
            Reward the selected action according to Eq.(8-1);
        else if (β==1) then//unfavorable response
            Penalize the selected action according to Eq.(8-2);
        end-if
    end-for
    apply exclusion operator Eq. (8-11);
end
```

Fig. 8.9 Pseudocode of FE-management in MP methods based on VSLA

Afterward, the corresponding sub-population is executed using the inner optimizer. Then, the VSLA updates its probability vector using the feedback received from the evolved sub-population.

Once an environmental change is detected, all sub-populations are re-evaluated, and the action probability vector of the automaton is restarted to its initial value. The structure of the VSLA-based FE management is shown in Fig. 8.9.

8.6.2 FSLA-Based FE Management Strategies

The main drawback of the above VSLA-based FE management is that VSLA chooses its action randomly based on the action probability vector kept over the action set. Therefore, it is probable that the automaton switches to another action even though it has received a reward for doing the current action until the selection probability of an action converges to one. With this in mind, in this sub-section, we propose two FE-management strategies based on FSLA.

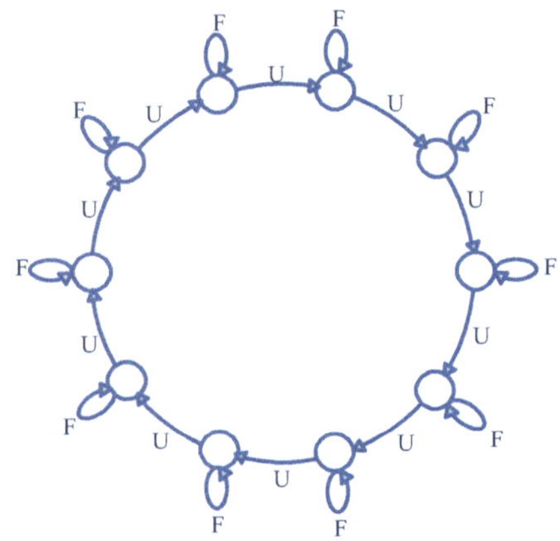

Fig. 8.10 The schematic of the FE management strategy in a typical MP method based on Tesetline automaton

8.6.2.1 FE Management with the Extension of Tsetline Automaton

The first FSLA strategy is an extension of the Tsetline $L_{2,2}$ automaton (Narendra and Thathachar 2012). The $L_{2,2}$ automaton has two states, Φ_1 and Φ_2 and two actions α_1 and α_2. The automaton accept input from a set of {0, 1} and switches its states upon encountering an input 1 (unfavorable response) and remains in the same state on receiving an input 0 (favorable response).

The proposed method is modeled with N states $\Phi = (\Phi_1, \Phi_2, \ldots, \Phi_N)$, the action set $\alpha = (\alpha_1, \cdots, \alpha_N)$, and the environment response $\beta = \{0, 1\}$ is related to reward and penalty. Figure 8.10 illustrates the state transition and action selection in the proposed model. In Fig. 8.10, each circle represents the state of the automaton. The symbols "U" and "F" correspond to unfavorable and favorable responses accordingly.

Initially, the automaton is in state $i = 1$. Therefore, the automaton selects the action α_1 which is related to evolving the sub-population n_1. Generally speaking, when the automaton is in state $i = 1, \ldots, N$, it performs the corresponding action α_i which is executing the sub-population n_i. Afterward, the FSLA receives an input β according to Eq. (8.10). If $\beta = 0$, FSLA stays in the same state, otherwise it goes to the next state.

Qualitatively, the simple strategy used by FSLA-based FE management implies that the corresponding MP algorithm continues to execute whatever sub-population it was evolving earlier as long as the response is favorable but changes to another sub-population as soon as the response is unfavorable. Figure 8.11 shows different steps of the FSLA model for FE management.

Algorithm 8-7. FE-management strategy based on the Tsetline automaton

set the parameters r_{excl}, N, *pop_size*; /*Set the exclusion radius, multi-population cardinality and sub-population size*/
randomly initialize N sub-populations of individuals in the search space; /*Initialize sub-populations into the search space*/
initialize the FSLA with $\Phi = (\Phi_1, \cdots, \Phi_N)$ and $\alpha = (\alpha_1, \cdots, \alpha_N)$
set the current state of *FSLA* to $i = 1$;
while termination criterion not met **do**
    if a change in the environment is detected **then**
        re-evaluate all sub-populations;
    end-if
    select action i and evolve the corresponding sub-population i according to the inner optimizer; /*e.g., PSO, FFA, JAYA etc.*/
    compute the reinforcement signal β according to Eq. (8-10);
    if (β==0) **then** //favorable response
        //do nothing and stay in the current state
    else if (β==1) **then** //unfavorable response
        $i = (i\%N) + 1$; //go to next state
    end-if
    apply exclusion operator according to Eq. (8-11);
end

Fig. 8.11 Pseudocode of FE-management in MP methods based on Tsetline automaton

8.6.2.2 FE Management with STAR Automaton

Another FSLA-based FE management strategy is based on the STAR with deterministic reward and deterministic penalty (Economides and Kehagias 2002). The automaton can be in any of $N + 1$ states $\Phi = (\Phi_0, \Phi_1, \Phi_2, \ldots, \Phi_N)$, the action set is $\alpha = (\alpha_1, \cdots, \alpha_N)$, and the environment response is $\beta = \{0, 1\}$ which is related to reward and penalty. The state transition and action selection are illustrated in Fig. 8.12. Again, in Fig. 8.12, each circle represents a state of the automaton. The symbols "U" and "F" correspond to unfavorable and favorable responses accordingly.

When the automaton is in any state $i = 1, \ldots, N$, it performs the corresponding action α_i which is related to evolving the sub-population n_i. On the other hand, the state Φ_0 is called "neutral" state: when in that state, the automaton chooses any of the r actions with equal probability $\frac{1}{N}$.

Both reward and penalty cause deterministic state transitions, according to the following rules (Economides and Kehagias 2002):

1. when in state Φ_0 and chosen action is $i (i = 1, \ldots, N)$, if rewarded go to state Φ_i with probability 1. if punished, stay in state Φ_0 with probability 1.
2. when in state Φ_i, $i \neq 0$ and chosen action is $i (i = 1, \ldots, N)$, if rewarded stay in state Φ_i with probability 1. if punished, go to state Φ_0 with probability 1.

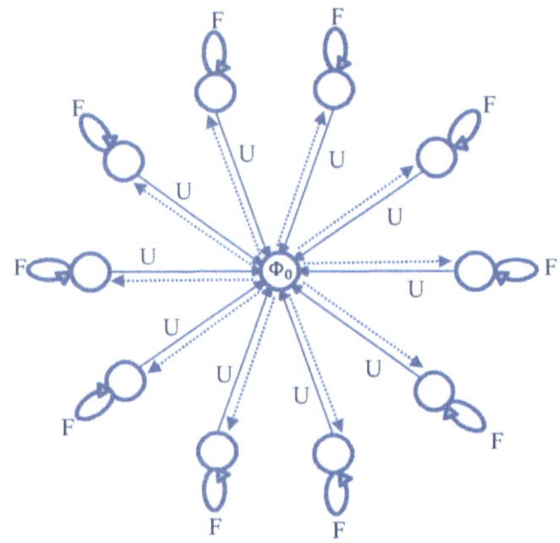

Fig. 8.12 The schematic of the STAR-based FE management strategy in a typical MP method

With these in mind, the STAR-based FE management can be described as follows. At the commencement of the run, the automaton is in state Φ_0. Therefore, the automaton chooses one of its actions, say α_r, randomly. The action chosen by the automaton is applied to the environment by evolving the corresponding sub-population r, which emits a reinforcement signal β according to Eq. (8.10). The automaton moves to state Φ_r if it has been rewarded and it stays in state Φ_0 in the case of punishment. On the other hand, when in state $\Phi_{i \neq 0}$, the automaton chooses an action based on the current state it resides in. That is, action α_i is chosen by the automaton if it is in the state $\Phi_{i \neq 0}$.

Consequently, sub-population i is evolved. Again, the environment responds to the automaton by generating a reinforcement signal according to Eq. (8.10). Based on the received signal β, the state of the automaton is updated. If $\beta = 0$, STAR stays in the same state, otherwise it moves to Φ_0. The algorithm repeats this process until the terminating condition is satisfied. Figure 8.13 shows the details of the proposed scheduling method.

Bearing in mind the proposed model, we can have different MP methods with FE-management schemes. Specifically, in this chapter, we embed the EC methods reviewed in Sect. 8.4 into the proposed model. The resulting MP methods are PSO+VSLA, PSO+FSLA, PSO+STAR, FFA+VSLA, FFA+FSLA, FFA+STAR, JAYA+VSLA, JAYA+FSLA, JAYA+STAR.

Algorithm 8-8. FE-management strategy based on STAR automaton

set the parameters r_{excl}, N, *pop_size*; /*Set the exclusion radius, multi-population cardinality and sub-population size*/
randomly initialize N sub-populations of individuals in the search space; /*Initialize sub-populations into the search space*/
initialize the STAR with $\Phi = (\Phi_0, \Phi_1, \cdots, \Phi_n)$ and $\alpha = (\alpha_1, \cdots, \alpha_N)$
set the current state of STAR[1] to $i = 0$;
while termination criterion not met **do**
    if a change in the environment is detected **then**
        re-evaluate all sub-populations;
    end-if
    if the current state is Φ_0 **then** //select an action randomly
        randomly choose an action from $\alpha_r \in \{\alpha_1, \cdots, \alpha_N\}$ and evolve the corresponding sub-population r according to the inner optimizer; /*e.g. PSO, FFA, JAYA etc.*/
        compute the reinforcement signal β according to Eq. (8-10);
        if (β==0) **then** //favorable response
            $i = r$; //go to state Φ_r
        else if (β==1) **then** //unfavorable response
            //do nothing and stay in the state Φ_0
        end-if
    else if current state is $\Phi_{i \neq 0}$ **then**//select action i
        select action α_i and evolve the corresponding subpopulation i according to the inner optimizer; /*e.g. PSO, FFA, JAYA etc.*/
        compute the reinforcement signal β according to Eq. (8-10);
        if (β==0) **then** //favorable response
            //do nothing and stay in the current state
        else if (β==1) //unfavorable response
            $i = 0$; //go to state Φ_0
        end-if
    end-if
    apply exclusion operator according to Eq. (8-11);
end

Fig. 8.13 Pseudocode of FE-management in MP methods based on STAR automaton

8.7 Experimental Study

8.7.1 Experimental Setup

8.7.1.1 Dynamic Test Function

One of the most widely used synthetic dynamic optimization test suites in the literature is the MPB problem proposed by Branke (Branke 1999b), which is highly regarded due to its configurability. MPB is a real-valued dynamic environment with a D-dimensional landscape consisting of m peaks, where the height, the width, and the position of each peak are changed slightly every time a change occurs in the environment (Branke 2002). Different landscapes can be defined by specifying the shape of the peaks. A typical peak shape is conical, which is defined as follows:

$$f(\vec{x},t) = \max_{i=1,\dots,m} H_t(i) - W_t(i)\sqrt{\sum\nolimits_{j=1}^{D} (x_t(j) - X_t(i,j))^2} \tag{8.12}$$

where $H_t(i)$ and $W_t(i)$ are the height and the width of peak i at time t, respectively. The coordinates of each dimension $j \in [1, D]$ related to the location of peak i at time t, is expressed by $X_t(i, j)$, Furthermore, D is the problem dimensionality. A typical change of a single peak can be modeled as follows:

$$H_{t+1}(i) = H_t(i) + height_{severity}.\sigma_h \tag{8.13}$$

$$W_{t+1}(i) = W_t(i) + width_{severity}.\sigma_w \tag{8.14}$$

$$\vec{X}_{t+1}(i) = \vec{X}_t(i) + \vec{v}_{t+1}(i) \tag{8.15}$$

$$\vec{v}_{t+1}(i) = \frac{s}{|\vec{r} + \vec{v}_t(i)|}((1-\lambda)\vec{r} + \lambda\vec{v}_t(i)) \tag{8.16}$$

where σ_h and σ_w are two random Gaussian numbers with zero mean and standard deviation one. Moreover, the shift vector $\vec{v}_{t+1}(i)$ is a combination of a random vector $\vec{r}$, which is created by drawing random numbers in [–0.5, 0.5] for each dimension, and the current shift vector $\vec{v}_t(i)$, and normalized to the length s. Parameter $\lambda \in [0.0, 1.0]$ specifies the correlation of each peak's changes to the previous one. This parameter determines the trajectory of changes, where $\lambda = 0$ means that the peaks are shifted in completely random directions and $= 1$ means that the peaks always follow the same direction until they hit the boundaries where they bounce off.

Different instances of the MPB can be obtained by changing the environmental parameters. Three sets of configurations have been introduced to provide a unified testbed for the researchers to investigate their approaches' performance in the same condition. The second configuration (Scenario 2) is the most widely used configuration, which was also used as the base configuration for the experiments conducted in this chapter. Unless stated otherwise, environmental parameters are set according to the values listed in Table 8.1 (default values). Besides, to investigate the effect of the environmental parameters (i.e., number of peaks, change period, number of dimensions) on the proposed methods' performance, various experiments were carried out with different combinations of other tested values listed in Table 8.1.

Table 8.1 Parameter settings for the moving peaks benchmark

Parameter	Default values	Other tested values
Number of peaks (m)	10	1
Height severity	7.0	
Width severity	1.0	
Peak function	Cone	
Number of dimensions (D)	5	25, 50, 100, 200
Height range (H)	$\in [30, 70]$	
Width range (W)	$\in [1, 12]$	
Standard height (I)	50.0	
Search space range (A)	$[0, 100]^D$	
Frequency of change (f)	5000	200, 300, 400, 500, 500
Shift severity (s)	1	
Correlation coefficient (λ)	0.0	
Basic function	No	

8.7.1.2 Performance Measure

To measure the optimization algorithms' efficiency on DOPs, we use the best error before changes that were first proposed in Trojanowski and Michalewicz (1999) as to the accuracy and later named the best error before the change (Nguyen et al. 2012). The measure is calculated as the average of the minimum fitness error achieved by the algorithm at the end of each period right before the moment of change, as follows:

$$E_{bbc} = \frac{1}{K}\sum\nolimits_{k=1}^{K}(h_k - f_k) \tag{8.17}$$

where f^k is the fitness value of the best solution obtained by the algorithm just before the kth change occurs, h^k is the optimum value of the kth environment, and K is the total number of environments. For more details about the above measure, interested users can refer to (Ranginkaman et al. 2014; Kazemi Kordestani et al. 2019d).

Table 8.2 Default parameter settings for the proposed MP with various FE management schemes

Parameter	Default value
Number of sub-populations	10
Number of individuals in each sub-population	10
ω	0.729843788
c_1	2.05
c_2	2.05
r_{excl}	Calculated according to Eq. (8.11)
Randomization parameter α	0.25
Attractiveness β_0	1
ϕ	0.000001
Reward parameter (a)	0.15
Penalty parameter (b)	0.05

8.7.1.3 Experimental Settings

For each experiment of the proposed algorithms on a specific DOP, 31 independent runs were executed with different random seeds for the dynamic environment and algorithm. For each run of the algorithm, $f \times 100$ FEs were considered as the termination condition. The experimental results are reported in terms of average best error before the change and standard error, calculated as the standard deviation divided by the number of runs' squared root.

To determine the significant differences among the tested algorithms, we applied Friedman's nonparametric test (with $P < 0.05$). The results of the best-performing algorithm in each experiment are highlighted or printed in boldface.

Finally, the parameters setting of different EC methods, the reward and penalty factors for the proposed approach, are shown in Table 8.2.

8.7.2 Experimental Results and Discussion

8.7.2.1 Extrema Tracking

This experiment's main goal is to compare the optimum tracking ability of different FE-management strategies with the basic method. To this end, the tested DOP contains a single moving peak. Figures 8.14, 8.15, and 8.16 show the performance of different MP methods in tracking a 1-peak DOP.

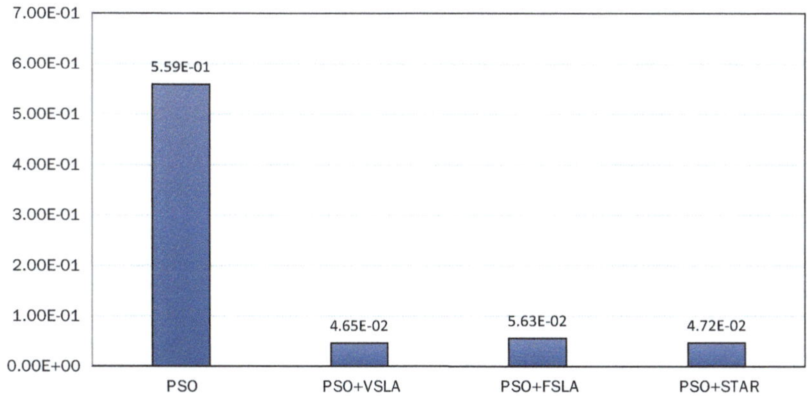

Fig. 8.14 Numerical results of different PSO variants on 1-peak DOP generated by MPB

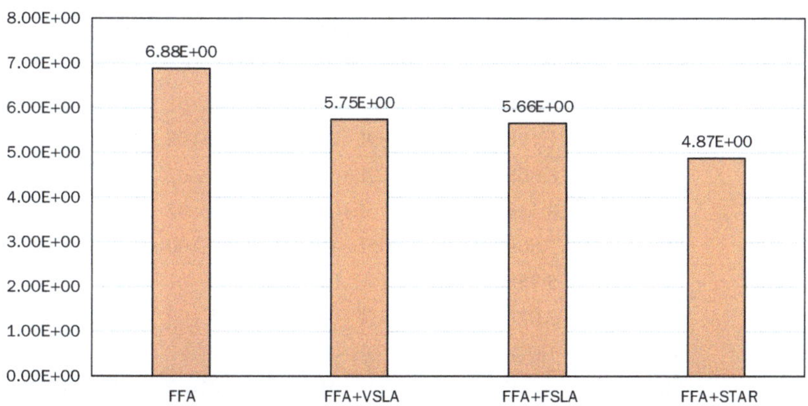

Fig. 8.15 Numerical results of different FFA variants on 1-peak DOP generated by MPB

Several conclusions can be drawn from Figs. 8.14, 8.15, and 8.16. Here we highlight several important conclusions as follows:

- The performance of MP methods with the FE-management strategy is better than the basic methods in all experiments.
- There is no one omnipotent FE-management strategy to outperform the others on all MP methods.

Regarding the reported results, we can conclude that the FE-management strategy can improve the optimum tracking ability of MP methods in DOPs.

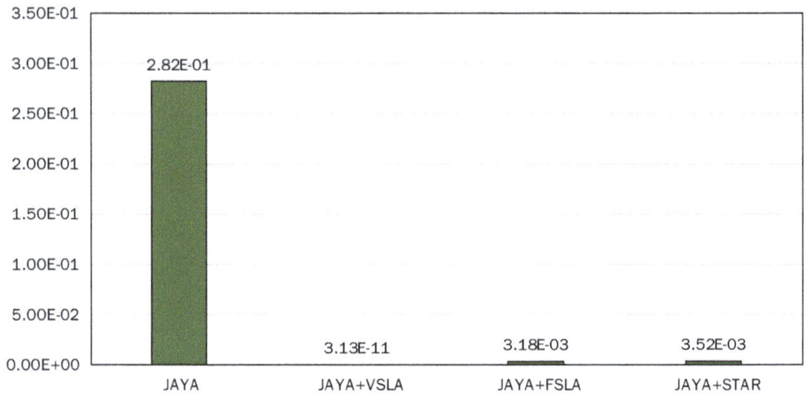

Fig. 8.16 Numerical results of different JAYA variants on 1-peak DOP generated by MPB

Table 8.3 Obtained results by different PSO variant on DOPs with varying number of dimensions

Method	Stats.	D = 25	D = 50	D = 100	D = 200
PSO	Best	5.68E+00	2.04E+01	7.94E+01	2.48E+02
	Worst	1.19E+01	3.89E+01	1.77E+02	7.40E+02
	Median	8.60E+00	2.77E+01	1.02E+02	4.13E+02
	Mean	8.41E+00	2.76E+01	1.11E+02	4.32E+02
	StdErr	2.86E−01	7.45E−01	5.06E+00	2.35E+01
PSO+VSLA	Best	5.75E+00	1.19E+01	1.89E+01	1.11E+02
	Worst	2.02E+01	8.37E+01	1.77E+02	4.66E+02
	Median	1.07E+01	3.50E+01	1.15E+02	2.85E+02
	Mean	1.16E+01	3.43E+01	1.13E+02	2.93E+02
	StdErr	6.66E−01	2.90E+00	7.53E+00	1.60E+01
PSO+FSLA	Best	3.36E+00	1.04E+01	4.53E+01	1.42E+02
	Worst	9.09E+00	2.77E+01	1.32E+02	6.13E+02
	Median	4.95E+00	1.86E+00	8.26E+01	3.00E+02
	Mean	**5.56E+00**	**1.88E+01**	**7.91E+01**	3.28E+02
	StdErr	2.77E−01	6.78E−01	3.50E+00	2.05E+01
PSO+STAR	Best	3.11E+00	1.19E+01	3.45E+01	1.60E+02
	Worst	1.06E+01	3.28E+01	1.25E+02	4.81E+02
	Median	5.84E+00	1.78E+01	8.28E+01	2.77E+02
	Mean	5.87E+00	**1.89E+01**	8.73E+01	**2.91E+02**
	StdErr	2.54E−01	9.06E−01	3.45E+00	1.55E+01

8.7.2.2 Effect of a Varying Number of Dimensions

Many real-world problems consist of optimizing a large number of decision variables. Therefore, in this experiment, the performance of different MP algorithms is investigated on DOPs with different dimensionalities D ∈ {25,50,15,100,200} and the other dynamic and complexity parameters are the same as the Scenario 2 of the MPB listed in Table 8.1. The experimental results of different MP methods are reported in Tables 8.3, 8.4, and 8.5.

As shown in Tables 8.3, 8.4, and 8.5, FE-management significantly influences different MP methods' performance. Specifically, we can make the following remarks concerning the experimental results shown in Tables 8.3, 8.4, and 8.5:

- The first thing that stands out is the applicability of FE management in enhancing MP methods' performance when solving DOPs with larger scales. Moreover, as the problem's dimension increases, the best-performing FE-management strategy over the basic algorithm also increases.

Table 8.4 Obtained results by different FFA variant on DOPs with varying number of dimensions

Method	Stats.	D = 25	D = 50	D = 100	D = 200
FFA	Best	5.89E+01	1.57E+02	5.16E+02	7.53E+02
	Worst	1.45E+02	5.11E+02	8.42E+02	1.68E+03
	Median	9.43E+01	3.28E+02	6.33E+02	9.07E+02
	Mean	9.78E+01	3.35E+02	6.54E+02	9.42E+02
	StdErr	4.29E+00	1.25E+01	1.56E+01	3.36E+01
FFA+VSLA	Best	5.41E+01	2.07E+02	5.67E+02	7.33E+02
	Worst	1.74E+02	5.40E+02	9.13E+02	1.48E+03
	Median	9.56E+01	3.47E+02	6.71E+02	9.83E+02
	Mean	1.02E+02	3.50E+02	7.14E+02	9.91E+02
	StdErr	5.26E+00	1.18E+01	2.33E+01	2.71E+01
FFA+FSLA	Best	6.27E+01	2.36E+02	4.62E+02	6.92E+02
	Worst	1.34E+02	5.36E+02	7.88E+02	1.40E+03
	Median	8.76E+01	3.40E+02	5.71E+02	8.91E+02
	Mean	9.22E+01	3.54E+02	5.89E+02	8.99E+02
	StdErr	3.36E+00	1.20E+01	1.53E+01	2.85E+01
FFA+STAR	Best	4.91E+01	1.93E+02	4.68E+02	6.89E+02
	Worst	1.38E+02	5.39E+02	6.87E+02	1.11E+03
	Median	8.48E+01	3.19E+02	5.55E+02	8.36E+02
	Mean	**9.06E+01**	**3.30E+02**	**5.51E+02**	**8.81E+02**
	StdErr	4.01E+00	1.24E+01	9.90E+00	2.10E+01

Table 8.5 Obtained results by different JAYA variant on DOPs with varying number of dimensions

Method	Stats.	D = 25	D = 50	D = 100	D = 200
JAYA	Best	8.29E+00	2.87E+01	1.05E+02	3.11E+02
	Worst	1.56E+01	6.41E+01	2.29E+02	7.26E+02
	Median	1.11E+01	4.42E+01	1.42E+02	4.96E+02
	Mean	1.13E+01	4.35E+01	1.49E+02	4.93E+02
	StdErr	3.22E−01	1.13E+00	5.91E+00	1.99E+01
JAYA+VSLA	Best	5.76E+00	1.00E+01	1.47E+01	2.46E+01
	Worst	1.44E+01	2.18E+01	2.95E+01	7.24E+01
	Median	9.10E+00	1.34E+01	1.87E+01	4.00E+01
	Mean	9.19E+00	**1.39E+01**	**1.93E+01**	**3.95E+01**
	StdErr	3.73E−01	4.92E−01	5.37E−01	1.74E+00
JAYA+FSLA	Best	4.74E+00	1.20E+01	3.52E+01	6.24E+01
	Worst	1.11E+01	2.63E+01	6.74E+01	1.81E+02
	Median	6.87E+00	1.87E+01	4.49E+01	1.12E+02
	Mean	7.55E+00	1.89E+01	4.58E+01	1.20E+02
	StdErr	3.08E−01	6.63E−01	1.26E+00	5.79E+00
JAYA+STAR	Best	4.25E+00	1.29E+01	3.58E+01	6.66E+01
	Worst	1.09E+01	2.46E+01	7.06E+01	2.10E+02
	Median	7.08E+00	1.92E+01	4.47E+01	1.14E+02
	Mean	**7.23E+00**	1.85E+01	4.67E+01	1.20E+02
	StdErr	3.48E−01	5.26E−01	1.27E+00	6.19E+00

- For PSO and FFA, unlike FSLA and STAR methods, the performance of the VSLA-based FE-management is worse than the corresponding base method. The reasons can be linked to the following two factors: (1) Different EC methods has different characteristics which can affect their convergence which in turn have deleterious effects on the functionality of the VSLA FE-management strategy, (2) The performance of VSLA is sensitive to the value of its reward and penalty parameters.

To support the above supposition, here, we reconfigure the parameters and to check whether their values can help to rectify the above situation. To this end, we conduct an experiment to explore further the effect of a different combination of and on the performance of PSO on DOPs generated by MPB with ten peaks. The obtained results in terms of average best error before the change is illustrated in Fig. 8.17.

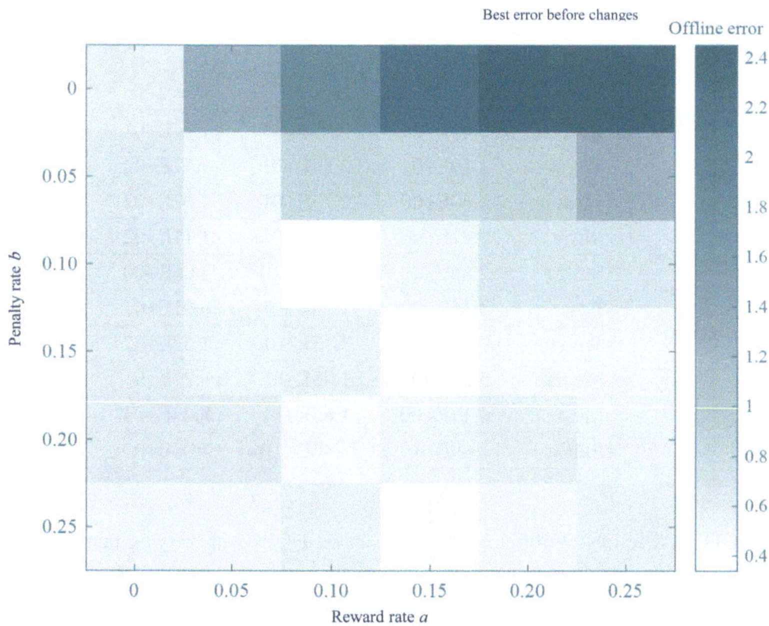

Fig. 8.17 Average best error before the change of PSO+VSLA with different reward and penalty parameters in MPB's Scenario 2

As can be observed in Fig. 8.17, the combination of and has an important effect on the performance of the PSO+VSLA. See for example that the worst result is obtained when $a = 0.25$ and $b = 0.00$. The reason can be attributed to the fact that when $b = 0.00$ and a has large values (i.e., 0.20 and 0.25), the proposed approach quickly converges toward the non-optimal actions. Similarly, it is also possible to note that the best result is obtained with $a = 0.10$ and $b = 0.10$, which is related to learning scheme. In the rest of this subsection, we compare the performance of PSO+VSLA with the new set of values for parameters a and b against the PSO. Table 8.6 shows the results of such a comparison.

From Table 8.6, it is clear that the performance of PSO+VSLA with $a = 0.10$ and $b = 0.10$ is better than that of PSO.

8.7.2.3 Effect of Varying Change Intervals

This experiment aims to investigate the performance comparison between different MP methods on fast-changing DOPs (i.e., change frequencies $f \in \{200, 300, 400, 500\}$). The numerical results of different algorithms are reported in Tables 8.7,8.8, and 8.9.

Table 8.6 Comparison between PSO and PSO + VSLA with different parameters for reward and penalty rates

Method	Stats.	D = 25	D = 50	D = 100	D = 200
PSO	Best	5.68E+00	2.04E+01	7.94E+01	2.48E+02
	Worst	1.19E+01	3.89E+01	1.77E+02	7.40E+02
	Median	8.60E+00	2.77E+01	1.02E+02	4.13E+02
	Mean	8.41E+00	2.76E+01	1.11E+02	4.32E+02
	StdErr	2.86E−01	7.45E−01	5.06E+00	2.35E+01
PSO+VSLA ($a = 0.10$, $b = 0.10$)	Best	3.16E+00	1.20E+01	6.01E+01	1.51E+02
	Worst	9.09E+00	3.15E+01	1.32E+02	5.27E+02
	Median	5.84E+00	1.78E+01	8.27E+01	2.90E+02
	Mean	**6.19E+00**	**1.86E+01**	**8.49E+01**	**3.12E+02**
	StdErr	2.89E−01	8.46E−01	3.08E+00	1.73E+01

Table 8.7 Obtained results by different PSO variant on DOPs with varying number of change intervals

Method	Stats.	f = 200	f = 300	f = 400	f = 500
PSO	Best	1.06E+01	6.56E+00	5.21E+00	4.60E+00
	Worst	2.70E+01	1.28E+01	1.01E+01	9.45E+00
	Median	1.81E+01	9.42E+00	8.29E+00	6.68E+00
	Mean	1.90E+01	9.35E+00	8.25E+00	6.89E+00
	StdErr	7.36E−01	2.83E−01	2.25E−01	2.06E−01
PSO+VSLA ($a = 0.10$, $b = 0.10$)	Best	1.01E+01	5.93E+00	4.17E+00	3.20E+00
	Worst	2.51E+01	1.01E+01	9.77E+00	7.19E+00
	Median	1.46E+01	8.02E+00	6.04E+00	4.74E+00
	Mean	1.60E+01	**8.00E+00**	6.21E+00	**4.99E+00**
	StdErr	7.73E−01	2.22E−01	2.23E−01	1.68E−01
PSO+STAR	Best	7.73E+00	5.72E+00	4.50E+00	3.17E+00
	Worst	2.35E+01	1.12E+01	9.53E+00	6.74E+00
	Median	1.47E+01	7.97E+00	6.17E+00	5.17E+00
	Mean	**1.50E+01**	8.13E+00	6.34E+00	5.20E+00
	StdErr	7.17E−01	2.74E−01	2.28E−01	1.71E−01
PSO+FSLA	Best	1.04E+01	5.50E+00	4.30E+00	3.55E+00
	Worst	2.63E+01	1.05E+01	9.71E+00	8.01E+00
	Median	1.73E+01	8.55E+00	6.12E+00	5.22E+00
	Mean	1.83E+01	8.37E+00	**6.10E+00**	5.28E+00
	StdErr	7.69E−01	2.52E−01	2.01E−01	1.96E−01

Table 8.8 Obtained results by different FFA variant on DOPs with varying number of change intervals

Method	Stats.	f = 200	f = 300	f = 400	f = 500
FFA	Best	1.00E+01	8.16E+00	6.95E+00	6.25E+00
	Worst	3.02E+01	2.13E+01	1.57E+01	1.81E+01
	Median	1.77E+01	1.22E+01	1.09E+01	9.80E+00
	Mean	**1.78E+01**	1.27E+01	1.05E+01	1.00E+01
	StdErr	7.74E−01	6.08E−01	3.80E−01	3.59E−01
FFA+VSLA	Best	1.73E+01	9.34E+00	6.36E+00	5.26E+00
	Worst	6.15E+01	2.36E+01	1.49E+01	1.22E+01
	Median	2.70E+01	1.22E+01	1.11E+01	9.59E+00
	Mean	3.07E+01	1.33E+01	1.08E+01	9.48E+00
	StdErr	2.06E+00	5.87E−01	4.23E−01	3.15E−01
FFA+FSLA	Best	8.37E+00	5.61E+00	6.42E+00	4.14E+00
	Worst	3.41E+01	2.58E+01	1.51E+01	1.17E+01
	Median	2.06E+01	1.09E+01	9.51E+00	8.96E+00
	Mean	2.04E+01	**1.16E+01**	**9.82E+00**	**8.72E+00**
	StdErr	8.88E−01	7.70E−01	3.90E−01	3.33E−01
FFA+STAR	Best	1.58E+01	8.51E+00	5.95E+00	5.01E+00
	Worst	6.52E+01	2.49E+01	1.62E+01	1.36E+01
	Median	2.59E+01	1.43E+01	9.53E+00	8.85E+00
	Mean	2.80E+01	1.44E+01	9.88E+00	8.90E+00
	StdErr	1.95E+00	6.38E−01	4.39E−01	3.35E−01

This experiment's results also confirm the FE management's effectiveness for improving MP methods' performance for DOPs. In the case of VSLA, as the DOP perturbation frequency increases, the automaton loses its ability to track the optimal action. Therefore, FSLA and STAR are better options for fast-changing non-stationary environments.

Table 8.9 Obtained results by different JAYA variant on DOPs with varying number of change intervals

Method	Stats	f = 200	f = 300	f = 400	f = 500
JAYA	Best	1.08E+01	7.69E+00	5.97E+00	5.19E+00
	Worst	2.64E+01	1.38E+01	1.18E+01	9.47E+00
	Median	1.61E+01	9.88E+00	8.14E+00	7.82E+00
	Mean	1.69E+01	9.87E+00	8.39E+00	7.59E+00
	StdErr	7.28E−01	2.68E−01	2.34E−01	2.37E−01
JAYA+VSLA	Best	8.92E+00	6.37E+00	5.15E+00	3.33E+00
	Worst	2.48E+01	1.31E+01	9.93E+00	8.92E+00
	Median	1.48E+01	8.76E+01	7.37E+00	6.08E+00
	Mean	1.55E+01	8.74E+01	7.36E+00	5.99E+00
	StdErr	6.72E−01	2.85E−01	1.90E−01	2.15E−01
JAYA+FSLA	Best	9.56E+00	6.16E+00	5.78E+00	4.24E+00
	Worst	2.60E+01	1.31E+01	9.05E+00	7.96E+00
	Median	1.55E+01	8.30E+00	7.35E+00	6.07E+00
	Mean	1.64E+01	8.51E+00	7.23E+00	5.97E+00
	StdErr	7.54E−01	2.94E−01	1.39E−01	1.53E−01
JAYA+ssSTAR	Best	8.04E+00	5.87E+00	4.31E+00	4.05E+00
	Worst	2.35E+01	1.12E+01	1.03E+01	8.39E+00
	Median	1.46E+01	8.53E+00	6.66E+00	5.81E+00
	Mean	**1.46E+01**	**8.46E+00**	**6.94E+00**	**5.71E+00**
	StdErr	7.59E−01	2.30E−01	2.82E−01	1.52E−01

8.8 Conclusion

This chapter develops a framework for applying function evaluation management to multi-population methods when solving dynamic optimization problems. Using the proposed framework, any population-based stochastic optimization method can be converted into a multi-population version with a function management strategy. Three learning automata-based population selection strategies were used, of which one is based on variable structure learning automaton, and two are based on fixed structure learning automaton. These strategies were embedded into multi-population algorithms based on particle swarm optimization, firefly algorithm, and JAYA. Experiments were carried on different dynamic instances generated by moving peaks benchmark problem. From the obtained results, one can conclude that function evaluation management is an effective way to enhance multi-population methods in dynamic environments. The work studied in this chapter can be further extended in several directions. Developing other mechanisms for function evaluation management would be an interesting direction for future work. Studying the effect of different feedback parameters is also interesting. Proposing other multi-population frameworks with function evaluation management is a valuable future work to pursue. Evaluating the function evaluation management concept for noisy dynamic environments is another future work. In the end, it is also very valuable to apply the proposed approach to several real-world dynamic problems.

References

Blackwell, T.: Particle swarm optimization in dynamic environments. In: Yang, S., Ong, Y.-S., Jin, Y. (eds.) Evolutionary Computation in Dynamic and Uncertain Environments, pp. 29–49. Springer, Heidelberg (2007)

Blackwell, T., Branke, J.: Multiswarms, exclusion, and anti-convergence in dynamic environments. IEEE Trans. Evol. Comput. **10**, 459–472 (2006). https://doi.org/10.1109/TEVC.2005.857074

Branke, J.: Memory enhanced evolutionary algorithms for changing optimization problems. In: Proceedings of the IEEE Congress on Evolutionary Computation, pp. 1875–1882. IEEE (1999)

Branke, J.: Evolutionary Optimization in Dynamic Environments. Springer, Heidelberg (2002)

du Plessis, M.C., Engelbrecht, A.P.: Using competitive population evaluation in a differential evolution algorithm for dynamic environments. Eur. J. Oper. Res. **218**, 7–20 (2012). https://doi.org/10.1016/j.ejor.2011.08.031

du Plessis, M.C., Engelbrecht, A.P.: Differential evolution for dynamic environments with unknown numbers of optima. J. Glob. Optim. **55**, 73–99 (2013). https://doi.org/10.1007/s10898-012-9864-9

Economides, A., Kehagias, A.: The STAR automaton: expediency and optimality properties. IEEE Trans. Syst. Man Cybern. Part B (Cybern.) **32**, 723–737 (2002). https://doi.org/10.1109/TSMCB.2002.1049607

Hashemi, A.B., Meybodi, M.R.: A multi-role cellular PSO for dynamic environments. In: Proceedings of the 14th International CSI Computer Conference, pp. 412–417. IEEE (2009a)

Hashemi, A.B., Meybodi, M.R.: Cellular PSO: A PSO for dynamic environments. In: Cai, Z., Li, Z., Kang, Z., Liu, Y. (eds.) Advances in Computation and Intelligence. ISICA 2009, pp. 422–433. Springer, Heidelberg (2009b)

Kamosi, M., Hashemi, A.B., Meybodi, M.R.: A new particle swarm optimization algorithm for dynamic environments. In: Proceedings of the First International Conference on Swarm, Evolutionary, and Memetic Computing, pp. 129–138. Springer, Heidelberg (2010b)

Kamosi, M., Hashemi, A.B., Meybodi, M.R.: A hibernating multi-swarm optimization algorithm for dynamic environments. In: Proceedings of the Second World Congress on Nature and Biologically Inspired Computing, pp. 363–369. IEEE (2010a)

Kazemi Kordestani, J., Meybodi, M.R.: Application of sub-population scheduling algorithm in multi-population evolutionary dynamic optimization. In: Gandomi, A.H., Emrouznejad, A., Jamshidi, M.M., Deb, K., Rahimi, I. (eds.) Evolutionary Computation in Scheduling, pp. 150–192. Wiley (2020)

Kazemi Kordestani, J., Ahmadi, A., Meybodi, M.R.: An improved differential evolution algorithm using learning automata and population topologies. Appl. Intell. **41**, 1150–1169 (2014). https://doi.org/10.1007/s10489-014-0585-2

Kazemi Kordestani, J., Abedi Firouzjaee, H., Meybodi, M.R.: An adaptive bi-flight cuckoo search with variable nests for continuous dynamic optimization problems. Appl. Intell. **48**, 97–117 (2018). https://doi.org/10.1007/s10489-017-0963-7

Kazemi Kordestani, J., Rezvanian, A., Meybodi, M.R.: New measures for comparing optimization algorithms on dynamic optimization problems. Nat. Comput. **18**, 705–720 (2019b). https://doi.org/10.1007/s11047-016-9596-8

Kazemi Kordestani, J., Ranginkaman, A.E., Meybodi, M.R., Novoa-Hernández, P.: A novel framework for improving multi-population algorithms for dynamic optimization problems: a scheduling approach. Swarm Evol. Comput. **44**, 788–805 (2019a). https://doi.org/10.1016/j.swevo.2018.09.002

Kordestani, J.K., Rezvanian, A., Meybodi, M.R.: An efficient oscillating inertia weight of particle swarm optimisation for tracking optima in dynamic environments. J. Exp. Theor. Artif. Intell. **28**, 137–149 (2016). https://doi.org/10.1080/0952813X.2015.1020521

Kordestani, J.K., Meybodi, M.R., Rahmani, A.M.: A note on the exclusion operator in multi-swarm PSO algorithms for dynamic environments. Connection Sci. **0**, 1–25 (2019). https://doi.org/10.1080/09540091.2019.1700912

Kordestani, J.K., Meybodi, M.R., Rahmani, A.M.: A two-level function evaluation management model for multi-population methods in dynamic environments: hierarchical learning automata approach. J. Exp. Theor. Artif. Intell. **0**, 1–26 (2020). https://doi.org/10.1080/0952813X.2020.1721568

Mahdaviani, M., Kazemi Kordestani, J., Rezvanian, A., Meybodi, M.R.: LADE: learning automata based differential evolution. Int. J. Artif. Intell. Tools **24**, 1550023 (2015). https://doi.org/10.1142/S0218213015500232

Mendes, R., Mohais, A.S.: DynDE: a differential evolution for dynamic optimization problems. In: IEEE Congress on Evolutionary Computation, vol. 3, pp. 2808–2815 (2005)

Nabizadeh, S., Rezvanian, A., Meybodi, M.R.: A multi-swarm cellular PSO based on clonal selection algorithm in dynamic environments. In: 2012 International Conference on Informatics, Electronics & Vision (ICIEV), Dhaka, Bangladesh, pp. 482–486. IEEE (2012a)

Narendra, K.S., Thathachar, M.A.: Learning Automata: An Introduction. Prentice-Hall, Hoboken (1989)

Narendra, K.S., Thathachar, M.A.L.: Learning Automata: An Introduction. Courier Corporation, North Chelmsford (2012b)

Nguyen, T.T., Yang, S., Branke, J.: Evolutionary dynamic optimization: a survey of the state of the art. Swarm Evol. Comput. **6**, 1–24 (2012). https://doi.org/10.1016/j.swevo.2012.05.001

Nickabadi, A., Ebadzadeh, M.M., Safabakhsh, R.: A novel particle swarm optimization algorithm with adaptive inertia weight. Appl. Soft Comput. **11**, 3658–3670 (2011). https://doi.org/10.1016/j.asoc.2011.01.037

Noroozi, V., AliB, Hashemi, Meybodi, M.: CellularDE: a cellular based differential evolution for dynamic optimization problems. In: Dobnikar, A., Lotrič, U., Šter, B. (eds.) Adaptive and Natural Computing Algorithms, pp. 340–349. Springer, Heidelberg (2011)

Noroozi, V., Hashemi, A.B., Meybodi, M.R.: Alpinist CellularDE: a cellular based optimization algorithm for dynamic environments. In: Proceedings of the 14th Annual Conference Companion on Genetic and Evolutionary Computation, Philadelphia, Pennsylvania, USA, pp. 1519–1520. ACM Press (2012)

Novoa-Hernández, P., Corona, C.C., Pelta, D.A.: Efficient multi-swarm PSO algorithms for dynamic environments. Memetic Comput. **3**, 163–174 (2011)

Ozsoydan, F.B., Baykasoglu, A.: A multi-population firefly algorithm for dynamic optimization problems. In: 2015 IEEE International Conference on Evolving and Adaptive Intelligent Systems (EAIS), pp. 1–7 (2015)

Ozsoydan, F.B., Baykasoğlu, A.: Quantum firefly swarms for multimodal dynamic optimization problems. Expert Syst. Appl. **115**, 189–199 (2019). https://doi.org/10.1016/j.eswa.2018.08.007

Ranginkaman, A.E., Kazemi Kordestani, J., Rezvanian, A., Meybodi, M.R.: A note on the paper "a multi-population harmony search algorithm with external archive for dynamic optimization problems" by Turky and Abdullah. Inf. Sci. **288**, 12–14 (2014). https://doi.org/10.1016/j.ins.2014.07.049

Rezvanian, A., Saghiri, A.M., Vahidipour, S.M., Esnaashari, M., Meybodi, M.R.: Recent Advances in Learning Automata. Springer, Berlin (2018a)

Rezvanian, A., Saghiri, A.M., Vahidipour, S.M., Esnaashari, M., Meybodi, M.R.: Learning automata theory. In: Recent Advances in Learning Automata, pp. 3–19. Springer, Heidelberg (2018c)

Rezvanian, A., Moradabadi, B., Ghavipour, M., Daliri Khomami, M.M., Meybodi, M.R.: Introduction to learning automata models. In: Learning Automata Approach for Social Networks, pp. 1–49. Springer, Heidelberg (2019)
Sharifi, A., Kazemi Kordestani, J., Mahdaviani, M., Meybodi, M.R.: A novel hybrid adaptive collaborative approach based on particle swarm optimization and local search for dynamic optimization problems. Appl. Soft Comput. **32**, 432–448 (2015). https://doi.org/10.1016/j.asoc.2015.04.001
Sharifi, A., Noroozi, V., Bashiri, M., Hashemi, A.B., Meybodi, M.R.: Two phased cellular PSO: a new collaborative cellular algorithm for optimization in dynamic environments. In: Proceedings of the IEEE Congress on Evolutionary Computation, pp. 1–8 (2012)
Trojanowski, K., Michalewicz, Z.: Searching for optima in non-stationary environments. In: Proceedings of the IEEE Congress on Evolutionary Computation, pp. 1843–1850. IEEE (1999)
Vafashoar, R., Morshedlou, H., Rezvanian, A., Meybodi, M.R.: Cellular Learning Automata: Theory and Applications. Springer, Heidelberg (2021)
Venkata Rao, R.: Jaya: a simple and new optimization algorithm for solving constrained and unconstrained optimization problems. Int. J. Industr. Eng. Comput. **7**, 19–34 (2016)
Yang, S., Li, C.: A clustering particle swarm optimizer for locating and tracking multiple optima in dynamic environment. IEEE Trans. Evol. Comput. **14**, 959–974 (2010). https://doi.org/10.1109/TEVC.2010.2046667

Chapter 9
Function Management in Multi-population Methods with a Variable Number of Populations: A Variable Action Learning Automaton Approach

Javidan Kazemi Kordestani, Mehdi Razapoor Mirsaleh, Alireza Rezvanian, and Mohammad Reza Meybodi

Abstract The fitness evaluation management (FEM) has been successfully applied to improve the performance of multi-population (MP) methods with a fixed number of populations for dynamic optimization problems (DOPs). Along with the benefits they offered, the proper number of populations in MP with a fixed number of populations is difficult to determine. The number of populations in this approach is specified according to the number of local optima and before the commencement of the optimization process. However, the number of optima in real-world problems is mostly unknown. Therefore, it is valuable to study the usefulness of FEM for MP with a varying number of populations. In this chapter, the concept of FEM is extended to the MP approach with a variable number of populations. Since the number of sub-populations in this method is varied during the run, the FEM system should adapt to this new emerging challenge. To do so, we use a variable-action set learning automaton (VALA) to change the action-set of the learning automaton (LA) accordingly. As a starting point, we first need to use an MP method with a variable number of populations as the base method to develop our proposed FEM scheme.

J. Kazemi Kordestani
Department of Computer Engineering, Science and Research Branch, Islamic Azad University, Tehran, Iran

M. Razapoor Mirsaleh
Department of Computer Engineering and Information Technology, Payame Noor University (PNU), P.O. BOX 19395-3697, Tehran, Iran
e-mail: mrezapoorm@aut.ac.ir

A. Rezvanian (✉)
Department of Computer Engineering, University of Science and Culture, Tehran, Iran
e-mail: rezvanian@usc.ac.ir

M. R. Meybodi
Computer Engineering Department, Amirkabir University of Technology (Tehran Polytechnic), Tehran, Iran
e-mail: mmeybodi@aut.ac.ir

J. Kazemi Kordestani et al. (eds.), *Advances in Learning Automata and Intelligent Optimization*, Intelligent Systems Reference Library 208, https://doi.org/10.1007/978-3-030-76291-9_9

Although several frameworks exist to create the MP with a variable number of sub-populations, we have considered the framework for locating and tracking multiple optima in dynamic environments using clustering particle swarm optimizer (CPSO). CPSO has been extensively studied in the past, showing very promising results in challenging DOPs. Therefore, in the next, we review the CPSO in brief. Then we describe our proposal in detail. Finally, the effectiveness of the proposed FEM is evaluated through numerical experiments.

9.1 Introduction

The fitness evaluation management (FEM) has been successfully applied to improve the performance of multi-population (MP) methods with a fixed number of populations for dynamic optimization problems (DOPs) (Kordestani et al. 2019; Kordestani and Meybodi 2020). Along with the benefits they offered, the proper number of populations in MP with a fixed number of populations is difficult to determine. As mentioned in Chapter 7, the number of populations in this approach is specified according to the number of local optima and before the commencement of the optimization process. However, the number of optima in real-world problems is mostly unknown. Therefore, it is valuable to study the usefulness of FEM for MP with a varying number of populations. In this chapter, the concept of FEM, which was introduced in the previous chapter, is extended to the MP approach with a variable number of populations. Since the number of sub-populations in this method is varied during the run, the FEM system should adapt to this new emerging challenge. To do so, we use a variable-action set learning automaton (VALA) to change the action-set of the learning automaton (LA) accordingly (Rezvanian et al. 2018, 2019). As a starting point, we first need to use an MP method with a variable number of populations as the base method to develop our proposed FEM scheme. Although several frameworks exist to create the MP with a variable number of sub-populations, we have considered the framework for locating and tracking multiple optima in dynamic environments using clustering particle swarm optimizer (CPSO), the algorithm proposed in Yang and Li (2010). CPSO has been extensively studied in the past, showing very promising results in challenging DOPs. Therefore, in the next section, we review the CPSO in brief. Then we describe our proposal in detail. Finally, the effectiveness of the proposed FEM is evaluated by means of numerical experiments.

9.2 Main Framework of Clustering Particle Swarm Optimization

CPSO (Yang and Li 2010) applies a hierarchical clustering method to divide an initial cradle swarm into various sub-swarms, each of which covers a different sub-area of the landscape. Each sub-swarm is evolved using a modified particle swarm optimizer (PSO). Afterward, each sub-swarm status (i.e., overlapping, overcrowding, and convergence) is checked, and an operator is executed on each sub-accordingly. Upon detecting a change in the environment, a new cradle swarm will be regenerated. The pseudocode for CPSO is shown in Fig. 9.1.

9.2.1 Creating Multiple Sub-swarms from the Cradle Swarm

CPSO uses a single linkage hierarchical clustering method to split off the sub-swarms from the cradle swarm to generate sub-populations. Single-linkage clustering is one of the earliest methods for clustering multi-dimensional vectors based

Algorithm 9-1. Main framework of CPSO

Setting the parameters R_{conv}, $R_{overlap}$, *M*, *max_subsize*; /*Set the convergence radius, overlapping radius, population size, and the maximum size of the child swarm particles*/
Randomly initialize the main swarm *parentlist* with *M* particles in the search space; /*Randomize the initial population into the search space*/
sublist = Clustering (*parentlist*);
while the termination condition is not met **do**
    for each sub-swarm *i* in sublist **do** //movement of sub-swarm *i*
        for each particle *j* in sub-swarm *i* **do**
            Evolve particles $\vec{x}_j$ of sub-swarm *i* according to Eq. (9-2) & Eq.(9-3);
            CheckBoundaries $(\vec{x}_j, \vec{v}_j)$;
            if $(\text{fitness}(\vec{p}_j) < \text{fitness}(\vec{x}_j))$ **then**
                $\vec{p}_j := \vec{x}_j$;
                if $(\text{fitness}(\vec{p}_g) < \text{fitness}(\vec{x}_j))$ **then**
                    $\vec{p}_g := \vec{x}_j$
                end-if
            end-if
        end-for
    end-for
    CheckOverlaping (*sublist*); //merge overlapped sub-swarms
    CheckRedundancy (*sublist*); //remove redundant particles
    CheckConvergence (*sublist*); //deactivate the converged sub-swarms
    if a change is detected **then**
        Save the *gbest* of each sub-swarm in a temporary list memory;
        Randomly generate a new *parentlist* with *M*-\|memory\| particles;
        Add memory list particles to the newly generated *parentlist*; /**M* particles exist in the *parentlist* now*/
        sublist = Clustering (*parentlist*);
    end
end

Fig. 9.1 General framework for CPSO

Algorithm 9-2. Clustering (parentlist)

```
Let M(r,s) = min_{i∈r, j∈s} d(i,j) be the distance of the two closest particles i and j that belong to clusters r and s respectively;
Create a temporary cluster list G of size |parentlist|;
for each particle i in parentlist do
    G[i] = parentlist[i];//each single particle forms one cluster in G
end-for
Calculate the distance between all clusters (i.e., particles) in G and construct a distance matrix M of size |G|×|G|;
while TRUE do
  | if FindNearestPair(G, r, s) = FALSE then
  |   | Break;
  | end-if
  | r:=r+s;//merge r and s into r
  | Delete the cluster s from G;
  | Re-calculate all distance elements in M which have been affected by merging the
  | clusters r and s;
  | if each cluster in G has more than one particle then
  |   | Break;
  | end-if
end-while
Empty parentlist;
return G
```

Fig. 9.2 General framework for CPSO

on the concept of nearest-neighbor. Given a dataset consisting of n real-valued D-dimensional vectors $v_1, \ldots, v_n \in R^D$ d the goal of single-linkage clustering is to construct a partition of these vectors into k clusters $C_1, \ldots, C_k$ such that the smallest distance between two vectors in different clusters is maximized.

The details of the single-linkage clustering algorithm in CPSO are shown in Fig. 9.2.

In Algorithm 9-2, the distance between particle i and j is computed using the Euclidean distance between them as follows:

$$d(i, j) = \sqrt{\sum\nolimits_{d=1}^{D} \left(x_i^d - x_j^d\right)^2} \tag{9.1}$$

Moreover, Fig. 9.3 shows the algorithm for finding the two nearest pairs of clusters in which the total number of particles in the two clusters is not greater than the parameter *max_subsize*.

Using the above clustering method, sub-swarms will be created based on the distribution of initial particles in the fitness landscape. Moreover, the total number of sub-swarms and each one's size are also determined with respect to parameter *max_subsize*.

Algorithm 9-3. FindNearestPair (*G,r,s*)

```
found = FALSE;
min_dist := sqrt(Σ_{d=1}^{D} (ub_d − lb_d)^2);//ub_d and lb_d are the upper and lower bounds of the d-th dimension of the
    search space;
for i := 0 to |G| do
    for j := i+1 to |G| do
        if (|G[i]| + |G[j]| > max_subsize) then
            continue;
        end-if
        if (min_dist > M(G[i], G[j])) then
            min_dist := M(G[i], G[j]);
            r := G[i];
            s := G[j];
            found := TRUE;
        end-if
    end-for
end-for
return found
```

Fig. 9.3 Pseudo-code for finding the closest pair of clusters

9.2.2 Local Search by PSO

PSO is a versatile population-based stochastic optimization method that was first proposed by Kennedy and Eberhart (Kennedy and Eberhart 1995) in 1995. PSO begins with a population of randomly generated particles in a *D*-dimensional search space. Each particle *i* of the swarm has three features: $\vec{x}_i$ that shows the current position of the particle i in the search space, $\vec{v}_i$ which is the velocity of the particle *i* and $\vec{p}_i$ which denotes the best position found so far by the particle *i*. Each particle *i* updates its position in the search space, at every time step *t*, according to the following equations:

$$\vec{v}_i(t+1) = \omega\vec{v}_i(t) + c_1r_1[\vec{p}_i(t) - \vec{x}_i(t)] + c_2r_2[\vec{p}_g(t) - \vec{x}_i(t)] \tag{9.2}$$

$$\vec{x}_i(t+1) = \vec{x}_i(t) + \vec{v}_i(t+1), \tag{9.3}$$

where ω is an inertia weight that governs the amount of speed preserved from the previous iteration. c_1 and c_2 denote the cognitive and social learning factors used to adjust the degree particles' movements toward their personal best position and the global best position of the swarm. r_1 and r_2 are two independent random variables drawn with uniform probability from [0, 1]. Finally, $\vec{p}_g$ is the globally best position found so far by the swarm. The pseudocode of the PSO is shown in Fig. 9.4.

Once a sub-swarm is created using the clustering method, it starts searching its respective sub-region using the PSO algorithm. For a sub-swarm to locate a local peak quickly, the authors used the PSO with the *gbest* model. After a subswarm updates, its position using Eq. (9.2) and Eq. (9.3), a repair operator is used according to Fig. 9.5 to prevent particles from going beyond the search space boundaries.

Algorithm 9-4. Pseudocode for canonical PSO

```
Setting parameters ω, c1 and c2 ;
Generate the initial swarm of the particles with random positions and velocities
Evaluate the fitness of each particle of the swarm
repeat
|  for each particle i in the swarm do
|  |  update particle i according to Eq. (9-2) and Eq. (9-3);
|  |  if(fitness(p⃗_i)<fitness(x⃗_i)) then
|  |  |  p⃗_i:= x⃗_i;
|  |  |  if (fitness(p⃗_g)<fitness(x⃗_i)) then
|  |  |  |     p⃗_g:= x⃗_i;
|  |  |  end-if
|  |  end-if
|  end-for
until a termination condition is met
```

Fig. 9.4 Canonical particle swarm optimization

Algorithm 9-5. CheckBoundaries $(\vec{x}.\vec{v})$

```
for each dimension d ∈ [1 ... D] do
|  if x_d > ub_d then //ub_d is the upper-bound of the search space in d-th dimension
|  |  x_d = ub_d;
|  |  v_d = 0;
|  else if x_d < lb_d then //lb_d is the lower-bound of the search space in d-th dimension
|  |  x_d = lb_d;
|  |  v_d = 0;
|  end-if
end-for
```

Fig. 9.5 Pseudocode for boundary check

Besides, at each iteration, the authors used a local search operator (Yang and Li 2010) to learn useful information from successful particles' dimensions. Moreover, to speed up the convergence of the sub-swarms, the authors also used a linearly decreasing strategy to change the value of inertia weight ω for each sub-swarm.

9.2.3 Status of Sub-swarms

Afterward, the status of the sub-swarms is checked to stop them from wasting computing resources. These checks are: (a) *overlapping check* which prevents two sub-swarms from searching the same sub-area of the search space, (b) *redundancy check,* which keeps the number of particles in each sub-swarm within a predefined interval and (c) *convergence check,* which removes the converged sub-swarms from the landscape. These checks are performed using algorithms in Figs. 9.6, 9.7, and 9.8.

Algorithm 9-6. CheckOverlaping (sublist)

```
for each pair of sub-swarms (r,s) in the sublist do
    if r_overlap > R_overlap then
        Merge r and s into r;
        Remove s from sublist;
    end-if
end-for
```

Fig. 9.6 Pseudocode for overlapping check

Algorithm 9-7. CheckRedundancy (sublist)

```
for each sub-swarms r in sublist do
    if |r|>max_subsize then
        Remove worst (|r|-maxsub_size) particles from r;
    end-if
end-for
```

Fig. 9.7 Pseudocode for redundancy check

Algorithm 9-8. CheckConvergence (sublist)

```
for each sub-swarms s in sublist do
    if radius(s)< R_conv then
        Add gbest sub-swarm s into memory;
        Remove s from sublist;
    end-if
end-for
```

Fig. 9.8 Pseudocode for convergence check

9.2.4 Detection and Response to Environmental Changes

Several methods have been suggested for detecting changes in dynamic environments. See, for example (Nguyen et al. 2012) for a good survey on the topic. The most frequently used method consists of reevaluating the algorithm's memories for detecting inconsistencies in their corresponding fitness values. In the proposed framework, the best solution for each sub-population was considered. At the beginning of each iteration, the algorithm reevaluates these solutions and compares their current fitness values against those from the previous iteration. If at least one inconsistency is found from these comparisons, one can assume that the environment has changed. Once an environmental change is detected: Firstly, the best position *gbest* of each sub-swarm in *sublist* before the change is stored into a temporarily memory list. Secondly, all particles are re-initialized to create a new cradle swarm, and the particles stored in the memory list are added to the new cradle swarm by replacing the worst particles in it. Finally, *sublist* and memory list are re-set to be empty. Again, the clustering method will be performed to generate the new sub-swarm list.

9.3 Variable Action-Set Learning Automata

A variable action-set learning automaton is an automaton in which the number of actions available at each instant changes with time (Narendra and Thathachar 2012). Such an automaton has a finite set of n actions, $\alpha = \{\alpha_1, \alpha_2, \ldots, \alpha_n\}$. $A = \{A_1, A_2, \ldots, A_n\}$ denotes the set of action subsets and $A(k) \subseteq \alpha$ is the subset of all the actions that can be chosen by the learning automaton, at each instant$k$. The selection of the particular action subsets is randomly made by an external agency according to the probability distribution $q(k) = \{q_1(k), q_2(k), \ldots, q_m(k)\}$ defined over the possible subsets of the actions, where $q_i(k) = prob[A(k) = A_i | A_i \in A, 1 \leq i \leq 2^n - 1].\widehat{p}_i(k) = prob[\alpha(k) = \alpha_i | A(k), \alpha_i \in A(k)]$ is the probability of choosing actionα_i, conditioned on the event that the action subset $A(k)$ has already been selected and also$\alpha_i \in A(k)$. The scaled probability $\widehat{p}_i(k)$ is defined as:

$$\widehat{p}_i(k) = p_i(k)/K(k) \tag{9.4}$$

where $K(k) = \sum_{\alpha_i \in A(k)} p_i(k)$ is the sum of the probabilities of the actions in subset $A(k)$, and $p_i(k) = prob[\alpha(k) = \alpha_i]$.

The procedure of choosing an action and updating the action probabilities in a variable action-set learning automaton can be described. Let $A(k)$ be the action subset selected at instant k. Before choosing an action, the probabilities of all the actions in the selected subset are scaled as defined in Eq. (9.3). The automaton then randomly selects one of its possible actions according to the scaled action probability vector $\widehat{p}(k)$. Depending on the response received from the environment, the learning automaton updates its scaled action probability vector. Note that the probability of the available actions is only updated. Finally, the probability vector of the actions of the chosen subset is rescaled as $p_i(k+1) = \widehat{p}_i(k+1) \cdot K(k)$, for all $\alpha_i \in A(k)$. The absolute expediency and ε-optimality of the method described above have been proved (Rezvanian et al. 2018, 2019; Vafashoar et al. 2021).

9.4 FEM in MP Methods with a Variable Number of Populations

This chapter uses a modified version of CPSO, called BCPSO, as a base algorithm to test our VALA-based FEM strategy. The CPSO and BCPSO are different in the following ways: Firstly, CPSO uses a local search operator around the best particle of each sub-swarm to improve the overall performance of the CPSO as much as possible. In BCPSO, we have removed this operator. The reason is that the local search operator consumes a considerable amount of FEs and makes it impossible to study the effect of the proposed FEM strategy. Secondly, CPSO uses a linearly decreasing inertia weight which its calculation depends on the change frequency of the DOP. However, the information about the change frequency of the DOP is not at hand in real-life scenarios. Instead, we use the constriction model proposed by Clerc (Clerc 1999) to guarantee the convergence of each sub-swarm. Using constriction mode, w, c_1 and c_2 is chosen in a way to ensure the convergence of the population. One of such constriction models is:

$$\vec{v}_i(t+1) = \chi\left[\vec{v}_i(t) + c_1 r_1(\vec{p}_i(t) - \vec{x}_i(t)) + c_2 r_2(\vec{p}_g(t) - \vec{x}_i(t))\right] \quad (9.5)$$

where χ is computed as follows:

$$\chi = \frac{2}{\left|2 - \phi - \sqrt{\phi^2 - 4\phi}\right|}, \phi = c_1 + c_2, \phi > 4 \quad (9.6)$$

In what follows, we describe the BCPSO+VALA in detail. BCPSO+VALA method is modeled with a VALA. At the beginning of each environment, the number of actions for VALA (i.e., $\alpha_1, \alpha_2, \ldots, \alpha_N$) is set according to the number of sub-populations obtained by the hierarchical clustering procedure. In order to select a new sub-population to be executed, the VALA selects one of its actions (e.g., α_i) according to its action probability vector $\widehat{p}$. The pseudocode for the action selection procedure in VSLA is shown in Fig. 9.9.

Afterward, the corresponding sub-population is executed using the principles of PSO in Eqs. (9.4), (9.5), and (9.6).

After evolving the selected sub-population, it is checked to see whether the selected sub-population's execution improved the algorithm's overall performance. To this end, we compare the fitness value of the global best-found solution against that from before the execution of the selected sub-population. Then a reinforcement signal is generated as follows:

$$Reinforcement\ signal\beta = \begin{cases} 0\ if\ f\left(\vec{X}_{G,t}\right) < f\left(\vec{X}_{G,t+1}\right) \\ 1 \qquad otherwise \end{cases} \quad (9.7)$$

Algorithm 9-9. Action selection in VALA

```
for each action jεA(k) of VALA do
| Calculate p̂j according to Eq. (9-3);
end
rnd := rand();
sum := 0;
selectedPopulation := Null;
for each action jεA(k) of VALA do
| if rnd < sum + p̂j then
| | selectedPopulation := j;
| | break;
| end-if
| sum := sum + p̂j;
end
```

Fig. 9.9 Pseudocode for action selection in VALA

Algorithm 9-10. Updating the probability vector $\hat{p}$ for actions in VALA using the environment feedback

for each action $j\epsilon A(k)$ of VALA **do**
  if ($\beta = 0$) **then** //favorable response
    $\hat{p}_j(k+1) = \begin{cases} \hat{p}_j(k) + \alpha \cdot (1 - \hat{p}_j(k)) & if\ i = j \\ \hat{p}_j(k) \cdot (1 - \alpha) & if\ i \neq j \end{cases}$
  else if ($\beta = 1$) //unfavorable response
    $\hat{p}_j(k+1) = \begin{cases} \hat{p}_j(k) \cdot (1 - \beta) & if\ i = j \\ \frac{\beta}{r-1} + (1 - \beta) \cdot \hat{p}_j(k) & if\ i \neq j \end{cases}$
  end-if
end-for

Fig. 9.10 Pseudocode for updating the probability vector of VALA

where f is the fitness function, and $\overrightarrow{X}_{G,t}$ is the global best particle of the algorithm at time step t. The above equation implies that if the quality of the best-found solution by the BCPSO algorithm improves by executing the selected sub-population, a positive signal is generated. The generated reinforcement signal is used to determine whether the selected action was right or wrong.

Then, the VALA updates its probability vector using the feedback β according to Algorithm 9–10 in Fig. 9.10.

Then, overlapping checks, redundancy checks, and convergence checks are performed according to (Yang and Li 2010) for each sub-swarm, and corresponding operations are executed for each subs-warm, accordingly. Once an environmental change is detected, a new cradle swarm will be regenerated, and the best memory of each sub-swarms is transferred to the new cradle swarm. The pseudocode for BCPSO+VALA is shown in Fig. 9.11.

The number of active sub-swarms in BCPSO is varied due to overlapping and convergence of sub-swarms. Figure 9.12 shows the evolution of converged sub-swarms and active sub-swarms during 1000 iteration.

9.5 Experimental Study

9.5.1 Dynamic Test Function

One of the most widely used synthetic dynamic optimization test suites in the literature is the MPB problem proposed by Branke (Branke 1999), which is highly regarded due to its configurability. MPB is a real-valued dynamic environment with a D-dimensional landscape consisting of m peaks, where the height, the width, and the position of each peak are changed slightly every time a change occurs in the environment (Branke 2002). Different landscapes can be defined by specifying the shape of the peaks. A typical peak shape is conical, which is defined as follows:

$$f(\overrightarrow{x},t) = \max_{i=1,\ldots,m} H_t(i) - W_t(i)\sqrt{\sum\nolimits_{j=1}^{D} (x_t(j) - X_t(i,j))^2} \tag{9.8}$$

Algorithm 9-11. BCPSO+VALA

```
setting the parameters R_conv, R_overlap, M, max_subsize; /*Set the convergence radius, overlapping radius,
population size and the maximum size of the child swarm particles*/
randomly initialize the main swarm parentlist with M particles in the search space; /*Randomize the initial
population into the search space*/
set the reward and penalty parameters a = 0.15 and b = 0.05; /* Reward and penalty of the VALA*/
sublist = Clustering (parentlist);
initialize the VALA with action set α = {α_1,···,α_|sublist|} and action probability vector p = [(1/|sublist|),
(1/|sublist|),..., (1/|sublist|)].
let A be the set of active actions to at each stage;
A = α;
while termination condition is not met do
    select an action α_r ∈ {α_1,···,α_|sublist|} using Algorithm 9;
    evolve the corresponding sub-swarm α_r according to the PSO principles using Eq. (9-5) and Eq.(9-6);
    compute the reinforcement signal β according to Eq. (9-12);
    update the probability vector p̂ for active actions in A using Algorithm 9-10;
    //overlapping check
    for each pair of sub-swarms (r,s) in the sublist do
        if r_overlap > R_overlap then
            merge r and s into r;
            remove s from sublist;
            remove the corresponding α_s from A;
        end-if
    end-for
    //redundancy check
    for each sub-swarms r in sublist do
        if |r|>max_subsize then
            remove worst (|r|-maxsub_size) particles from r;
        end-if
    end-for
    //convergence check
    for each sub-swarms s in sublist do
        if radius(s)< R_conv then
            add gbest sub-swarm s into memory;
            remove s from sublist;
            remove the corresponding α_s from A;
        end-if
    end-for
    if a change is detected then
        save the gbest of each sub-swarm in a temporary list memory;
        randomly generate a new parentlist with M-|memory| particles;
        add memory list particles to the newly generated parentlist; /*M particles exist in
        the parentlist now*/
        sublist = Clustering (parentlist);
        re-initialize the VALA with action set α = {1, 2,..., |sublist|} and action probability
        vector p = [(1/|sublist|), (1/|sublist|),..., (1/|sublist|)].
        A = α;
    end
end
```

Fig. 9.11 Pseudocode for BCPSO+VALA

where $H_t(i)$ and $W_t(i)$ are the height and the width of peak i at time t, respectively. The coordinates of each dimension $j \in [1, D]$ related to the location of peak i at time t, is expressed by $X_t(i, j)$, and D is the problem dimensionality. A typical change of a single peak can be modeled as follows:

$$H_{t+1}(i) = H_t(i) + height_{severity}.\sigma_h \tag{9.9}$$

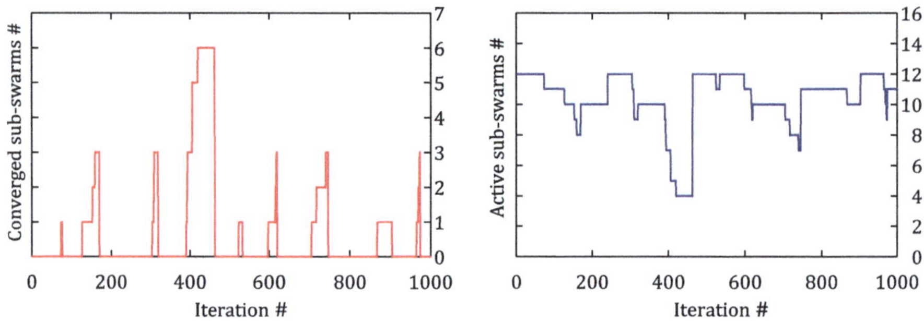

Fig. 9.12 Evolution of the number of converged and active sub-swarms over time in BCPSO

$$W_{t+1}(i) = W_t(i) + width_{severity}.\sigma_w \tag{9.10}$$

$$\overrightarrow{X}_{t+1}(i) = \overrightarrow{X}_t(i) + \overrightarrow{v}_{t+1}(i) \tag{9.11}$$

$$\overrightarrow{v}_{t+1}(i) = \frac{s}{\left|\overrightarrow{r} + \overrightarrow{v}_t(i)\right|}\left((1-\lambda)\overrightarrow{r} + \lambda\overrightarrow{v}_t(i)\right) \tag{9.12}$$

where σ_h and σ_w are two random Gaussian numbers with zero mean and standard deviation one. Moreover, the shift vector $\overrightarrow{v}_{t+1}(i)$ is a combination of a random vector $\overrightarrow{r}$, which is created by drawing random numbers in $[-0.5, 0.5]$ for each dimension, and the current shift vector $\overrightarrow{v}_t(i)$, and normalized to the length s. Parameter $\lambda \in [0.0, 1.0]$ specifies the correlation of each peak's changes to the previous one. This parameter determines the trajectory of changes, where $\lambda = 0$ means that the peaks are shifted in completely random directions and $= 1$ means that the peaks always follow the same direction until they hit the boundaries where they bounce off.

Different instances of the MPB can be obtained by changing the environmental parameters. Three sets of configurations have been introduced to provide a unified testbed for the researchers to investigate their approaches' performance in the same condition. The second configuration (Scenario 2) is the most widely used configuration, which was also used as the base configuration for the experiments conducted in this chapter. Unless stated otherwise, environmental parameters are set according to the values listed in Table 9.1 (default values). Besides, to investigate the effect of the environmental parameters (i.e., number of peaks, change period, number of dimensions) on the proposed methods' performance, various experiments were carried out with different combinations of other tested values listed in Table 9.1.

Table 9.1 Parameter settings for the moving peaks benchmark

Parameter	Default values	Other tested values
Number of peaks (m)	10	1, 5, 20, 30, 50, 100, 200
Height severity	7	
Width severity	1	
Peak function	Cone	
Number of dimensions (D)	5	
Height range (H)	$\in [30, 70]$	
Width range (W)	$\in [1, 12]$	
Standard height (I)	50	
Search space range (A)	$[0, 100]^D$	
Frequency of change (f)	5000	500, 1000, 2500
Shift severity (s)	1	0, 2, 3, 4, 5
Correlation coefficient (λ)	0.0	
Basic function	No	

9.5.2 Performance Measure

To measure the efficiency of the optimization algorithms on DOPs, we use the best error before changes which was first proposed in Trojanowski and Michalewicz (1999) as to accuracy and later named as the best error before the change in Nguyen et al. (2012), is calculated as the average of the minimum fitness error achieved by the algorithm at the end of each period right before the moment of change, as follows:

$$E_{bbc} = \frac{1}{K} \sum\nolimits_{k=1}^{K} (h_k - f_k) \tag{9.13}$$

where f_k is the fitness value of the best solution obtained by the algorithm just before the kth change occurs, h_k is the optimum value of the kth environment, and K is the total number of environments. For more details about this measure, interested users can refer to (Ranginkaman et al. 2014; Kazemi Kordestani et al. 2019).

9.5.3 Experimental Settings

For each experiment of the proposed algorithm on a specific DOP, 31 independent runs were executed with different random seeds for the dynamic environment and

Table 9.2 Default parameter settings for the proposed MP with FE management schemes

Parameter	Default value
Parent population size	50
max_subsize	5
$R_{overlap}$	0.7
R_{conv}	0.0001
χ	0.7298438
c_1	2.05
c_2	2.05
Reward rate (a)	0.15
Penalty rate (b)	0.05

algorithm. For each run of the algorithm, $f \times 100$ FEs were considered as the termination condition. The experimental results are reported in terms of average best error before the change and standard error, calculated as the standard deviation divided by the number of runs' squared root.

Finally, the parameters set for BCPSO and the reward and penalty factors for the proposed approach are shown in Table 9.2.

To show the significant statistical difference between our BCPSO and BCPSO+VALA, a non-parametric statistical test called the Wilcoxon rank-sum test is conducted for independent samples at the 0.05 significance. In situations where the Wilcoxon rank-sum test outcomes reveal that the algorithms' difference is not statically significant, an asterisk symbol is printed after the corresponding p-value. Also, when reporting the results, the last column(row) of the table includes the improvement rate (%Imp) of the BCPSO+VALA vs. BCPSO. This measure is computed for every problem instance i as follows:

$$\%Imp_i = 100 \times \left(1 - \frac{e_{i,BCPSO+VALA}}{e_{i,BCPSO}}\right) \tag{9.14}$$

where $e_{i,BCPSO+VALA}$ is the error value obtained by BCPSO+VALA for problem i. Similarly, $e_{i,BCPSO}$ is the best (minimum) offline error obtained by BCPSO.

9.5.4 Experimental Results

9.5.4.1 Extrema Tracking

In this experiment, a single moving peak is defined in the fitness landscape, and its position in the environment is changed after every 5000 FEs. Over time, the evolution of the current fitness error for BCPSO and BCPSO+VALA is depicted in Fig. 9.13.

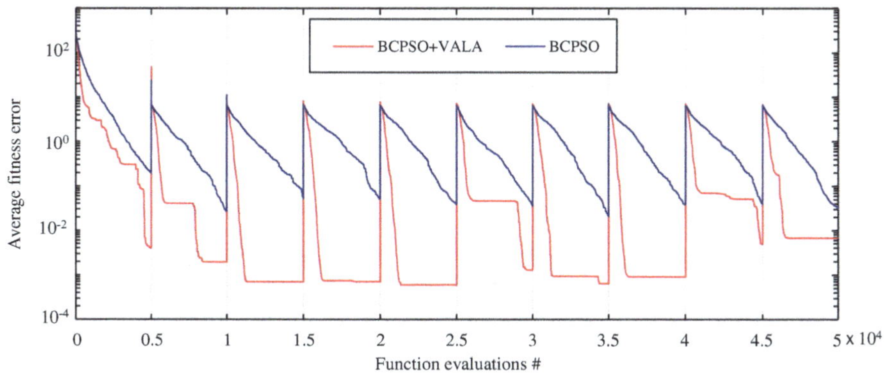

Fig. 9.13 Evolution of the current fitness error for different algorithms in a single-peak DOP with $f = 5000$. The vertical dotted lines indicate the beginning of a new environment

The results perfectly verify the effectiveness of FEM on the performance of BCPSO+VALA given in Sect. 9.4. As the convergence curves of BCPSO and BCPSO+VALA depict, in the initial periods (i.e., FEs less than 5000), BCPSO+VALA can locate the optimum position faster. This is because BCPSO+VALA allocates more FEs to the most promising sub-swarms instead of sequential execution of sub-populations. Moreover, after each environment change, the BCPSO+VALA has a better capability to track the new position of the optimum.

9.5.4.2 Experiment on DOPs with Different Change Intervals

Table 9.3 shows the performance comparison between BCPSO and BCPSO+VALA on DOPs with different change intervals, $f \in \{500, 1000, 2500, 5000\}$.

We can see from Table 9.3 that both algorithms' performance improved with an increase in the change interval. We can also see that the results from BCPSO+VALA are much better than the base algorithm. These results indicate that the proposed algorithm can quickly adapt to environmental changes.

Table 9.3 Comparison between BCPSO and BCPSO+VALA on MBP with various change intervals

MPB	Method		% Imp
	BCPSO	BCPSO+VALA	
f = 500	11.12 ± 0.36	**10.36 ± 0.27**	6.83
f = 1000	7.72 ± 0.24	**6.65 ± 0.21**	13.86
f = 2500	3.45 ± 0.14	**2.81 ± 0.14**	18.55
f = 5000	2.64 ± 0.17	**2.18 ± 0.14**	17.42

Table 9.4 Performance of BCPSO+VALA vs. BCPSO on DOPs with different number of peaks generated by MPB

MPB	Method		%Imp
	BCPSO	BCPSO+VALA	
$m = 1$	0.04 ± 0.00	**0.001 ± 0.00**	99
$m = 5$	1.75 ± 0.15	**1.37 ± 0.16**	21.71
$m = 10$	2.64 ± 0.17	**2.18 ± 0.14**	17.42
$m = 20$	2.41 ± 0.10	**2.20 ± 0.09**	8.71
$m = 30$	2.42 ± 0.07	**2.34 ± 0.09**	3.30*
$m = 50$	2.28 ± 0.07	**2.09 ± 0.07**	8.33
$m = 100$	1.87 ± 0.05	**1.70 ± 0.04**	9.09
$m = 200$	1.64 ± 0.03	**1.43 ± 0.03**	12.8

9.5.4.3 Experiment on DOPs with Various Number of Peaks

In this experiment, the performance of BCPSO and BCPSO+VALA was investigated using MPB with a different number of peaks, m ∈ {1, 5, 10, 20, 30, 50, 100, 200}. Table 9.4 presents the numerical results of BCPSO and BCPSO+VALA in terms of the E_{bbc} and standard error.

Regarding Table 9.4, the first thing that stands out is that the number of peaks affects both algorithms' performance. When the number of peaks in the environment is not very large (i.e., 1, 5, 10), BCPSO+VALA shows its best performance where the rate of improvement values of the BCPSO+VALA over the BCPSO is ranging from 17.42% to 99%. The reason can be contributed to the fact that in BCPSO+VALA, the FEs are allocated to more promising sub-populations. For DOPs with a larger number of peaks, although the improvement rate values become smaller, the BCPSO+VALA can still perform better.

9.5.4.4 Experiment on DOPs with Various Shift Lengths

In this experiment, we investigate the performance of BCPSO and BCPSO+VALA using MPB with different shift lengths, s ∈ {0, 1, 2, 3, 4, 5}. The obtained results in terms of average best error before changes are illustrated in Fig. 9.14.

From Fig. 14, it is observed that it became more difficult for the algorithms to track the global optima as the severity of the environmental changes was increased.

As can be observed in Fig. 9.14, it is clear that the VALA-based FEM strategy has a significant influence on the performance of BCPSO+VALA. Compared to BCPSO, BCPSO+VALA reached better results in all tested DOPs.

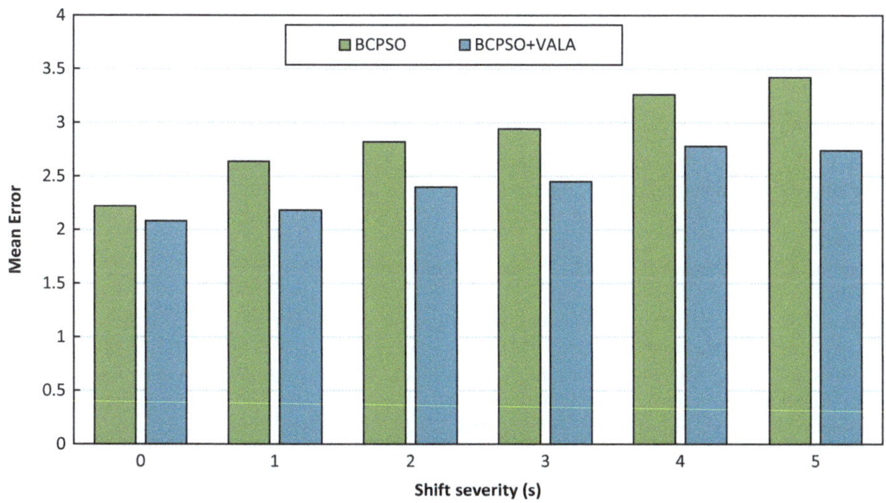

Fig. 9.14 Simulation results for MPB with different shift lengths

9.6 Conclusions

This chapter has extended the concept of function evaluation management for multi-population methods with various populations. To this end, a variable action-set learning automaton is used to allocate function evaluations to sub-populations in situations where the number of sub-populations is changing over time by factors like the convergence of sub-populations or merging the collided sub-populations. The proposed algorithm was integrated into the clustering particle swarm optimizer (CPSO), a well-known algorithm from the literature. Experiments were conducted to compare the proposed algorithm with the base algorithm. Results show the applicability of the proposed function evaluation management for multi-population with a changing population over time.

References

Branke, J.: Memory enhanced evolutionary algorithms for changing optimization problems. In: Proceedings of the IEEE Congress on Evolutionary Computation. IEEE, pp. 1875–1882 (1999)

Branke, J.: Evolutionary Optimization in Dynamic Environments. Springer (2002)

Clerc, M.: The swarm and the queen: towards a deterministic and adaptive particle swarm optimization. In: Proceedings of the 1999 Congress on Evolutionary Computation-CEC99 (Cat. No. 99TH8406), vol. 3, pp. 1951–1957 (1999)

Kazemi Kordestani, J., Rezvanian, A., Meybodi, M.R.: New measures for comparing optimization algorithms on dynamic optimization problems. Nat. Comput. **18**, 705–720 (2019). https://doi.org/10.1007/s11047-016-9596-8

Kennedy, J., Eberhart, R.: Particle swarm optimization. In: IEEE International Conference on Neural Networks, pp 1942–1948 (1995)

Kordestani, J.K., Meybodi, M.R.: Application of sub-population scheduling algorithm in multi-population evolutionary dynamic optimization. In: Evolutionary Computation in Scheduling, pp. 169–211. Wiley (2020)

Kordestani, J.K., Ranginkaman, A.E., Meybodi, M.R., Novoa-Hernández, P.: A novel framework for improving multi-population algorithms for dynamic optimization problems: a scheduling approach. Swarm Evol. Comput. **44**, 788–805 (2019). https://doi.org/10.1016/j.swevo.2018.09.002

Narendra, K.S., Thathachar, M.A.L.: Learning automata: an introduction. Courier Corporation (2012)

Nguyen, T.T., Yang, S., Branke, J.: Evolutionary dynamic optimization: a survey of the state of the art. Swarm Evol. Comput. **6**, 1–24 (2012). https://doi.org/10.1016/j.swevo.2012.05.001

Ranginkaman, A.E., Kazemi Kordestani, J., Rezvanian, A., Meybodi, M.R.: A note on the paper "A multi-population harmony search algorithm with external archive for dynamic optimization problems" by Turky and Abdullah. Inf. Sci. **288**, 12–14 (2014). https://doi.org/10.1016/j.ins.2014.07.049

Rezvanian, A., Saghiri, A.M., Vahidipour, S.M., Esnaashari, M., Meybodi, M.R.: Learning automata theory. In: Recent Advances in Learning Automata. Springer, pp. 3–19 (2018)

Rezvanian, A., Moradabadi, B., Ghavipour, M., Daliri Khomami, M.M., Meybodi, M.R.: Introduction to Learning Automata Models. In: Learning Automata Approach for Social Networks. Springer, pp 1–49 (2019)

Trojanowski, K., Michalewicz, Z.: Searching for optima in non-stationary environments. In: Proceedings of the IEEE Congress on Evolutionary Computation. IEEE, pp 1843–1850 (1999)

Vafashoar, R., Morshedlou, H., Rezvanian, A., Meybodi, M.R.: Cellular Learning Automata: Theory and Applications. Springer (2021)

Yang, S., Li, C.: A clustering particle swarm optimizer for locating and tracking multiple optima in dynamic environments. IEEE Trans. Evol. Comput. **14**, 959–974 (2010). https://doi.org/10.1109/TEVC.2010.2046667